PATIENT H.M.

PATIENT H.M.

A STORY OF MEMORY, MADNESS, AND FAMILY SECRETS

LUKE DITTRICH

Chatto & Windus
LONDON

1 3 5 7 9 10 8 6 4 2

Chatto & Windus, an imprint of Vintage,
20 Vauxhall Bridge Road,
London SW1V 2SA

Chatto & Windus is part of the Penguin Random House group
of companies whose addresses can be found at
global.penguinrandomhouse.com

First published by Chatto & Windus in 2016

www.vintage-books.co.uk

A CIP catalogue record for this book is available from the British Library

ISBN 9780701187132

Printed and bound by Clays Ltd, St Ives plc

For Bambam, Lolo, Laska, and Anwyn

Man is certainly no poorer as an experimental animal merely because he can talk.

PAUL BUCY

Every day is alone in itself. Whatever enjoyment I've had, and whatever sorrow I've had.

HENRY MOLAISON

CONTENTS

PROLOGUE xi

PART I · ORIGINS

1 *The Fall* 3
2 *Crumpled Lead and Rippled Copper* 12
3 *Dream Jobs* 20
4 *The Bridge* 27
5 *Arline* 38

PART II · MADNESS

6 *Pomander Walk* 51
7 *Water, Fire, Electricity* 63
8 *Melius Anceps Remedium Quam Nullum* 75
9 *The Broken* 89
10 *Room 2200* 98
11 *Sunset Hill* 105
12 *Experiment Successful, but the Patient Died* 110
13 *Unlimited Access* 122
14 *Ecphory* 134
15 *The Vacuum and the Ice Pick* 144

PART III · THE HUNT

16 *It Was Brought into the Sea* 159
17 *Proust on the Operating Table* 179
18 *Fortunate Misfortunes* 190
19 *Henry Gustave Molaison (1926–1953)* 201

PART IV · DISCOVERY

20 *Where Angels Fear to Tread* 219
21 *Monkeys and Men* 234
22 *Interpreting the Stars* 250
23 *The Son-of-a-Bitch Center* 259
24 *The MIT Research Project Known as the Amnesic Patient H.M.* 267

PART V · SECRET WARS

25 *Dewey Defeats Truman* 293
26 *A Sweet, Tractable Man* 316
27 *It Is Necessary to Go to Niagara to See Niagara Falls* 322
28 *Patient H.M. (1953–2008)* 336
29 *The Smell of Bone Dust* 347
30 *Every Day Is Alone in Itself* 357
31 *Postmortem* 379

EPILOGUE 407
ACKNOWLEDGMENTS 413
INDEX 417

PROLOGUE

The laboratory at night, the lights down low. An iMac streams a Pat Metheny version of an Ennio Morricone tune while Dr. Jacopo Annese, sitting in front of his ventilated biosafety cabinet, a small paintbrush in his hand, teases apart a crumpled slice of brain. The slice floats in saline solution in a shallow black plastic tray, looking exactly like a piece of ginger at a good sushi restaurant, one where they don't dye the ginger but leave it pale. Annese takes his brush and, with practiced dabs and tugs, gently unfurls it. The slice becomes a silhouette, recognizable for what it is, what organ it comes from, even if you are not, as Annese is, a neuroanatomist.

He loves quiet nights like these, when his lab assistants set him up with everything he needs—the numbered specimen containers, the paintbrushes, the empty glass slides—and then leave him alone with his music and his work.

Annese coaxes the slice into position on the slide that lies half submerged in the tray, cocking his head, peering at it from different angles, checking to see that he has the orientation right. When you're looking directly at the slide, the left hemisphere must be on the right side of your field of view, just as it would be if you were you staring into the eyes of the brain's owner. Although brains are roughly symmetrical, they are not entirely so, and Annese has become familiar with the topography of this one, all its subtly asymmetrical sulci. At the very center of this slice, in an area that would normally contain a buttressing framework of neural tissue, there are instead two gaping holes, one in each hemisphere. Annese takes extra care not to tear the edges of the holes or distort them, dabbing painstakingly at their frayed perimeters with the tip of his brush. The holes are historic, precious in their own way. Annese does not want to become famous as the second doctor to desecrate this particular brain.

A few more prods and Annese begins to pull the glass out of the tray. Before he trained as a scientist, he worked as a cook, and he often uses cooking analogies to explain his techniques. The art of histology is a lot like baking, he says, since in both everything must be finely calibrated, with little room for improvisation. Soon the slide, with its burden perfectly positioned, is resting safely on the tepid surface of a warmer, where it will be left to dry overnight.

Annese reaches for another cryogenic vial, number 451, screws off the lid. Just before he tips the next slice into the tray, he turns to me and smiles.

"See how much work I have to do to clean up the mess your grandfather made?" he says.

There were things Henry loved to do.

He loved to pet the animals. Bickford Health Care Center was one of the first Eden Alternative facilities in Connecticut, which meant that along with its forty-eight or so patients, the center housed three cats, four or five birds, a bunch of fish, a rabbit, and a dog named Sadie. Henry would spend hours sitting in his wheelchair in the courtyard with the rabbit on his lap and Sadie by his side.

He loved to watch the trains go by. His room, 133, was on the far side of the center, and from his window, several times a day, he could watch the Amtrak rumble past the abandoned redbrick husk of the old paper mill across the street.

He loved word games. He'd sit for hours and hours and work through books full of them. Many of the scientific papers that have been written about Henry over the past six decades describe his avidity for crossword puzzles, though in his later years he found them too great a challenge and started doing simple find-a-word puzzles instead.

He loved old movies. Bogart and Bacall, that era. *The African Queen. Gone with the Wind. North by Northwest.* We call them classics, though of course they were not classics to him. He'd ask to see one of these movies, and a nurse or attendant would pop in a videocassette. Television sets were no shock to him, TV being a technology that developed during his time. But he never did figure out how to operate a remote control.

He loved talking to people. He'd tell them stories. He told the same stories, over and over, but he always told them with equal enthusiasm. When people asked him if he remembered meeting them before, he'd often tell them that yes, he thought they'd once been friends. Hadn't they gone to high school together? Even when his uncertainty about these sorts of things frustrated him, he usually remained courteous and cheerful. Compliant, too. When the scientists would come to pick him up and take him to the laboratory, he never objected. And he almost always took his meds when the nurses asked him to. On the rare occasions that he refused, the nurses knew of an easy way to get him to cooperate. It was a trick passed down over decades, from one nurse to another.

"Henry," a nurse would say, "Dr. Scoville insists that you take your meds right now!"

Invariably he would comply.

This strategy worked right up to the end, until Henry died. The fact that Scoville had died decades before then, and that they'd had no contact for decades before that, made no difference. Scoville remained an authority figure in Henry's life because Henry's life never progressed beyond the day in 1953 when Dr. William Beecher Scoville, my grandfather, removed some small but important pieces of Henry's brain.

• • •

I remember following my grandfather up a snowy hill during his last winter.

I think he was wearing a light blue parka, and in my mind the parka is worn and threadbare, though that would have been uncharacteristic of him. This is a man who was once described by a *New York Times* reporter as "almost unreal in his dashing appearance." But there it is in my memory: a threadbare blue parka. Maybe he even had a woolen cap, one with a pom-pom top, pulled down over his pomaded hair. He always combed his hair with olive oil, that's what my mom says.

We were going sledding.

I remember snow, a white sky, some trees. Cold. Tramping up the hill together.

He was dragging an old-fashioned wooden toboggan behind him, big enough for the two of us. When he reached the top of the hill, he stopped, looked back toward me, and waited.

Why do I remember any of this?

I remember because when the cascade of impressions from my eyes, ears, and skin bombarded me with sights and sounds and textures, with leafless trees and my grandfather's hat and the crunch of our boots in the fresh snow, those impressions were channeled to some small but important parts of my ten-year-old brain. Then my brain went to work, processing raw sensation into something else: a memory, one that still resides inside me, three decades later, to be called up on occasion and dragged, blinking and uncertain, into the light.

I'm getting ahead of myself.

Memories make us. Everything we are is everything we were. This has always been true and is so obvious that it hardly needs to be said. But though memories make us, we've only recently begun to understand how we make memories. The story of how we've gained this understanding is the story I'm telling in this book. It's a story with heroes and villains, tragedy and romance, violence and tenderness. My grandfather plays a part, but it's much bigger than my grandfather.

It's a story about science, and about nature, human and otherwise. And it begins, like a lot of stories do, with a fall.

PART I
ORIGINS

ONE

THE FALL

In the northwest corner of Colt Park, in downtown Hartford, Connecticut, a ten-foot-tall bronze statue of the park's namesake rose from a granite pedestal. Engraved tributes to Samuel Colt, inventor of the Colt .45, covered one side of the pedestal, but the boy trudging toward it wouldn't have been able to read them even if he'd wanted to, since he wasn't wearing his glasses. It was dinnertime, July 3, and it was probably 1933 or 1934, though the exact year would be one of the things that scientists would argue about in the decades to come. His family's second-floor walk-up apartment was about a quarter mile away. He was seven or eight years old and already he'd moved at least three times. His father was an electrician, didn't make much money, had to go wherever the work was. It must have been confusing sometimes for the boy, all these homes flashing by, all those fresh starts. He had blond hair and bright blue eyes and a sweet, uncertain smile.

A steep road skirted the northern edge of the park, and if the boy cut across it and down some backstreets, he could shave a little time off his walk home. The boy's eyesight may have been bad, but there was nothing wrong with his ears. He didn't hear any cars coming. He stepped off the sidewalk and started crossing the road.

The bicyclist, coasting down the hill, didn't see Henry until it was too late.

• • •

Hippocrates Asclepiades, a Greek physician born on the island of Cos in the fourth century B.C.E., is widely regarded as the father of modern medicine. Although his last name indicates a claimed family connection to Asclepius, the revered doctor-god of Greek myth, Hippocrates became famous by advancing the revolutionary argument that the gods had no place in medicine. Healers of one sort or another have existed for as long as humans have, but Hippocrates was one of the first to reject the magic and spiritualism and religion that most who came before him relied on. Instead he attempted to localize the sources of our ailments in our physical environment and inside our bodies themselves.

That approach was well illustrated in an essay he wrote called "On the Sacred Disease." The title was a little misleading, since Hippocrates preferred to call the disease in question by a different name: epilepsy, from the Greek *epilambanein,* which means "to seize." And the disease of epilepsy, he wrote, was "no more divine than others; but it has its nature such as other diseases have, and a cause whence it originates." He criticized the "conjurors, purificators, mountebanks, and charlatans" who used "divinity as a pretext and screen of their own inability to afford any assistance," and he ridiculed them for blaming the gods for the various ways epilepsy manifested itself in their patients: "For, if they imitate a goat, or grind their teeth, or if their right side be convulsed, they say that the mother of the gods is the cause. But if they speak in a sharper and more intense tone, they resemble this state to a horse, and say that Poseidon is the cause. Or if any excrement be passed, which is often the case, owing to the violence of the disease, the appellation of Enodia is adhibited; or if it be passed in smaller and denser masses, like a bird's, it is said to be from Apollo Nomius. But if foam be emitted by the mouth, and the patient kick with his feet, Ares then gets the blame."

After rejecting all the sacred explanations, Hippocrates presented a startling explanation of his own: "The brain is the cause of this affection," he wrote, "as it is of other very great diseases, and in what manner and from what cause it is formed, I will now plainly declare."

The details of Hippocrates's subsequent explanation of the aetiology of epilepsy, of course, haven't stood the test of time. In his view, the brain was a pneumatic organ, alternately pulsing with phlegm and bile. It was delicately attuned to the winds, and the wrong wind blowing on the wrong person at the wrong time could wreak havoc. If the west wind buffeted a constitutionally phlegmatic child, for example, it might cause the child's brain to temporarily "melt," at which point epileptic fits would occur. Hippocrates's prescription for such children would be to shield them from the west wind and expose them instead to the north wind, which would, presumably, recongeal their brains and set them right.

What's important about Hippocrates isn't that he figured out epilepsy's origins or its treatment—he did neither—but that he began looking in the right place: not up to the heavens or Mount Olympus but into the even more mysterious terrain inside our skulls.

In the years since, many doctors grappling with the problem of epilepsy followed Hippocrates's lead, venturing deeper and deeper into the brain, seeking a secular understanding of the "sacred disease."

By the early 1930s, when a bicyclist knocked down a young boy on a street in Hartford, Connecticut, they'd begun to find some answers.

Let's imagine ourselves inside Henry's skull.

Let's imagine the moment after the bicycle hit him and before he hit the ground, when he was neither standing nor lying down but was instead floating through the air.

His brain was floating, too. It was nestled in a warm pool of cerebrospinal fluid, while vivid sensations of every sort coursed through it. The pain from wherever the bicycle impacted him, the shards of scenery as he was knocked off his feet, the view of the fast-approaching ground, the sound of his own involuntary gasp, the feel of his wavy hair ruffling as he fell through the air—all of these sensations and more were relaying from the nerves in his retinas, his auditory canals, his skin, his vestibular balance system, and buffeting his brain, which processed them into the multidimensional stew we experience as in-the-moment consciousness.

Now let's imagine the impact.

Henry landed on the left side of his head, hard enough to tear a deep inch-long gash in his forehead just above his eyebrow. His brain then experienced what are known as torsional forces—that is, forces that caused it to twist inside his skull, in this case from left to right. At the same time, it sloshed forward in its watery womb, pushing up against the thin membrane of the pia mater and the thicker membranes of the arachnoid and dura mater, its weight compressing them all until it crashed against the unyielding barrier of his skull. His brain deformed. It changed shape exactly like a rubber ball does when it hits a hard surface, and then rebounded. If it was moving fast enough, if the rebound was strong enough, it again compressed the various layers of insulation that usually kept it safe, this time on the opposite side of his skull. This second impact would have been somewhat less violent than the first. And if it rebounded again, to make a third transit, it would be moving even more slowly. Within a second, it stopped its bouncing. The force of the impact dissipated, and Henry's brain was again floating serenely in its warm pool of cerebrospinal fluid.

But the damage was already done.

During that first concussive impact and its immediate aftermath, as Henry's brain twisted and compressed and rebounded, various things happened. Some of these things were physical and easy to understand. Neurons and glial cells—the stuff our brains are made of—were torn and ruptured. Other things that happened inside Henry's brain, in that violent moment, were chemical and electrical and harder to explain. For reasons that are still poorly understood, when a brain experiences a combination of torsional forces and blunt-force impact, like Henry's brain experienced, local clusters of neurons open up their floodgates in lockstep synchrony. Bursts of electricity surge down axons—the slender filaments that stretch out from each neuron—and trigger the release of neurotransmitters at their tips. These neurotransmitters bridge the synapses between the ends of the axons and the waiting dendrites of other nearby neurons, causing those neurons to trigger their own bursts of electricity. Eventually, the growing tsunami of neurotransmitters creates an overwhelming surge of brain activity. Whatever sensations and thoughts were inhabiting Henry's brain prior to this moment—the fear, the pain, the confusion—were wiped out by this burst of activity. Which means that, much like a power surge knocks out a computer, it knocked Henry out.

For five minutes, nothing. Henry's brain carried on with its usual autonomic, life-regulating tasks, but wherever his consciousness resided was temporarily shut down.

Then, slowly, he came back online.

He opened his eyes. The world came flooding in again, the bustle and noise of downtown Hartford, the voices of a gathering crowd, the pain from the gash in his forehead, the sticky warmth of the blood flowing down his face: The steady march of experience and sensation resumed.

He was back, but he was not the same.

The next day was the Fourth of July, and Henry went to a picnic with his family. It was perfect weather for it: warm, no rain. His forehead had been stitched up, and there was a bandage above his left eye. People joked with him about it, asking if he'd been playing with firecrackers.

"You must have been up early and got at it," somebody said.

Henry laughed.

He seemed fine.

He felt fine.

Soon, though, the seizures began.

While the exact origins of Henry's epilepsy can never be known for sure, many scientists believe that it was related to his fall. It could have been the direct physical damage: When brain injuries heal, the scars left behind have a tendency to become epileptogenic, meaning they can generate epileptic seizures. There's also a theory known as the kindling effect, which holds that the sort of short-circuiting Henry's brain underwent leaves a new circuit in its wake, a dangerously convulsive circuit, one that grows more active over time, kindling a fiercer and fiercer blaze.

The seizures were minor at first. Little instants of inattention. Dazed moments, small absences.

Still, the seed had been planted, and Henry's transformation into Patient H.M., the most studied individual in the history of neuroscience, had begun.

That's his real name: Henry.

I can even give it to you complete: Henry Gustave Molaison.

There was a time I couldn't. It was a secret.

For almost six decades, the scientists who studied Henry kept his name hidden away. When they wrote about him they were always careful not to reveal too much, for fear that outsiders might find him, and they were successful. There wasn't a single paper, out of the hundreds that chronicled in great detail the experiments performed on Henry during the fifty-five years between his operation and his death, that contained anything but the vaguest biographical information about Henry himself.

If you happened to read a lot of these papers, you could have pieced together a fragmentary portrait: One might have mentioned that he had relatives in Louisiana. Another that he was born in 1926. A third that his father's name was Gustave. A fourth that he was an only child.

And so on.

But most of his story, starting with that most basic fact of his name, was a tightly guarded mystery to the outside world.

. . .

Henry Gustave Molaison was born in Manchester, Connecticut, on February 26, 1926.

Two twenty-six, twenty-six.

"'Least it's easy enough to remember," he often told the scientists with a smile.

They prodded him for his birth date over and over, sometimes five, six, seven times during a single session, and though he never remembered the previous time they'd asked him, the correct answer always came tumbling out intact: two twenty-six, twenty-six.

Other questions had less consistent answers.

"Henry," a scientist asked him one afternoon, about fifteen years after the experiments began, "could you once more describe a little your earliest memory, very early in your life, when you were very small, the very first thing?"

"Well, gee," Henry said. "There is a jumble right there."

He paused. He was sitting in a laboratory at the Massachusetts Institute of Technology, though he didn't know that, and when the scientists had earlier asked where he thought he was, he guessed that he might be in Canada.

"Sort of," he continued, "to pinpoint, put them right down in a . . ."

Henry paused again. He was smoking a cigarette.

"Find the one that comes before or after," he said. He had a soft, gentle voice with a thick New England accent. You could almost hear the thoughts whirring inside as he reached back, deep into his childhood. That time, his earliest memory was of a place. A little blue house the Molaison family once lived in.

Another time, during the same session, responding to the same question, he described a person.

"I can think of my grandfather," he said. "Walking with him. I was very, very small. I think of, uh, well, right off I thought of a tall man, but he isn't, wasn't, tall. Medium-size. Not heavy-built. I always think of him in a gray suit. . . . He looked entirely different than my father did, of course. . . . He was, uh, I think of about five-eight."

"Your father?" the scientist asked.

"Grand," Henry corrected. "Grandfather. Because my father was almost exactly six foot, just had, oh, a quarter part of an inch or so to go, and he'd be six foot."

"How tall are you?" the scientist asked.

"I think of six-two right off."

"Pretty tall," the scientist said.

"Yes, I know I'm taller than my father," Henry said.

"Is your father still alive?" the scientist asked.

Henry thought about the question for a few moments before answering. "There I have an argument with myself. Right off, I think that he is. And then I have the argument, of course, that I think that he has been called."

"You're not sure?" the scientist said.

"I'm not sure," Henry said. "Can't put my finger, well, definitely on it." He paused again before continuing. "He is and he isn't."

The scientist made a note of this—Henry's father had died three years before—and then asked once again for his earliest memory.

"Now, Henry, I want you to go back as far as you can, and I want you to try to tell me what you think is your very first, earliest childhood memory, the memory which you think comes before any other."

"Well, I can go back to, uh, taking a sleigh ride for the first time. . . ."

He described being on Spruce Street, in Manchester, Connecticut, midwinter. He remembered the sleigh being pulled by a single horse. He thought the sleigh and horse belonged to the father of playmates of his, two brothers, Frankie and Jimmie. As he told the story, he picked up the pace, added more details, lost himself in the memory. The horse was on its way to a stable to be reshod. Frankie and Jimmie and Henry were nestled warmly in the back. Some other local kids, seeing them go by, threw snowballs, but the walls of the sleigh kept them safe.

Henry chuckled.

"It was good," he said. "I liked that."

The scientist nodded.

"You remember things from before the operation quite well, don't you?"

Later, when a graduate student transcribed the tape of this session, she noted in parentheses that Henry's response to this last question arrived in a hushed voice, and that he was almost in tears.

"Yes," Henry whispered. "Before that, yes. I do remember."

TWO

CRUMPLED LEAD AND RIPPLED COPPER

I remember midway through one Christmas dinner, when I was about eight, my grandfather pushing himself up from his chair at the head of the table, wandering off to his study, and returning a few minutes later with something in his hand. He placed the object beside his plate: It was a crumpled wad of dark metal, not much bigger than a pencil eraser. I looked at it, wondering what it was. Then he sat back down and told us a story.

Stamford, Connecticut, turn of the century. A burglar broke into the home of a young bachelor, and the bachelor woke up. He reached for the pistol he kept in his nightstand, aimed it at the intruder, but the pistol jammed. The burglar's didn't. A bullet entered the bachelor's chest, where it encountered a rib, deflecting it away from his heart. The bachelor survived and kept the bullet as a memento. He eventually passed it down to his son, my grandfather.

The bullet sat there for the rest of the dinner, and I found it both fascinating and terrible to contemplate. Had it found its target, had its aim been true, then my grandfather, his children, his children's children, most of the people sitting around the table, would have never existed. It was a matter of centimeters—a fluke of aim, bone, ballistics—and it had made all the difference, its repercussions rippling down through generations.

There were other artifacts in my grandfather's home, many equally fascinating—like the bleached human skull that sat on a shelf in his study—and some equally terrible. Each had its own story.

A carved wooden totem hung on one wall of the dining room, a representation of some sort of pagan king or god. It was maybe three feet tall, with a mournful look on its face. He'd received it during a trip to South America, an expression of gratitude for an operation he'd performed. The carving had apparently once been an object of worship, owing mainly to the fact that it would sometimes cry, drops of water trickling from the corners of its eyes. Did seasonal moisture variations and the way the wood responded to them cause the tears? Probably. That or magic. My grandfather had appreciated the totem's beauty but was unsentimental about its emotions. When he brought it home, he had someone shellac it before he hung it on the wall. It never cried again.

Hanging on a wall near the front door was something that at first glance looked like another piece of tribal art. It was made of metal, had a green patina, was about eight inches tall. Its top and bottom both had similar half-moon shapes, though the top had a face carved into it, and the bottom, which was sharpened to a razor's edge, did not. It was part of my grandfather's collection of ancient Inca neurosurgical instruments. The top was a handle, the bottom a blade. It used to mesmerize me. It wasn't just its age, it was its purpose. Somewhere, hundreds of years in the past and thousands of miles to the south, in a time and place incredibly far from my grandfather's warm New England home, this relic had been used to do the same sort of work he did. I would imagine that half-moon sweep of metal slicing through flesh, exposing the bone beneath, then cutting even deeper. I used to wonder if it was still flecked with old blood.

Neurosurgery, whatever the era, always requires at least two frightening qualities in its practitioners: the will to make forcible entry into another human's brain, and the hubris to believe you can fix the problems inside.

. . .

The early history of neurosurgery is written in skulls, not words.

Hundreds of skulls, thousands of skulls, all over the world. In Europe, Africa, South America, Asia. Skulls from different races, different societies, different millennia. All these different skulls, all telling variants of the same story.

The skulls have holes in them. Man-made holes. More than ten thousand years ago, people began cutting holes into the skulls of other people.

Medical historians have noted that ancient Inca skulls in Peru, when they had holes in them, almost always had those holes on the upper left side, the so-called left frontoparietal area. The Incans were a martial culture, fought hard at close quarters with maces and clubs. They were also, like modern humans, predominantly right-handed. When a right-handed man swings a club at the head of an adversary, it tends to land on the left frontoparietal area. So, the theory goes, these surgical holes in the Inca skulls were probably part of the treatment for head wounds they'd received in battle. Perhaps they were made to relieve intracranial pressure or were cut around smaller, brute-force holes and cracks made by the impact of the weapons to better reach and remove the bony shrapnel inside.

In other parts of the world—more specifically, in a trove of skulls discovered at a seven-thousand-year-old grave site in Ensisheim, France—the holes seemed to be evenly divided between the left and the right sides of the head. This was taken as evidence that not all the skull openings had been made to treat war wounds. But if not, then what were they for? To release evil spirits? To cure headaches? To accelerate enlightenment? Nobody knows for sure.

One thing we do know: Having a hole cut in your skull, even seven thousand years ago, didn't necessarily kill you. A close examination of the edges of the holes in those ancient skulls revealed that in most of them, stretching inward from the serrated or punched or smooth edge where surgeons made their marks, new bone had grown, the beginning of an attempt by the skull to reseal itself. The bones in our heads grow slowly, and stop growing as soon as we die. Which means that the owners of those skulls, with their indications of postoperative growth, had survived their surgeries.

In some skulls, in some cultures, the holes weren't so much cut as they were scraped away by surgeons wielding tools more like Brillo pads or sanders than scalpels or drills. Paul Broca, a pioneering French nineteenth-century neuroanatomist, was fascinated by these scraped skulls. He noted that they predated anesthesia—which originated in its crudest form around 400 B.C.E., when Assyrian surgeons would induce unconsciousness by compressing the carotid arteries of their patients—by at least 3,500 years, and speculated that the surgeries must have been performed when the patients were young children, because a child's thinner skull wouldn't take so agonizingly long to rub through. To prove his point, Broca obtained a number of corpses of all ages and demonstrated that while it took him almost an hour to rub through the skull of an adult, he could do the same to a two-year-old child's in less than five minutes. Others took exception to Broca's theory, noting that although dying from these operations was evidently rare—see again: the evidence of postoperative bone growth—it did sometimes happen, and if ancient brain surgeons had been scraping holes in ancient infants, you would expect to find at least a few ancient infant skulls with holes in them, victims of unsuccessful operations. No such skulls had been found.

These sorts of debates will go on and on. People continue to study these silent skulls, trying to read the stories their preliterate former owners couldn't document.

Eventually, of course, humans did gain the ability to record their own lives. We began to write. And among the first things we wrote about?

Brain injuries and how to treat them.

• • •

In 1862, an American collector of antiquities named Edwin Smith bought a scroll of papyrus from a dealer in Luxor, Egypt. The papyrus was fifteen feet long, and an unknown ancient had used a reed brush and inks derived from clay and burnt oils to cover it in a thicket of hieratic script. Hieratic was the less formal and ornate descendant of the Egyptian hieroglyphs, their version of shorthand. For almost all of the thousand years prior to Smith's purchase, both forms of writing—hieroglyphs and hieratic—had been relics of a dead language, unused and untranslatable. The Egyptians themselves debated whether the two scripts even represented a language at all or whether their ancestors had just enjoyed covering scrolls and tombs with meaningless decorative symbols. In one often-repeated tale, an Italian merchant visiting the Giza pyramids in the 1700s was offered a wooden chest containing forty ancient papyrus rolls. He purchased only one of them, and the villagers supposedly "burned the rest in order to enjoy the smell they gave off."

But by 1862, things had changed. The 1822 translation of the Rosetta Stone—which contains versions of the same text in hieroglyphs, hieratic, *and* Greek—provided sudden access to an entire epoch of the ancient world that had been previously sealed off. For people interested in history, or in profiting from it, this was the equivalent of the gold rush: Egypt swarmed with tomb raiders and treasure hunters, such as Edwin Smith, who gathered up as many of these formerly inscrutable documents as they could.

Smith took the scroll home to Connecticut and spent much of the rest of his life trying to make sense of it. Even the best linguist might spend years translating a single passage of hieratic. Edwin Smith was not the best linguist. He'd obtained an objectively beautiful piece of writing—the papyrus was incredibly well preserved, and its ink changed intriguingly from a deep black to, for certain words and lines, a crimson made out of ground ochre—but it refused to give up its secrets. When Smith died in 1906, his daughter donated the scroll to the New-York Historical Society, where it was found by James Henry Breasted, a professor of Egyptology at the University of Chicago.

Breasted spent nearly ten years working on his own translation of the scroll, and when he published it, in 1930, he declared the so-called Edwin Smith Papyrus "the oldest nucleus of really scientific knowledge in the world."

The scroll, Breasted revealed, was a medical textbook. And in its formatting, it was a strikingly modern one: The passages written in red, for example, were done to highlight the key parts of the text that the author wanted the reader to remember. But much more surprising was how modern its contents were, this despite the fact that the scroll itself was at least 3,600 years old and contained archaic turns of phrase that indicated it may have been a transcription of a text eight hundred years older than that.

Up until Breasted's translation, the prevailing view of medicine in ancient Egypt was that it was based in magic, not science. Previously discovered papyri about medical topics limited their prescriptions to incantations and dubious potions. And the Edwin Smith Papyrus had some of that: The scroll was organized as a series of forty-eight case studies of battlefield injuries, each with its own treatment recommendation. In case nine, a man with a bashed-in forehead, the would-be attending physician was advised to chant the following spell while standing over his patient: "Repelled is the enemy that is in the wound / Cast out is the evil that is in the blood / The adversary of Horus, on every side of the mouth of Isis / This temple does not fall down / There is no enemy of the vessel therein / I am under the protection of Isis / My rescue is the son of Osiris!"

Most of the prescriptions, however, were secular. And most of the case studies—twenty-seven out of forty-eight—involved head trauma.

For example:

> Case 6: Medical instructions for an oozing gash/cutting wound in his head that penetrates to the bone, smashing in his cranium and exposing the brain in his cranium.

—You have to probe his wound.

—Should you find [in] that smash fracture that is in his cranium, ripples [like] those that occur [in] copper in the smelting process,

—and something within that throbs and flutters under your fingers like the weak spot of the crown of a child not yet fused and made "whole."

You should daub that wound of his with oil.

—Do not bandage it.

—Do not put dressings on it until you know that he has passed the crisis!

The smash fracture is large, opening to the inside of his cranium and the membrane enveloping his brain is ruptured and its fluid falls from the interior of his head.

Although the treatment in this case by modern standards is conservative—clean the wound and hope for the best—it's about as promising an approach as you could hope to find in a four-thousand-year-old hospital. Sealing the wound, for example, would have probably done more harm than good, causing death through swelling or infection. The Egyptians were apparently not only restrained physicians but accomplished neuroanatomists as well: Up until the translation of the Smith papyrus, nobody had found the word *brain* in any prior hieratic or hieroglyphic documents. In this one, however, they didn't just name the brain, they described it in vivid poetic detail—with its "ripples [like] those that occur [in] copper in the smelting process"—along with the membranes and cerebrospinal fluid that swaddled it. The brain, the ancients clearly understood, was a delicate and important organ. It was also an organ that, in general, should be protected and not actively messed with. In the case of a patient with a fractured skull, you might "clean it for him with a swab of linen until you see its fragments of bone," but the brain itself should remain untouched.

After the revelations of the Edwin Smith Papyrus, some Egyptologists argued that the Egyptian ankh—which represents the human spinal column—would make a more historically accurate symbol for the medical arts than the traditional snake-twined staff. (That symbol derived from myths about the staff-wielding, snake-revering Greek doctor-god, Asclepius, who was so good at keeping people alive that Zeus killed him to prevent overpopulation.) In any event, the papyrus seemed to prove that modern medicine had begun far earlier than was previously thought, and the cautious treatments it prescribed indicated that at least some ancient doctors were adhering to the central tenet of the Hippocratic oath more than a thousand years before Hippocrates's birth.

"Abstain from doing harm. . . ."

A simple principle, and an enduring one.

For most of the long history of the healing arts, that principle has guided the care and treatment of our most mysterious and delicate organ.

Protect it when possible, keep it clean, don't muck about inside.

That was the status quo for thousands of years.

Until suddenly it wasn't.

THREE

DREAM JOBS

In the lab at MIT, Henry was explaining, again, the many moves his family had made when he was a child. Even the scientists found his odyssey confusing. Dr. William Marslen-Wilson, a British psychologist who was interviewing him, worked hard to follow Henry's story.

"I see," Marslen-Wilson said at one point. "Right. It's all becoming clearer now. A lot of schools, a lot of houses, difficult to sort out."

"And from there," Henry continued, "we moved from Franklin Avenue out to South Coventry, Connecticut. And I had to take a school bus, which stopped—I was the last one to get on it in the morning—and take me home, take me from South Coventry to Willimantic, and I think it was exactly five miles, right from our house to Willimantic."

"And you were in . . . what sort of school were you in?"

"That was in a high school. Windham High."

"Windham?"

"Windham High School."

"Do you remember how to spell Windham?"

"Well, it's W-I-N-D-H-A-M."

"And what grade were you in there?"

"Second year of high . . . and well, half of the third year."

"Why only half of the third year?"

"Because we moved from South Coventry back to Hartford, and I quit school."

"Yes."

"And it was then . . . well, then after that we moved from light-housekeeping rooms that we had. . . ."

"Lighthouse-keeping?"

"Yes."

"Uh, I don't really understand about this lighthouse-keeping. You mean your parents were working in a lighthouse?"

Henry's parents had not worked in a lighthouse. His father was an electrician, his mother a housekeeper. They didn't make much money. Their savings, small to begin with, had been hammered down to almost nothing in the stock market crash of 1929. A light-housekeeping room was mid-twentieth-century American shorthand for a partly furnished tenement apartment. When Henry was a teenager, his family put their furniture into storage and lived in a string of light-housekeeping rooms in and around Hartford.

To supplement the family income, Henry took on part-time jobs. He worked as an usher in a movie theater, a stock boy in the shoe department of the G. Fox & Co. department store, and a scrap-metal salvager at a junkyard. After dropping out of high school, he started learning a trade, training as a motor winder at Ace Electric Motors. The job involved taking small electric motors apart, eyeballing their individual parts for problems or defects, and then rewinding the copper wires that coiled tightly around each motor's magnetic core. It was a good job, one with a future, but eventually he had to give it up. By that time, his epilepsy had become much more severe. He had his first grand mal seizure on his fifteenth birthday. He remembered driving with his father. He didn't remember if he was in the front or the backseat, but he thought that it was probably the backseat, because when he began convulsing and fell forward his father didn't notice at first and just kept driving. Then he had another seizure, in which he fell down someone's front steps and awoke on a sidewalk. He began having them more and more often. Afterward he would typically remember what he'd been doing immediately prior to seizing up, but the seizures themselves were inaccessible, a blank spot in his mind. When they struck, they sent him to oblivion, exactly like that bicycle in Colt

Park had done. Motor winding required delicacy and precision, and even when Henry wasn't having a full-blown grand mal, the petit mals were enough to compromise his work, causing him to strip too much insulation from wires or to leave out some essential part of the motor when he put it back together.

He eventually went back to high school, enrolling at East Hartford High and struggling his way through to a diploma. By the time of his graduation, his seizures were so frequent that the decision was made to keep him from collecting the diploma onstage. In the laboratory, he explained to Dr. Marslen-Wilson about what might have happened otherwise.

"Well, in a way that was more protection of themselves," he said, "so if I was to have an attack or something like that, I wouldn't fall or something and disturb the others that were there, that were graduating, and the people in the audience." He explained that even a petit mal would have been enough to cause a disruption. "You black out. You could be walking to get your diploma or something, and you walk right by the person that was handing out the diploma instead of stopping and getting the diploma and *then* walking off."

"That would have been a bit difficult," Marslen-Wilson said.

"Yes, it would," Henry said.

After high school, Henry got the last job he remembers, working on the assembly line at the Underwood Typewriter factory. Again, his illness became an issue. He was in the middle of the line, helping put together the frame of the typewriter before another worker attached the keys. This required less skill than motor winding, but still, sometimes the parts of the typewriter might be laid out in front of him, ready for assembly, and he would suddenly go absent, frozen, eyes open but looking at nothing, and the line would back up behind him until he came to.

I recently found a moldering cardboard box in my mother's basement. Inside was a stack of letters wrapped in twine. The letters had all been sent by either my grandfather or his older brother, Gurdon, to their mother in the late 1920s. Most were in their original envelopes, and on a few of these envelopes my great-grandmother had scrawled little notes to herself. Sometimes the notes were just hints about the contents—"About graduation"; "My birthday"—but sometimes they were more like mini-reviews: "To be kept always in my lifetime, this letter of Gurdon's"; "About our beloved home, Treetop"; or "A beautiful letter to be kept from Gurdon."

Usually, she would only add those extra heartfelt little notes to Gurdon's letters. And reading the letters themselves, it was easy to see why. Gurdon was so effusive in his love for his mother that it was almost creepy. Here's a typical passage, written when he was several years into seminary school and already sounding like the minister he would become:

> I have never felt before what I feel at this moment—the union of our hearts and souls in a way that has broken all the barriers of distance and brought me into a sudden new understanding of what love between two persons really can be and how it lifts them into the realm of the eternal beyond the mile posts and clock ticks of our little earth. You must be thinking of me, Mother darling, and praying for me this morning—you have given me just as I have been writing this letter something more than I ever before had—a feeling of how we cannot ever be separated that I will always count one of my sacred experiences.

The letters from my grandfather were different. They, too, were loving—both sons clearly adored their mother—but my grandfather's letters tended to be maudlin, full of apologies.

"My mother darling," he wrote toward the end of his senior year at Yale University, "I have just written father about 42 pages, but seem to be just as discouraged, so will continue with you. After reviewing all of your last letters and his, the world has become darker and darker, letter by letter. I am sorry about everything I have done, for I don't mean to be so selfish and mean. I have been awfully wrought about money matters, and am trying to work it out all right." He laid out the various ways he was trying to supplement his "allowance," ranging from tutoring to selling jewelry he had imported from China. "I am working my head off with this jewelry in an effort to be self-supporting," he wrote, then added a clause that said a lot about the privileges of his upbringing, "and am giving the Rolls back to Pen without using it, as you wished." Toward the end of the letter, he unleashed a fusillade of self-pity and self-recrimination. "I am more sorry than I can say that I disappoint you both with my carefree-ness and thoughtless selfishness. I have worried too much to be exactly carefree. I love you so much and think of you so much and your goodness to me and in you—as I grow older—that I feel very miserable when I hurt you so—please forgive me."

The painful insecurity on display in this letter was present in several others from that same year as well. In one he sent to his mother on her thirty-ninth birthday, he wrote, "I want all of my friends to see what a mother I have—then they will realize that even if I am not going to be a success in life, it was not my mother's fault."

It was hard to know what to make of this. People I'd spoken with over the years had applied a lot of adjectives to my grandfather—brilliant, arrogant, dashing, reckless—but insecure wasn't one of them. The man berating himself in these letters didn't match the image I had of Dr. William Beecher Scoville. But maybe that was the point. The person writing these letters was still just Bill Scoville, a rich kid from Philadelphia, struggling through college, trying to decide what to do with his life. He was smart and ambitious, and possessed certain talents—he was, for example, good with his hands, loved tinkering with cars, taking them apart and putting them back together to understand their inner workings—but he didn't yet know what he wanted to do with these skills.

Deeper in the same box the letters had been hiding in there was a brown-jacketed photo album with yellowing pages, and in the album was a photograph of my grandfather. He was probably two or three, wearing one of those frocks that toddler boys wore in the early 1900s, reaching a tiny hand up toward the mouth of a live rattlesnake that had wrapped itself around the staff of his father's arm. (His father was an eccentric polymath—a lawyer, a writer of children's adventure books, and an amateur naturalist—and used to keep a variety of venomous serpents as pets until his wife forced him to stop.) Seeing that photo, I got a cheap little jolt, since the obvious symbolism of it—the snake and the staff—seemed to portend my grandfather's eventual choice of career.

Within a year of writing those letters to his mother, he had moved back to Philadelphia, enrolled in the University of Pennsylvania's medical school, and thrown himself into his studies.

Sometime around 1969, in the middle of taking some tests, Henry stopped and looked up at his examiners.

"Right now," he said, "I'm wondering, have I done or said something amiss? You see, at this moment, everything looks clear to me, but what happened just before? That's what worries me. It's like waking from a dream."

Henry often described his inner state that way: a constant feeling of having just emerged from a dream. There's a psychological term used to describe that feeling: hypnopompia. It derives from the Greek *hypnos* (sleep) and *pompe* (sending away), and it's a common sensation, experienced as our minds dispel our dreams and snap us back to reality. In Henry's case this feeling never went away.

As it happens, Henry's dreams, like so many other aspects of his life, were also a subject of intense scientific inquiry. He once spent several nights in a sleep laboratory at MIT, hooked up to sensors. Whenever he entered REM sleep, a researcher would shake him until his eyes opened and then ask him what he'd been dreaming about. In the end, the researchers never published their dream studies, in part because nobody could decide whether Henry was actually dreaming at all, in the normal sense, or whether he was even capable of dreaming, since many scientists consider dreams to be patchworks constructed out of recent memories. But some of the transcripts survived. In them, he usually just talked about the same sorts of things he liked to talk about when he was awake—boyhood recollections of a road trip to Florida with his parents, of shooting targets in his backyard, of fishing with his dad. Occasionally, he would talk about what he claimed were old childhood ambitions.

"Henry, Henry, Henry?"

"Oh!"

"Were you dreaming?"

"Yeah."

"What were you dreaming about?"

"I was having an argument with myself. . . ."

"About what?"

"What I could have been . . . I dreamed of Pennsylvania. I dreamed of being a doctor. A brain surgeon. And it was all quick. Flashlike, being successful. And living down that way . . . tall straight trees."

It's doubtful that Henry ever shared my grandfather's aspirations. Instead he probably just imagined himself into his shoes. Even from the depths of his perpetual hypnopompic murk, some part of Henry saw things clearly enough to know that the nearsighted child of an electrician and a housekeeper would have had trouble pursuing that particular goal. A decade after the dream study, fully awake, Henry explained to another scientist that he had eventually decided against becoming a neurosurgeon due to his poor eyesight.

"Because I know, in brain surgery"—maybe Henry gestured with one of his hands as though he were holding a scalpel—"that wearing glasses, these little . . ."

Maybe he made a slight twitch with his hand then, pantomiming the scalpel going a little too far, cutting a little too deeply.

"That person is gone," Henry said of his imaginary patient.

FOUR

THE BRIDGE

On September 21, 1930, just before dawn, a twenty-four-year-old man from Pennsylvania named Norman J. Terry snuck past the watchmen for the still-under-construction George Washington Bridge, then climbed to the top of the enormous steel tower on the Manhattan side, bypassing the elevator for the stairs. Once at the summit, six hundred feet above the Hudson River, he edged out onto one of the four steel cables that hung between the tower and its twin in New Jersey. The distance between the two towers was precisely 3,600 feet, but the cables were far longer than that; they hung down in gentle parabolic arcs, creating the bottom halves of enormous bowls that, if extended into complete circles, could have swallowed the Empire State Building whole. Each cable contained 26,474 pencil-thin strands of steel, forged at a factory near Trenton and woven together into one mammoth braid three feet in diameter. If you unspooled the individual wires in all the cables, you would have a length of metal rope long enough to circle Earth four times or reach halfway to the moon, depending on your taste in superlatives.

As Terry crept down the cable, holding on to the two guide ropes at waist height for balance, the intricacies of the braid beneath his feet would have been invisible to him. Workmen had slathered the whole thing with a thick, hard layer of zinc paste to protect it from the elements. But still: The sheer mass of the cable must have been obvious, sufficient to hold up a quarter of what was then the world's largest bridge. A foot-long slice of the cable weighed nearly two tons. And not only was it massive, it was supple. The bridge had been designed to give. If a gust had stormed down the Hudson from upstate New York at that moment, Terry would have had the unpleasant sensation of standing atop the world's thickest braid of steel as it swung gently toward the sea.

Hundreds of feet below, a handful of people in two small speedboats peered up toward the cable. Terry and his manager had contracted with the New York *Daily News* to allow exclusive newspaper coverage of the event now unfolding. They'd also contracted with an independent motion picture company, so there were both still cameras and movie cameras on the boats. A couple of the young man's friends were out on the water as well, though they were only there to watch. At first nobody could see much of anything. They craned their necks, gazing upward, dwarfed by the monumental tower with its ten million pounds of exposed girders rising like God's own Erector set from the murky river. Then eventually a small gray dot presented itself as a silhouette against the morning sky, inching slowly down the cable.

When he reached the center point between the two towers, the convex of the bowl, Terry stopped. The bridge was designed to sag in the middle, and engineers believed it would come within approximately 196 feet of the surface of the water under a full load of cars. On this morning, however, more than a year before opening day, with no traffic at all, the center of the bridge was as high as it ever would be, at least 207 feet above the river. Terry looked down at the water below. The cameras in the boats were snapping and whirring, but he couldn't hear them. He removed his street clothes and stood there for a few moments in nothing but a swimsuit.

And then he dove.

Norman J. Terry was a daring man and had already performed all sorts of remarkable feats during his short life. He had learned how to jump between the wings of biplanes and had once tightroped between the Carbide & Carbon and Mather skyscrapers in Chicago, more than five hundred feet above the pavement. Just a few months before this particular morning, he'd hung from the undercarriage of an airplane, waited till it swooped ten feet above the ground, then let go, landing safely.

His dive, at first, was perfect. He swanned forward, stretched his arms above his head, pressed his legs together, and began a smooth downward arc toward the water. Soon he was approaching the river straight and true, at an accelerating rate of speed. The *Daily News* photographers fired off as many shots as they could and hoped the negatives wouldn't turn out too blurry. One hundred feet he'd fallen, 125, 150, 175.

Even today, eighty years later, the world record for the highest high dive is 177 feet. Terry passed through that mark and kept going, faster and faster: 180 feet, 190 feet. And then, according to the witnesses in the boats, something happened. His body, rigid, determined, confident, suddenly lost its composure. In midair, still fifteen feet or so above the water, he crumpled. Instead of staying straight, he bent in the middle and toppled forward, the beginning of a somersault.

What was going through his mind in that moment? What caused that loss of poise? Had he suddenly realized that he had gone too far? That his ambitions had finally oustripped his abilities?

He made it halfway through the somersault before landing on his back. The impact of the Hudson River against his skull knocked him unconscious in an instant, but worse was the impact against his spinal column, which ruptured his vertebrae and severed his spinal cord and prevented his unconsciously toiling brain stem from delivering all its usual life-sustaining imperatives, chief among them: Breathe.

By the time Norman J. Terry's friends had dragged him out of the water and delivered him to the hospital, he was already dead.

Strangely, considering its outcome, Terry's daring inspired copycat climbers. A few weeks afterward, *The New York Times* reported that a group of teenagers, hearing of the feat, had dared each other to climb to the top of the bridge and that two of them had done it. They were arrested on the way down. Surely there were others who managed to climb and descend undetected, but I only know of one for sure. As it happens, he was, like Norman J. Terry, another twenty-four-year-old man from Pennsylvania.

My grandfather was in his second year of medical school when he made the climb. There were no cameras, no police, no friends. He hadn't told anyone of his plans, and it's unclear whether those plans gestated for long before he acted on them. Perhaps he was just in the area, saw the opportunity, and seized it, hoisting himself onto one of the cables and beginning to climb. The climb would have been easy at first, almost parallel to the ground. And then the angle would have grown dramatically steeper. As the cable flared upward away from the island and the bridge extended out over the river, I imagine him climbing faster and faster, using both his legs and his arms to power himself up toward the tower's peak. It was nighttime, and the river below would have been a dark morass. I imagine he kept his mind on the climb, on placing each foot in front of the other, on gripping the guidelines hard enough to keep him steady but not so hard that his hands would cramp, on his breathing and his balance.

Eventually he reached the top of the cable and stepped off onto the flat roof of the tower. There were construction crates and bits of steel and coils of wire. Whenever the wind blew, he would have felt the tower sway beneath his feet. He had planned to climb up and then down right away, but now, up on that precipice, in the darkness alone, he couldn't bear to move. He sat down instead, placed his back up against something solid, and shivered in the cold.

One of the things our brains do, constantly, unconsciously, whether we like it or not, is make connections. They make connections in the literal sense, in that our neurons are promiscuous, always reaching out with their yearning axons to bond with other neurons. They also make connections in the figurative sense, in the way we're all familiar with, provoking endless little leaps of time travel during our daily lives. A few molecules of a certain burnt coffee bean adhere to the sensory neurons that project to your olfactory bulbs and build an instant and fleeting bridge to the last moment you smelled the same thing, in another town, city, year. A woman walking past you on a busy street talking on the phone suddenly laughs at something she hears, and her laugh—bracing and unself-conscious—conjures up an ex you've avoided thinking about for months.

When I think about my grandfather up on that tower, my mind yanks me back to a parallel night of my own. My memory of that night starts with me creeping along a narrow path through a lane of crypts, toward a breach in a high stone wall that I knew would give me access to the Giza Plateau, that otherwise well-guarded expanse of natural desert and man-made mountains on the outskirts of Cairo. Once I was out of the cemetery and onto the sand, however, there was nowhere to hide. About a quarter mile separated me from the Great Pyramid, and I started walking as fast as I could, passing tents that belonged to the families who rented scraggly horses and ludicrously pom-pommed camels to tourists in the daytime. A small clump of barking dogs materialized out of somewhere and started running toward me. *"Emshee!"* I shouted, using one of the too few Arabic words I'd learned during the year I'd so far spent in Egypt. "Go away!" They kept their distance and slunk off as soon as I got to the top of the plateau.

There are six pyramids in the Giza Necropolis, but only the three biggest ones are referred to as "great" and only the biggest of those three is known as the Great Pyramid. And it is. It's great in the way no photograph can prepare you for. Over the course of a life, you accumulate internal templates of what is and what can be. You know what a building is. You know what they look like, roughly, and you know their range, in size and spectacle. Then you see the Great Pyramid and your idea of what a building can be explodes in an instant.

Which isn't to say that we haven't done our best to diminish it. The diminishment began after the Mohammedan invasion of Egypt, in the seventh century, when its smooth white limestone carapace was stripped away for use in mosque building, leaving its flanks rough and terraced. Much of its subsequent uglification was at least well intentioned. Modern Egyptians no longer willfully destroyed their country's wonders; they just smothered them with unnecessary adornment. (See again: those dismal, pom-pommed camels.) In the case of the Great Pyramid, for example, a shabby museum squatted right at its base, twelve years old and already decrepit. Looking at the museum, and at the masterpiece that reared up behind it, I had a hard time not feeling that humanity's grasp of architecture and workmanship had declined steeply in the past four and a half thousand years.

And then there was the light show. Every evening, at eight P.M., the Egyptian Supreme Council of Antiquities mounted a psychedelic spectacle on the Giza Plateau. A few hundred foreigners sat in a hemicycle of seats in an outdoor amphitheater below the Sphinx. Then, for forty-five minutes, the tourists would be presented with a ridiculous display of Pink Floydian lasers and strobe lights projected against the side of the Great Pyramid, while an Attenborough clone intoned factoids and dry-ice smoke billowed out from somewhere.

But that night I was thankful for the garishness. I'd timed my visit for the beginning of that evening's show, and when it kicked in and I saw the colored lights and lasers begin to splash up against the southern end of the Great Pyramid, I ran across the remaining distance separating me from the western side. I had counted on the light show to distract and partially blind the guards who patrolled the plateau, and as soon as I reached the pyramid I hoisted myself onto the first block and began to climb. Each block was four or five feet high, and each ledge was between two and three feet wide. I climbed quickly, keeping my eyes up as much as I could. The pyramid itself filled most of my field of view, but at its edges the sky pulsed with migraine colors.

Within ten minutes I reached the final tier, just below the apex, 455 feet above the sand. Before it lost its limestone sheath, the Great Pyramid had ended in a point, one that was itself a miniature pyramid, rumored to have been made of solid gold. But now it ended in a square, flat summit, about fifteen by fifteen feet. In the center of the summit was one last bit of unnecessary ugly: a large metal tripod, bolted into the rock, aiming skyward, meant to remind people viewing the pyramid from below that, yes, it had once been pointy. I pulled myself up onto the summit, staying low so the light show's audience wouldn't see me. Then I lay back on the stone and looked up at the throbbing sky.

I'd been in Egypt for a little more than a year, and that evening was a sort of self-conscious commemoration, a moment to reminisce about where I'd been and to think about where I might be going. I was twenty-four years old, and my résumé was a joke. I'd graduated from college with a degree in U.S. history, mainly because the other degrees I considered—English, economics, sociology—seemed more theory than facts, and my brain tank was ridiculously low on the latter. Since college, I'd had a few different jobs. I'd hauled furniture for Cheap Date Moving in Watertown, Massachusetts, bussed Walter Cronkite's table at the Docksider restaurant in northeastern Maine, and for the past six months taught English to Egyptian oil rig workers at the Cairo offices of a Kuwaiti-owned, American-named company called the Santa Fe International Corporation. I'd moved to Egypt on a whim, inspired by Lawrence Durrell's *Alexandria Quartet,* an intricately interwoven series of novels that told a single story from four different perspectives. One of the quartet's protagonists was an English teacher who had all sorts of romantic adventures in Egypt, and I figured maybe I could do the same. Beyond that, I had no idea what I was going to do with my life. Prior to my move, when people asked, I'd fallen into the habit of saying that I thought I might apprentice myself to a furniture maker, even though I had zero interest in carpentry and even less skill.

But a few months before the pyramid climb, I'd started doing something that felt different from my previous jobs. I'd started writing. The Egyptian roustabouts I was supposed to be teaching during their vacation days had mounted a campaign of passive resistance, refusing to show up for class, which left me drawing a paycheck in an empty conference room overlooking the Nile. I had lots of free time and used it to draft long letters home. Eventually I decided to repurpose one of my letters to my grandmother—it was about a day I found a dead body while rowing on the Nile—and submitted it to a thrice-weekly newspaper called *The Middle East Times.* Then, armed with my single clip, I decided that my next target would be the *New Yorker* of Cairo's English-language publications, a glossy monthly called *Egypt Today.* I walked into the magazine's offices, asked to speak with an editor, and somehow walked out with an assignment: a feature about the statues in Cairo's public squares.

In retrospect, that sounds like an incredibly boring story idea, and later the editor told me she'd given the same assignment to dozens of other people, that it was a sort of hazing ritual. But I did my best with it. I roamed the city, notebook in hand, soaking up as much information as I could on a topic I'd previously known absolutely nothing about. I ended up filling the feature with mini-profiles—a man whose job it was to clean the statues, another who slept on the street near a monument to a famous Egyptian poet, a Spanish diplomat who oversaw the installation of the bust of José Martí near his embassy—and, to my surprise, the magazine accepted the story. Not only that, they were short-staffed, so they offered me a full-time job as a writer. My contract with the oil company was almost over, and I'd been planning to return to the States. I'd been thinking of going back to the moving company, working myself up from grunt to driver. The offer from *Egypt Today* had made those plans moot.

I decided to stay in Egypt, to take the job. For the first time in my life, I felt confident about my future, and that confidence, that sense of heading in the right direction, cried out for some sort of dramatic commemoration. So there I was, up on that pyramid, under that laser-beamed sky, riding that unfamiliar surge of optimism. Eventually the lasers blinked off, and the darkness descended. The tourists left their seats, and I stood and looked out over the smog-choked lights of Cairo. A light wind carried a fine grit of desert sand and the faint sound of car horns. The pockmarked Sphinx, 455 feet below, seemed as small as a kitten, and the stars above were nowhere to be seen. The Great Pyramid had been there for 4,600 years, and on that night, for a few hours, it was mine alone under my feet.

My grandfather sat shivering on the top of the George Washington Bridge. Later he would speak about the climb many times to many people. He would speak about the cold, the height, the fear, and the long night spent waiting for dawn, hiding in a packing crate. But he never, that I'm aware of, spoke about *why* he'd decided to climb in the first place.

I think I know.

He was twenty-four years old and in the process of burying his old self, the one who'd worried about always being a failure. He was halfway through medical school and discovering gifts he'd never known he had. He was learning how to make sick people better, and he was learning that this was something he could not only do but do well. He was newly confident that he was on the right path, and this feeling, this revelation, cried out for commemoration.

Of course this leap I'm making might be bullshit, but I'll defend it. There is only one way we tell stories about other people, and it's the only way we've ever told stories about other people. We find the connections between us and them, and then we use those connections as a bridge. Sometimes the connections are solid, buttressed by primary sources and interviews and every other sort of documentation you can imagine. Sometimes they're tenuous. Sometimes they're as ephemeral as two young men, divided by two generations, climbing two monuments for what the younger man believes must have been the same reasons.

My grandfather waited on top of the bridge until dawn. Maybe he slept at some point. I doubt it. I imagine him sitting there watching the lights of the city below until they were eclipsed by the light of the brightening sky. Eventually he stood and walked back to the cable and stepped out onto it. Like the man who'd jumped months before, my grandfather couldn't see the thousands of tiny filaments that united to form this one massive rope of steel beneath his feet. He didn't need to see them. He just needed them to sustain and hold him while his body coursed with adrenaline and endorphins and the sun began to warm him. Inside his skull, new neuronal connections were made, fresh impressions conveyed across yearning axons destined to imprint themselves as a memory trace that would reside there, inside him, until his last day.

One more thing: The length of all those thin strands that made up the thick cables of the George Washington Bridge—108,000 miles—not only could circle Earth four times or stretch halfway to the moon but also happened to match, almost exactly, the combined length of all the axons in the average adult human brain. Of course, they are very different, axons and steel. In the bridge, the wires came together in lockstep formation, pooling their strength in accordance with some civil engineer's formulas. In the brain, as my grandfather would have already learned during his first two years of medical school, axons came together and fell apart in infinite patterns that changed from microsecond to microsecond, conveying and creating everything we know and feel, believe and remember. He would have learned that there was no formula to describe what axons did, that they were the wiring of the brain, and that this wiring was embedded within small circuits, and those small circuits were embedded within larger ones, and the brain as a whole was made up of entire constellations of interlocking circuits. He would have studied these things and learned that a working human brain was so dauntingly complex that most of what we knew about it came from studying brains that *didn't* work, brains whose infinite circuitry had been disrupted and whose basic functions had been stifled or altered in a variety of interesting ways.

My grandfather held the guide ropes at either side of him and put one foot in front of the other. He must have felt proud. The bridge, even in its inchoate state, had already been extolled by *The New York Times* as a modern wonder of the world—"a symbol of man's mastery over matter, not less convincing than Cheops' pyramid itself"—and he had just conquered it. I imagine him, as he began his descent, freshly emboldened, ignoring the dark waters hundreds of feet below, ready to seize the opportunities that the world would soon tilt within his grasp.

FIVE

ARLINE

Two twenty-six, twenty-six.

February 26, 1926.

Henry's birthday.

There it was, right on his birth certificate. The certificate sat on a black velvet pedestal, inside a sealed rectangular box made of plexiglass. There were about a dozen other items on display, too: pictures, postcards, a pair of thick-lensed, thick-framed glasses, a Social Security card, a child's drawing, a tiny silver crucifix. A sign on the display said, PERSONAL BELONGINGS OF H.M. and gave thanks to the scientist who had lent the items for use in this exhibit. The exhibit was in the lobby of a building on the edge of the campus of MIT, one of the universities where scientists had spent decades studying Henry.

The largest photo in the display case was black-and-white, about eight and a half by eleven inches, and mounted in a chipped gold-painted wooden frame. It didn't have any sort of identifying label, but it was clearly a class photo. The kids were maybe in second or third grade. Thirty-seven of them, almost evenly divided between boys and girls. The girls all wore dresses, the boys shirts and ties.

I leaned in close and tried to spot him.

Between second and eighth grade, while the Molaison family lived in their light-housekeeping rooms in downtown Hartford, Henry attended St. Peter's, a Catholic school near Colt Park, where that bicyclist sent him crashing to the ground. The scientists who studied Henry often asked him about his time at St. Peter's, and he told them what he could.

"It was set off from Main Street," he said one afternoon in 1970, while sitting and smoking in an MIT laboratory, just a few blocks away from where that sampling of his personal relics would sit under plexiglass decades later. "Green grass in front of it. It was a brick building. Redbrick and white trim. White around the sill of the window. A basement and two stories . . . Had a window facing Main Street."

He remembered that there was so much noise from the traffic going by on Main Street that the nuns would often have to close those windows, and that at certain times of year this made things hot and stifling. They'd keep the classroom doors open so the air would circulate.

"There's one door who was forwards," Henry said, "and there was another door down in the back of the room, and both were open. And a draft could go through. Or a breeze."

He remembered that some of the younger students were scared of the nuns because of the imposing habits they wore.

"The black and the white," he said.

"Were *you* scared?" a scientist asked.

"No."

He told them that he'd been born a lefty but one of the nuns made him learn to write with his right hand.

"They changed me over," he said. "And my writing's never been good. But plain in a way . . . The sister . . . she was very . . . uh, wanted everyone to write one way."

"Did you complain about that?" the scientist asked.

He hadn't.

"Do you remember any of the kids? Any of the other children in the class?"

"Well," he said. "I think of one right off. I graduated with her. Her father was a cop at the time and went to become chief of police. And her name was Hallissey. Last name. First name was Arline."

A few minutes later, the scientist asked him the question again.

Did he remember any of the other children at the school?

"I think of one right off," he replied. "A girl. Her father is, uh, was, on the police force then. . . . Hallissey's the name."

"She was in the same class as you all the way through grade school?"

"Yes. Yes."

"And what was her first name?"

"Arline," he said.

Every time they asked him to recall another student at St. Peter's, he brought up Arline first. They had both attended St. Peter's from second through eighth grade, he said, though they hadn't always been in the same classroom. There were two classrooms for each grade, and school administrators would sometimes mix the kids as they rose through the grades. "Take some and just shift them around," Henry explained, "to make sure they wouldn't stay all together all the way through school." Still speaking about Arline, Henry recalled how he felt at the end of each school year, "hoping a certain one would stay with you . . . that they would be shifted, too."

Henry Molaison, it seemed, once had a crush on Arline Hallissey.

The walls of the classroom in the old photo were two-toned, darker on the bottom, lighter on top. There was a slate up high on the rear wall, and someone—a nun?—had written the ultimate imperative sentence across it in chalk: *Keep My Commandments.*

I studied the faces of the kids. The image had a shallow depth of field, and the faces of the students near the front of the room were in sharper focus than those in the back. In the upper left of the photo, his back against the wall, there was a boy, slightly blurred, with ramrod posture and thick-framed glasses. I thought that was probably Henry.

I looked at the girls and wondered which was Arline.

. . .

"Hello?"

"Hello. Is this Mrs. Pierce?"

"Yes."

"Hi, Mrs. Pierce. I'm calling because I'm working on a book about a man I believe was a former classmate of yours. Did you attend, back in the late 1930s, early 1940s, St. Peter's school?"

"Right, right."

"You did?"

"Yeah, I graduated from there."

She had married twice since then, had twice changed her name. And the people who transcribed Henry's laboratory interviews had usually misspelled her first and maiden names. Still, I was pretty sure.

"Your father was the police chief for a while, wasn't he?"

"Right," she said. "In Hartford."

"Okay, then I think you must be the Arline Pierce I'm looking for! The man I'm writing about, I don't know if you remember him or not. He passed away. But in interviews he recalled you as a childhood friend of his."

"I bet it was Bill Farrell!"

"It wasn't Bill Farrell."

"Oh, okay."

"It was a boy named Henry Molaison."

There was a long pause.

"I do not . . . see, that name . . . I do not remember."

Arline told me about the boys from St. Peter's that she *did* remember. Bill Farrell and William Brady. They both went on to become Catholic priests. One became a monsignor. "So they really made something of themselves. I remember the two priests because they were little devils. And I said, my God, how can you become priests?! You oughta be ashamed of yourselves. That's the only reason I remember them so vividly."

I told her more about Henry. I told her he'd worn glasses, had blond hair, and lived on Main Street, just a couple of blocks from the school.

"Now, wouldn't you think I'd remember?" she said. "That name, that boy . . . but I can't put my finger on it, you know? When you get to be eighty-seven, you forget."

"Henry Molaison," I repeated, first using the usual pronunciation—*Mo-luh-son*—and then trying an alternate one: "Or Henry *Mo-lie-ah-son*?"

"Honey," she said, after another long pause, "I don't remember that name. I truly don't."

One day in the summer of 1980, in a testing room at MIT, a scientist asked Henry to remove his shirt. Henry was fifty-four years old and getting a little flabby around the midsection. In the previous eight years, the scientist noted, he had begun to overeat, and his weight increased from 178.13 to 194.66 pounds. In fact, on a different visit to the lab, the scientist attempted to quantify Henry's increasing appetite and found that he was capable of eating two multicourse complete dinners, one after another, without reporting a feeling of fullness, though he would abstain from the salad on the second round. She thought his amnesia had something to do with his gluttony—it's easier to eat a second meal, after all, when you don't remember the first—but she wasn't sure. Now, during this visit, the scientist was intent on measuring something else: Henry's capacity to endure pain.

She applied six circles of india ink to his chest, each two centimeters in diameter. Then she picked up her dolorimeter. Although the device's name implies that it measures pain (*dolor* is Latin for *pain*), this is not exactly correct. What it does is cause pain. Its utility lies in the fact that the pain it causes is discrete, standardized, and adjustable. The dolorimeter looks like a gun, and there's a one-hundred-watt bulb inside it. The radiant heat energy from the bulb is directed and focused down the barrel through a series of lenses. Once the dolorimeter is activated, the heat at the tip of the barrel can be adjusted upward from zero in increments to a maximum of 370 millicalories, enough to cause second-degree burns.

She sat Henry in a chair and then applied the dolorimeter to his chest in the middle of one of the circles. During each application, she would hold the device to Henry's skin for three seconds and ask him to rate the intensity of the pain he was experiencing. During her tests with control subjects, she found that they tended to rate zero millicalories as "nothing," 90 millicalories as "warm," 180 millicalories as "very hot," 280 millicalories as "very faint pain," and 320 millicalories as "very painful." At the 370 millicalories setting, all of the control subjects found the pain to be so intense that they couldn't endure the three seconds of required contact and instead flinched away from the dolorimeter within fractions of a second.

Henry was different. He reported no pain at all through the dolorimeter's entire initial spectrum. Then the scientist cranked it all the way up to 370 millicalories and pushed the tip against his chest. Henry sat calmly as she held it there for the full three seconds, even as his skin began to burn and turn red.

"Unlike the control subjects," the scientist wrote later, "H.M. did not label any stimuli painful, no matter how intense it got."

The scientist speculated that Henry's high pain tolerance had something to do with the lesions to his brain, though again she couldn't say for sure. Other amnesic patients she had tested in similar fashion experienced normal pain sensitivity. What made Henry different?

Henry put his shirt back on and returned to the room he slept in during these visits to the laboratory. There, he likely pulled out a crossword book, picked up a pencil, and began quietly working one of his puzzles.

If Henry Molaison had a boyhood crush on Arline Hallissey, it was an unrequited one.

Later, after the operation, Henry never had a crush on anyone again.

The holes my grandfather dug in Henry's brain caused many deficits, some brutal and stark, some more subtle. Among the things he lost, according to the scientists who studied him, was a capacity for desire. As far as they could tell, in the six decades between his operation and his death he never had a girlfriend, or a boyfriend, never had sex, never even masturbated. The returning strangers who flitted in and out of his life, the movie stars who flickered on his television, he received them all with perfect neutrality, and they left behind neither traces of memory nor pangs of lust.

"The operation," one of the scientists who studied him concluded, "rendered him asexual."

There is no equivalent device to the dolorimeter to administer emotional as opposed to physical pain, but if there were, many of the people who studied Henry believed, he probably would have displayed a similar numbness. He was known for his outward placidity, for his equanimity, for the flatness of his demeanor. He tolerated whatever the scientists wanted to do to him without complaint. Patient H.M. was, above all else, patient. On the surface, at least, he rarely seemed troubled, even when confronted with troubling facts. After his parents had both died, for example, the scientists would often ask him about them to determine whether even such massive losses had made an imprint in his mind. In a characteristic exchange in 1986 two weeks before Henry's sixtieth birthday, five years after his mother died, and almost two decades after the death of his father, Henry had the following exchange with an MIT researcher:

Researcher: Where do you live now?

H.M.: In East Hartford.

Researcher: What sort of place do you live in?

H.M.: Well, I think of a house. A private house. But I can't think of the name of the street.

Researcher: Who lives with you there?

H.M.: Well, right off I think of my mother.

Researcher: Your mother?

H.M.: I'm not sure about Daddy.
RESEARCHER: You're not sure about your father?
H.M.: I know he was sick. And. But. I'm wondering if he has passed away?
RESEARCHER: I think he has passed away.
H.M.: Because he was sick before that and he was down to . . . He had to go to a hospital down in Niantic. Not Niantic. Uh. Mystic.
RESEARCHER: Mystic? He went to a hospital there?
H.M.: He went, well, uh. Like, TB.
RESEARCHER: That's not very nice, is it?
H.M.: No, because he was down there for, well, quite a spell.
RESEARCHER: But you don't know if he passed away?
H.M.: No.
RESEARCHER: I think he did.
H.M.: Gustave.
RESEARCHER: Hmm?
H.M.: His first name. Gustave.
RESEARCHER: That's your father's first name?
H.M.: Yeah.
RESEARCHER: I thought his name was Henry.
H.M.: No. It was Gustave Henry.
RESEARCHER: Oh, I see, what's your second name? Do you have a second name?
H.M.: Yeah, Gustave.
RESEARCHER: Your name's Henry Gustave. Ah. That's easy to remember!

Throughout the exchange, Henry's tone never changed. He spoke in the same soft, gentle, halting voice that he used when discussing almost anything and never betrayed any obvious sadness when the scientist confirmed to him, for both the thousandth and the first time, that his father was dead. One day during another similar session, Henry carefully wrote a note to himself on a scrap of paper that he began carrying with him in his shirt pocket wherever he went.

"Father's dead," the note read. "Mother's in a hospital, but she's well."

Arline Hallissey went to work, like her father, at the Hartford Police Department. She became one of the city's first meter maids. She remembers walking her beat, looking at the parked cars, noting the baby seats or the coffee cups or the paperwork or whatever else was visible inside and trying to imagine who the cars' owners were, what their lives were like, where they worked, how many children they had, their dreams and ambitions.

When I spoke with her she'd been retired a long time. She wasn't in great health. She sounded frail. She told me she had small seizures sometimes. It was something the doctors hadn't been able to pin down.

"I'm perfectly normal otherwise," she said. "But you know what the brain does. It does what it damn pleases."

Most of us have brains that work tirelessly, in dreams and in wakefulness, following their own mysterious routines. They absorb and process our experiences, they present and preserve our own personal worlds. Sometimes they grow ill, sometimes they grow cancerous, sometimes they simply grow old. But most of them do so according to their own mandates, far beyond our control. Most of us have brains, in other words, that do what they damn please.

Henry Molaison did not.

From the day of his operation and until the day of his death, Henry's brain could no longer follow its own natural inclinations. Many of its essential circuits had been disrupted or sealed off, and the impact of this intervention was persistent and ubiquitous. After its surgical transformation, Henry's brain, rather than doing what it damn pleased, was able to do only what it could, under radically altered and diminished circumstances.

That transformation of Henry's brain is what transformed a forgettable boy from St. Peter's school, whose own boyhood crush didn't even remember him, into the unforgettable amnesic Patient H.M. Scientists would end up studying almost every aspect and implication of Henry's brain, and what they learned revolutionized brain science.

The story of what they learned from Patient H.M. and his incomplete brain, however, is itself greatly diminished without first learning the story of what led my grandfather to make those devastating, enlightening cuts. That story is a dark one, full of the sort of emotional and physical pain, and fierce desires, that Patient H.M. himself couldn't experience. It's a story that's never been told, and even now, fingers on the keyboard, I hesitate.

PART II
MADNESS

SIX

POMANDER WALK

There were people in the cellar. My grandmother could hear them. She had thought she was alone in the house, except for her children, who were asleep in their bedrooms. Now it appeared she was wrong. The children were asleep and my grandmother was not and she could hear people in the cellar.

She was terrified.

It was late January 1944, in a comfortable single-family home on Frankland Street in Walla Walla, Washington. My grandfather, as usual, was at work, this time on an overnight shift at the U.S. Army's McCaw General Hospital, where he served as chief of neurosurgery. Just the day before, he had come back from a weeklong conference in Spokane. They had been married for ten years, and it had always been like that, his career constantly pulling him away.

Their wedding took place during the final year of his internship at Hartford Hospital, and in 1935 they moved from Hartford to Ithaca, New York, where my grandfather became an assistant resident in psychiatry at Cornell University. From there it was on to Manhattan, where he completed a residency in neurology at Bellevue Hospital. A neurosurgery residency at Massachusetts General Hospital followed soon after, and from there, in quick succession, he completed residencies and appointments in three other hospitals and clinics, from Baltimore to Boston, honing his skills under James Poppen and Walter Dandy, two of the best neurosurgeons of the day, before returning to

Hartford in 1939, where he founded Hartford Hospital's neurosurgical department. Neurosurgery quickly became a calling, not just a job. "Neurosurgery was my great desire," he'd write years later. "I couldn't stand the meditation in Neurology and Psychiatry in those days. The technical perfection of Poppen and a summer with Dandy thrilled and excited me. How could anyone choose any other specialty; cardiac surgery—only a pump; thoracic surgery—only a ventilator; orthopedic surgery—such crude instruments; urology—a sewage system appealing to the more morbid followers of Freud." He embraced brain surgery—that most difficult, demanding, and consequential of medical specialties—with all his heart.

He had been a practicing neurosurgeon for five years when the U.S. Army called him up for active duty and made him a major in the Medical Corps. He began his military career stationed at Walter Reed General Hospital in Washington, D.C., though he'd often make long-distance house calls, like when he was once flown on a bomber to the arctic airbase in Goose Bay, Labrador, to treat a head injury in the middle of winter. In 1943, the army ordered him to report to the opposite coast, in Walla Walla. There he spent his days treating the battered soldiers coming home from the South Pacific. He was good at his work, and his work energized and elevated him. The self-doubt of his college years had evaporated, replaced by a new self-confidence and a wry, upbeat outlook on life. That year, when Yale solicited all members of the class of 1928 to submit a one-sentence summation of their guiding philosophies, my grandfather wrote the following: "We are simply a fancy edition of the little flowers and fishes and should not get as wrought up as Hitler has about it all."

Meanwhile, my grandmother managed. As her husband found his passion and ascended professionally, she tried to adjust to the demands of raising her family three thousand miles from home. They had three kids, each two years apart. By 1944 the oldest, Barrett, was eight, and the youngest, Peter, was four. My mother, Lisa, was in the middle, six years old. My grandfather ran his operating rooms and my grandmother ran the household, attending to the day-in, day-out practicalities: school, meals, cleaning. She managed, but with each kid it got a little harder, and the past year had been the hardest. My grandfather had left for Walla Walla two months before she had, leaving her to handle most of the work of moving. She and the children ended up driving across the country in a Lincoln Continental convertible. Then she settled into her new life, in this new place, raising my mother and my uncles pretty much on her own as before, although now she was doing it without her support system, her family and friends. In the past several months, her weight had dropped from 108 pounds to 89. She'd always been thin, but it had gone beyond that.

And now there were people in the cellar. Or, she realized, there *had* been people in the cellar. They had migrated upstairs, and suddenly she could hear them behind one of the closed doors. She recognized the voices. They belonged to some of the only people she'd made friends with here. People who lived in the neighborhood, not far away. What were they doing there? She listened to the voices, tried to make out what they were saying. She caught only vague snatches, but it was enough to convince her that something terrible was looming. Some sort of plot against her and her husband. Her friends, the ones who snuck into her house, were whispering. She didn't dare open the door. The night passed slowly.

When she awoke the next morning, she wondered whether any of it had been real. Had she been dreaming? There hadn't really been people in the house, had there? At six A.M., before the children were awake, she slipped outside and rode her bicycle a few blocks to the house of the whispering friends. She rang the doorbell and waited for them to answer, to make sure they were there and not hiding somewhere in her own home. When they answered she hurriedly said she had to go, couldn't stay, and then she got back on the bicycle and rode back to start preparing breakfast. It was Saturday, and the children didn't have school, so she needed to tend to them.

As the morning unfolded, she called her husband three times. Each time she was told that he was busy performing an operation and couldn't call back till later, but something inside of her told her that this was a lie. They were lying to her. Her husband was not in surgery. Her husband was in jail. He had been court-martialed. They had him.

She did not know what to do, but she knew she had to do something. She tried to keep busy. She went upstairs, she went downstairs. She watched her children play. She noticed her youngest, four-year-old Peter, was not playing. He was not playing, and he was looking at her. He was looking at her and making motions with his hands. He was trying to tell her something without words. Just as she did the night before, when she tried to decipher the whispers behind the door, she tried to decipher Peter's hands. At first she didn't understand, but then suddenly she did, and she knew what she had to do.

She gave her children their lunch, then sent them off to play with some friends. Once they left, she walked to a neighbor's garage and went inside. There was a car in the garage, and she clambered up on top of it. She removed the shirt she was wearing. She began to tear the shirt into long strips of fabric, then tied the long strips of fabric together. This was what Peter had been pantomiming with his hands. This was what he had wanted her to do.

When she was done, she reached up and tied one end of the rope she had just made to a beam in the garage and placed the other end, the noose, over her head.

A neighbor found my grandmother standing there, half undressed, trying to kill herself with her handmade rope. The neighbor placed a phone call. My grandfather came home from the hospital at about three P.M., and my grandmother began telling him everything, about the plot and the court-martial and the whispering friends, about how what she had tried to do was for everyone's good, that it was the only way for her to stop the darkness from spreading. He listened to her, shocked. There had been no signs. None that he had seen, anyway. None that anyone had seen, for that matter. It was as though he and his wife had inhabited the same world all of their lives together, and then, with no warning, she departed for a new one. A world completely invisible to him.

They went to the hospital together. A psychiatrist interviewed her. They checked her in to an open ward. My grandfather stayed with her that night in her bed, while someone stayed home with the children. She was affectionate with him, imploring him to make love to her, but then became agitated, emotional, confused. In the middle of the night she got up out of bed and wandered into another room on the ward. There was a woman lying in a bed there. My grandmother stood at the bedside for a moment looking down at her.

"Why do you lie there in women's clothing?" she said, and slapped the patient across the face.

They restrained her. They took her to a closed ward, one where she was not free to wander. They kept her there for two months while my grandfather made arrangements for a longer-term solution. Her condition varied from day to day, week to week. She cried for hours and hours. She reported that someone had killed her children, that their heads were buried in the walls of her room. She reported that events too terrible to describe were taking place just beneath her floor. For the first ten days she refused to eat, claiming that the food they brought was contaminated with white lead. When my grandfather visited, she accused him of selfishness, of sleeping with other women, of giving her nothing but loneliness since the day they married.

On March 25, 1944, an ambulance took her from the closed ward to the train depot, and soon the entire family boarded a train for a three-day cross-country trip back home to Connecticut. The children were not aware of what was happening. They knew their mother had been somewhere else, and they knew they were now being tended to by a cartoonishly stern nanny, Mrs. Thornfeldt, but the details, the whys, were out of reach. A neighbor had given them a large bag filled with packages to take on their trip. They'd open a new one each morning, revealing a fresh gift for every day on the rails. They knew their mother was on the train with them, but they didn't know why she'd been confined to a locked drawing room. My grandfather visited her in the drawing room. He watched her chain-smoke. He watched her eat. Eventually he gave her several doses of Luminal and watched her sleep.

Three days after leaving Walla Walla, the train pulled into Hartford. A car was waiting for my grandmother at the station. It was just a short drive from there to the asylum.

In 1858, the president of the Connecticut State Medical Society gave a speech in which he recalled a scene from his childhood, fifty years earlier: "Travelers passing through the town of Farmington on the road to Hartford," he said, "would observe a little cage set in a bank near the turnpike, occupied by a raving maniac, staring and shouting to the passing travelers; subsequently he was removed to a barn nearby, where he sat crouched on his limbs till they inflamed and adhered together so that he could not be straightened. Here he sat year after year, covered over with an old blanket, and had his food given him as it was to the chickens in the barnyard. This is not mentioned as a reproach to the good people of Farmington, there being few towns whose inhabitants have a more enviable reputation for morality and true religion. Other cases perhaps as revolting existed in other towns."

His point was that those days had passed and that mentally ill men and women in Connecticut were no longer locked in roadside cages or, as had also once been common, bought and sold as cut-rate slaves, human damaged goods. Instead they now had a place to go. A welcoming place, one where they could experience a sort of treatment far removed from the hell on earth they'd once had to endure. That place was my grandmother's asylum, which was then known as the Hartford Retreat for the Insane. Ever since its founding in 1822, the institution had presented itself to the world as being among the most progressive asylums on earth. Its first superintendent, in a "Report of the Committee Respecting an Asylum for the Insane," which was delivered to the Connecticut General Assembly just prior to the opening of the institution, declared that the Hartford Retreat aspired to be "the reverse of everything which usually enters into our conceptions of a madhouse." Instead of just being a sort of jail where "the unfortunate maniacs are confined," his asylum would be dedicated to the "moral management" of the inmates.

In many ways, he succeeded. Patients at the Hartford Retreat, unlike the unfortunates who might have once wound up shackled in the dungeons of Bethlem or forgotten in the back ward of Bellevue, were given a chance to lead almost normal lives. They could walk the verdant grounds, eat communally in a gracious dining hall, attend religious services, and even participate in the drafting of the world's first newspaper ever produced in an asylum, *The Retreat Gazette*. The governor of Connecticut was a subscriber to the *Gazette* and remarked that the publication was "more rational than many papers I have lately seen, emanating from quarters much less suspicious so far as the seal of public reprobation is concerned."

The asylum didn't offer many treatments during those early years. Its founders believed that the best thing to do for the mentally ill was to simply provide them with a stimulating but pleasant environment, full of time-consuming but not too taxing activities. They hoped that if patients were kept busy while being sheltered from the stresses of their everyday lives, their problems would take care of themselves. Patients dressed in street clothes, as did their attendants, and unless residents were particularly violent or uncontrollable they were not confined with chains or anything else. Charles Dickens, who visited the asylum as a journalist in 1842, was struck by this. "I very much questioned within myself, as I walked through the Insane Asylum, whether I should have known the attendants from the patients, but for the few words which passed between the former, and the Doctor, in reference to the persons under their charge. Of course, I limit this remark merely to their looks; for the conversation of the mad people was mad enough." He goes on to recount one such conversation:

> There was one little, prim old lady, of very smiling and good-humoured appearance, who came sidling up to me from the end of

a long passage, and with a curtsey of inexpressible condescension propounded this unaccountable inquiry:

"Does Pontefract still flourish, sir, upon the soil of England?"

"He does, ma'am," I rejoined.

"When you last saw him, sir, he was—"

"Well, ma'am," said I, "extremely well. He begged me to present his compliments. I never saw him looking better."

At this the old lady was very much delighted. After glancing at me for a moment, as if to be quite sure that I was serious in my respectful air, she sidled back some paces; sidled forward again; made a sudden skip (at which I precipitately retreated a step or two); and said:

"I am an antediluvian, sir."

I thought the best thing to say was, that I had suspected as much from the first. Therefore I said so.

"It is an extremely proud and pleasant thing, sir, to be an antediluvian," said the old lady.

"I should think it was, ma'am," I rejoined.

The old lady kissed her hand, gave another skip, smirked and sidled down the gallery in a most extraordinary manner, and ambled gracefully into her own bed-chamber.

Dickens judged the asylum "admirably conducted" and cracked a joke when he learned that one of the inmates he met heard "voices in the air."

"Well!" thought I. "It would be well if we could shut up a few false prophets of these later times, who have professed to do the same; and I should like to try the experiment on a Mormonist or two to begin with."

Dickens was not the asylum's only admirer. One of the institution's later superintendents, writing of those early years, stated that "there apparently was no criticism of the hospital or the quality of care given to its patients," though he did add the following caveat: "There were old wives' tales about the supposed occult happenings in the hospital, of course, but that is not uncommon for mental institutions. People have difficulty understanding the nuances of mental disorder, so fantasy often takes the place of fact in their minds and all too often in their conversations." Another observer, a doctor who admired the gentle and largely hands-off "moral treatment" given to even the asylum's most hopeless cases, wrote that "it is only, as it were, twining fresh flowers on the graves of the dead, still it is a grateful sight to the humane, and a more certain indication of high civilization than the most refined taste in literature and the arts or the most fastidious of social etiquette."

By the time my grandmother passed through the gates of the asylum, a hundred years after Dickens, things had changed, and the moral treatment that was at the heart of the institution's original mandate had transformed into something very different.

It was beautiful, that much remained the same.

The car drove through the gate and up the gently sloped driveway that led to the asylum's main building. If she was looking out the window, my grandmother would have seen carefully tended flowers and bushes and lawns glide by. Frederick Law Olmsted, America's greatest landscape architect, the man behind Central Park, had overhauled the grounds in the late 1800s, seeding it with a harmonious assortment of ginkgoes, sugar maples, black walnuts, sweet gums, copper beeches, and the only living pecan tree this far north. The vehicle rolled past elegant cottages and a few larger dormitories, then pulled to a stop in front of the oldest structure on the campus, the palatial administration building. If my grandmother was feeling well enough to exit the car without assistance, Pete Souza, a Portuguese-born sailor who had worked as the asylum's head porter for almost a decade, would have been there to grab her bags and show her inside, where a receptionist sat behind a desk, waiting to check her in. From there, Souza would have escorted her to her room, which was tastefully appointed. On her bedside table, a member of the housekeeping staff had probably left a copy of a slim magazine called *The Chatterbox*. There was an Easter Bunny on the cover, peeking out through a thicket of lilies. If she picked the magazine up and flipped it open, she would have found a notice on the opposite page advising her that the publication was "prepared fortnightly for the Guests of THE INSTITUTE OF LIVING."

The name was new. One year earlier, the asylum's superintendent, Dr. C. Charles Burlingame, had successfully petitioned the Connecticut General Assembly to have it change the facility's name to the Institute of Living from the Neuro-Psychiatric Institute of the Hartford Retreat, which had itself replaced the original, the Hartford Retreat for the Insane, in 1931. Burlingame had also requested that in all future references to his institution the word *asylum* should be replaced with *mental hospital,* and that any use of the adjective *insane* should be replaced with the words *mentally ill.* Words, appearances: These things were important to Burlingame. As he once wrote, he strove to make the Institute of Living a "meticulously normal place."

On the surface, he more than succeeded: When my grandmother flipped through the pages of *The Chatterbox,* she would have found pictures of a place that looked more like a country club than anything else. There was an outdoor swimming pool, a small putting green next to the Golf House, a volleyball court, even an indoor ten-pin bowling alley. She'd see advertisements for some of the goods available at Vauxhall Row, the asylum's own shopping mall: She could buy artisanal Ajello candles and Easter greeting cards at the Here-It-Is Shop, purchase a variety of different corsages—starting at ninety cents apiece—at the Vauxhall Row Flower Shop, or browse a selection of newly arrived bestsellers at Ye Royale Booke. She would find photographs of some of the guest-created handicrafts on display at "our little arts and crafts colony in Center Building," a collection of "avocational shops"—Ye Glaziery, Ye Silver Smithy, Ye Bindery, and Costume à la Main—

known collectively as Pomander Walk. (A pomander is a perfumed ball that can be suspended from a necklace. It was a popular fashion accessory in the eighteenth century, when doctors prescribed it to help ward off plague.) One of the asylum's annual reports stated that Pomander Walk strove "to create an entirely non-institutional atmosphere," and that it was "designed to simulate an early English road, with the shop exteriors constructed of half-timbers in the Elizabethan manner."

In the entire magazine there was no indication of what the Institute of Living really was, except for one place, on the fourth page, where the veil was drawn back a bit. There, a gentle admonition was printed next to a drawing of a bunny standing on its hind legs: DON'T TELL YOUR TROUBLES, it began, then continued: "Easter time is an excellent opportunity to reflect upon the trials of another, to whose sacrifice it is dedicated. Your doctors and nurses always are willing listeners but others may be spared by trying to emulate the spirit of selflessness that marks this season." Keeping the guests from discussing their mental states with one another was a part of the institution's efforts to maintain its meticulously normal appearance, and some variation of this advice was contained in every issue of *The Chatterbox,* often in rhyming form, as in the previous issue, where the following lines ran under a cheery illustration of a horn-playing cherub:

To Our Guest

Welcome to our guest!
To you, we make this one request:
When your symptoms you would discuss,
Tell no other guest, but do tell us.

THE DOCTORS AND NURSES

My grandmother violated the spirit of that dictate within hours of her arrival. That evening, instead of going to sleep, she removed all her clothes, pounded incessantly on the locked door of her room, and ranted in a loud voice. It was, at the very least, a disturbance to the neighboring guests. Attendants soon arrived. They helped her get dressed, took her outside, then escorted her across the campus to a building known as South One. There, that first night, she received her first glimpse of the dark reality that lurked beneath the asylum's placid surface.

SEVEN

WATER, FIRE, ELECTRICITY

South One was a white-tiled, dimly lit, and sound-insulated room full of exposed plumbing and valves. There were five large tubs in the room, and it was quite likely that at least one of them was occupied by another of the 355 guests currently residing at the Institute of Living. The empty tubs had been scrubbed clean with yellow soap, and one of these tubs was now prepared for my grandmother. A nurse unlocked a control panel on the wall and used the valves and knobs behind it to fill the tub with water. Given my grandmother's behavior that evening, it is likely that the nurse chose to make the water very cold, as cold water, and its attendant hypothermia, was known to have sedative effects on mammals. Aides stripped off whatever clothes my grandmother still had on, coated her body with oil, then placed her in the tub. If she struggled, and there is no reason to believe she did not, attendants forced her down and strapped a large sheet of heavy fabric over the tub, leaving a small aperture for her head, trapping the rest of her body beneath.

It is not clear how long they kept her there. A treatment of "continuous hydrotherapy" at the Institute of Living typically ranged from hours to days to, in exceptional cases, weeks. If this session lasted a long time, a staff member would remove my grandmother every four hours and "relubricate" her before returning her to the tub. This was to prevent her skin from becoming excessively dry. Every twelve hours, she would receive an alcohol rub.

Eventually the staff decided she was calm enough to return to the residential area, where a clinician noted that she was, for the moment at least, much quieter and seemed to be enjoying herself playing the piano.

Guests at the Institute of Living underwent an extensive series of interviews over the course of their stays, during which psychiatrists probed their life stories for clues to their present conditions. Summations of these interviews were then typed up and placed in the patient's file. On the first page of my grandmother's clinical notes, under the heading "Personal History" and the subheading "Intellectual and Social Development," a psychiatrist wrote the following:

> A bright child and adult. Culturally, she is above the average, received high marks at school and at college, graduated from Vassar with a B average. She has an excellent mind, is a great reader of all types of literature, she majored in music and shows considerable ability, both with the piano and slightly on the cello. She has a true love of music, without any sentimental forced feelings towards arts, literature, or culture. Delinquency—none. Neither a model child or any immoral tendencies. Adult intellectual level—markedly above the average, but in a completely feminine fashion. A good mind, a logical mind, an extremely sensitive mind, especially towards the arts and music.

My grandmother had an unusually privileged upbringing. She grew up in Manchester, Connecticut, where her mother's side of the family owned a business—the Cheney Brothers Silk Manufacturing Company—that, during her teen years, was the largest producer of silk outside of China and one of the wealthiest corporations in the United States. Manchester was a company town—one in four residents worked for Cheney Brothers, and the company's campus, which sprawled over hundreds of acres and included its own hospital, school, and power utilities, was in many ways a city unto itself.

The company peaked when she was fifteen years old and then began a slow decline, brought on by the rise of rayon and other synthetic materials. No matter: She wasn't going to be tasked with saving the company. Women had never played important roles in the family business. Her mother had ten siblings; most of them lived on the Cheney Brothers compound, and my grandmother could have gauged the relative importance accorded the sexes with a quick glance at the comparative sizes of her aunts' and uncles' cottages and mansions, respectively.

She adored music, always. After graduating from Vassar in 1933, she moved to Vienna to study the piano and spent a gilded year in the city. She embarked on an affair with a married concert musician and found employment as a pianist in silent-movie theaters, improvising emotional soundtracks to the oversize characters flickering on the screens above her. But the year came to an end, and she boarded a ship back home to Connecticut. Shortly after her return, while she was riding in the passenger seat of a friend's Model T, a dashing young medical student jumped onto the running board and introduced himself as Bill Scoville.

The Cheney Brothers Silk Manufacturing Company declared bankruptcy in 1937, selling off, among other things, its private railroad. At the time of my grandmother's institutionalization, however, the company was experiencing a brief resurgence, having scored a contract to manufacture the silk parachutes used by most American aviators during World War II. But I imagine that the ups and downs of the silk industry were far from her mind as she sat at the piano in the residential hall, temporarily subdued by her hydrotherapy, perhaps playing from memory one of her beloved Chopin pieces.

It was a moment of respite, a brief interlude, before her next treatment began.

One morning during my grandmother's second week at the Institute of Living, attendants woke her at six A.M. and served her tea and toast before escorting her to the Pyretotherapy Room on the first floor of White Hall, a brick building on the eastern edge of the campus. The room was medium-size, and a mural of a desert oasis covered the walls, palm trees and sand dunes and birds silhouetted against a blue sky. Alongside one of the walls, pushed up against a painted dune, there was what appeared to be a copper coffin with a semicylindrical lid. A nurse opened the top of the device, revealing a thin mattress inside. Once my grandmother had disrobed, she lay down on the mattress, and a blanket was placed over her before the lid of the device closed. Her body, from her neck down, was entirely inside. Her head protruded from a hole at one end and rested on a pillow.

An infrared lamp inside the device radiated heat, while a machine called an inductotherm monitored the temperature, allowing staff members to adjust it as desired. A fan circulated hot air inside the device. A pan of water sat near the fan, maintaining high humidity to prevent excessive skin dryness. Every fifteen minutes, a nurse took my grandmother's temperature. Humans are warm-blooded and can generally maintain a uniform body temperature of approximately 98.6 degrees regardless of their environment. Eventually, however, when exposed to relentless heat, the body's thermostatic ability breaks down and its internal temperature begins to rise.

That first day, within the first two hours of treatment, the attendants managed to induce a fever of 102 degrees in my grandmother and then adjusted the inductotherm to maintain that approximate level of fever for the remainder of the session. Along with her temperature, they monitored my grandmother's pulse and respiration, looking for signs of acute distress. By the time they let her out of the cabinet, she had been inside for a total of eight hours. She was allowed to rest for an additional hour inside the Pyretotherapy Room and then was escorted back to her residence. The next morning attendants woke her at six A.M. again, and the process repeated itself, only this time they brought her temperature up to 103 degrees.

By the fourth day of treatment, they were able to induce a fever in my grandmother of somewhere between 105 and 106 degrees.

Pyretotherapy, or fever therapy, had been one of the treatments offered at the asylum for at least a decade, though the mechanisms by which the fevers were induced changed over the years. In the past, nurses would inject patients with a strain of malarial parasite, giving them a "benign malaria" that caused high fevers. The electropyrexia cabinet used on my grandmother was meant to achieve the same results, in a more modern, controllable way. It had been installed at the Institute of Living five years before her arrival, and an issue of the staff newsletter from around that time boasted that when compared to the biological method it "produces equally good results."

Toward the end of her first full week of pyretotherapy, a clinician noted that although my grandmother was still delusional, rambling about "transmigration" and falsely identifying asylum guests and staff as "various different friends she has known in the past," she was on the whole "quite pleasant" and "quieter."

On the second page of the asylum's clinical notes about my grandmother, under the subheading "Mental Make-Up and Type of Personality," a psychiatrist described her as "entirely unaggressive," "extremely sensitive," "gentle and kind," and "utterly feminine." The psychiatrist then spent some time discussing her relationship with my grandfather and judged it to be a healthy one: "The marriage actually has been extremely happy and congenial," he wrote, "with the two being together constantly for 10 years, making many trips, skiing abroad, and sharing a book, art, etc. The husband has been happy in his marriage and at no time has ever considered or wished he was married to another."

My grandfather would have agreed with this assessment. In a letter he wrote to his parents on January 29, two days after the breakdown, he wrote, "I have been so happily married, and am utterly heartbroken."

My grandmother, on the other hand, had a different, more negative opinion of the marriage. According to her psychiatrist, this view was itself a symptom of her mental problems. "She is too idealistic," he wrote, "demanding too much perfection in her husband." My grandfather, the psychiatrist continued, "is truly devoted and loyal to her, but has upset her by mild promiscuity." Her husband's infidelity, my grandmother's psychiatrist concluded, "has upset her to an exaggerated degree."

The psychiatrist was clearly comfortable making certain basic judgments about my grandmother's backstory and modes of thought, but he did not pretend to understand the precise biological or psychological causes of her breakdown. Just as it had been a century prior, mental illness remained largely a mystery. As Charles Burlingame, the superintendent of the Institute of Living, put it, "psychoses can hardly be called disease entities, even now, but are regarded as manifestations of a disease process, concerning the real sources of which we can do little more than speculate at the present." What had changed, however, was the asylum's attitude toward the treatment of these mysterious illnesses. Whereas in the institution's early days, treatments were conservative and minimalistic, they had by my grandmother's time become aggressive and prolific: In Burlingame's view, "in psychiatry there should be no conflict between the various therapies." Each type of treatment, he argued, "has merit and should not be discarded," and the best treatment plan for the asylum's guests was consequently almost always a multivalent one.

My grandmother had endured hydrotherapy and pyretotherapy. She was still not well. A third treatment was prescribed.

Some days she would wake up and they wouldn't give her any breakfast. Not even tea and toast. This had been going on for months now, about three times a week, and she had learned what it meant. She knew what was coming. Or she knew that something was coming, something she dreaded. It was a vague, indistinct dread, though, since what was coming always caused a short-term amnesia and she'd never quite remember it the next day.

The aide would escort her from her residence to the building called Butler One. They would follow a special route between the two buildings, one that minimized the exposure of my grandmother to other guests in case she caused a commotion. This was not because my grandmother was particularly volatile. The same instructions applied to any guest being escorted to Butler One. Conversation was also to be kept to a minimum. Burlingame stressed these points in an employment manual that was provided to the aide shortly after his hiring: "In escorting Guests, do not discuss the treatments; try to be reassuring, considerate, quiet and pleasant. Securing the Guests' cooperation aids the Guest in receiving the maximum benefit from the therapy."

Once the aide and my grandmother arrived at Butler One, the aide would escort her inside to the Therapy Room.

"You will not," the aide's manual read, "discuss anything which transpires in the Therapy Room with the Guest or anyone else, except members of the personnel to whom you are responsible in connection with the care of the Guest to whom you are assigned. Under no circumstances will you give any details to the Guests themselves."

This is what transpired in the Therapy Room.

The aide brought my grandmother to a chair, then checked to make sure that she did not have any glasses or hairpins or jewelry on her person. Had she been wearing false teeth, which she was not, the aide would have removed those as well. Had she been wearing a girdle, the aide would have loosened it. Same with a belt. The nurse in charge in the Therapy Room double-checked the aide's work, then the aide guided my grandmother to one of the beds. My grandmother lay down on a loose sheet atop the mattress, and the aide carefully placed a foam-rubber pillow just above her shoulder blades. The aide requested that my grandmother place her arms by her side and extend her legs. The sheet was then wrapped around my grandmother, pinioning her legs and arms in place. Her head remained exposed. The aide placed a rubber gag in her mouth, then held down her shoulders while a second aide held down her hips. It was important that she be immobilized as much as possible, as the therapy she was about to experience caused muscular contractions so violent that unrestrained limbs had been known to thrash around so hard that their bones broke.

Once the nurse in charge had checked again to make sure all was in order, she brushed out of the way any strands of hair that may have fallen across my grandmother's forehead and then applied the electrodes to her skin.

CLINICAL NOTES, MRS. WILLIAM B. SCOVILLE (EMILY LEARNED)

April 5, 1944: Mrs. Scoville has been started on a course of electric shock treatment.

April 12, 1944: Mrs. Scoville has been receiving electric shock treatments. She has shown some improvement in her behavior in that her relatively clear periods are of longer duration. However, there are many periods when she sings, laughs and dances about the hall; bangs loudly on the piano and is otherwise quite annoying to the other guests. She recognizes relatives when they come to visit, but her . . . interest appears to be very short-lived and she soon begins to ramble incoherently about herself and her operation, by which she means the removal of her head. She also talks about a trial she has undergone, and considers that she has been in prison since that time.

April 19, 1944: Mrs. Scoville has been receiving electric shock treatment since the preceding note. She shows some improvement in that she is more quiet and less disturbing to others. However, her psychotic ideation has shown little change. She still talks about being on trial and in prison following a trial, about having her [missing word] altered, about the way in which her boy talked to her, about migration of souls from one individual to another.

[The Clinical Notes for May through June are missing and don't pick up again until three months later, in July.]

July 12, 1944: Mrs. Scoville continues to show some improvement. She continues to be somewhat afraid of the shock treatment.

My grandmother was not the only asylum guest who was afraid of the shock treatment. The fear had become so pervasive, leading to so many disturbances and breakdowns among the patient population, that Burlingame had mounted a campaign to eliminate it.

Not the shock treatment. The fear.

The campaign was outlined in the asylum's counterpart to *The Chatterbox,* a for-staff-eyes-only newsletter called *Personews,* which ran a two-part series on the subject in successive issues in August 1944. The first of these articles, "Fear-Free Psychiatry," laid out the general problem. "The destructive force of fear," Burlingame wrote, "presents a special problem for the psychiatrist who is attempting to build up the confidence of his patient. Consequently, those assisting in this aim should be well acquainted with the psychology of fear, so that they may help rather than hinder the course of treatment." He recommended that nurses and aides should "not discuss treatments with guests but if the conversation reaches a point where it is impossible or awkward to avoid the topic, the positive side always should be stressed. It is never advisable to refer to or present these measures as punishment or threats."

By working toward "dispelling fear from the patient's mind," Burlingame believed his asylum staff would be "bolstering the foundation for life on leaving the hospital as well as facilitating the treatments here. Fear gets us nowhere since it stands in the way of a guest's progress, thus defeating our purpose. Bear this in mind and act accordingly so you will be contributing to, rather than obstructing, our collective goal—helping to get sick people well."

A second article, in the next issue of *Personews,* outlined one specific and disarmingly simple measure that Burlingame was taking to eliminate the fear infecting his institution. The article was called "'Sleep'—Not 'Shock,'" and it noted that there were three forms of shock treatment practiced at the asylum: insulin shock, Metrazol shock, and electric shock. The first, insulin shock, required injecting guests with massive amounts of insulin until their blood sugar levels collapsed and they fell into a hypoglycemic coma. (As with all three forms of shock treatment, the rationale for insulin shock was murky. Nobody claimed to have a clear understanding of *why* inducing unconsciousness through artificial means apparently helped alleviate, at least temporarily, the symptoms of mental illness. They just claimed that it did, for reasons unknown, and justified its continued use on this basis.) After one to two hours, attempts were made to revive the patient, which often involved artificial respiration or injections of adrenaline. As Burlingame himself noted in a journal article from 1938, the insulin treatment was dangerous, a "skillful sparring with death, where a few moments of neglect, inattention or inadvertence may cost a life." The second such treatment, Metrazol shock, required overdosing the guest with Metrazol, a chemical compound, which typically caused a few moments of primal panic followed by a daylong coma. The third treatment, electric shock therapy, was the latest of the three therapies to gain a foothold at the asylum and involved the induction of unconsciousness through the repeated application of high-voltage bursts of electricity to guests' brains.

Burlingame had latched on to all three shock treatments with a vengeance and, in a previous annual report, wrote a paragraph that perfectly captured the strange, two-sided nature of his institution, with its country club veneer and hidden underbelly: "The success of the indoor swimming pool as a therapeutic agent seemed to urge the construction of an outdoor swimming pool which was finished last spring. At the time this seemed to be somewhat of a new departure for a psychiatric hospital but a single season demonstrated its worth and its use has become an important feature on our program of normality. The continued development and increased importance of shock therapy demanded the construction of a specially equipped air-conditioned unit of twenty-five beds exclusively for this form of therapy. It has been used to capacity since its completion."

But now my grandmother and many other asylum guests were becoming increasingly afraid of these shock treatments, and this was posing a problem. Burlingame had come to the conclusion that the fear of his guests had less to do with the treatments themselves than with how the treatments were described. It was, in short, a problem of presentation, not substance. And so he ordered a change.

"Unfortunately," he wrote, "the misleading term 'shock' has come into common usage in reference to insulin, metrazol and electric treatments for mental and nervous illnesses and has proved the source of unwarranted fear. Rightly, since these three treatments produce states of unconsciousness akin to normal slumber, they should be known as insulin, metrazol and electric sleep—with 'sleep' replacing 'shock' in all three instances. . . . Because of these facts and for the benefit of those receiving treatment, we are adopting the names that are more truly descriptive of these treatments—INSULIN, METRAZOL and ELECTRIC SLEEP. By thus deleting the misnomer 'shock' from our psychiatric vocabulary, we demonstrate that the Institute of Living is not inhibited by traditions and morals that harm rather than further our aims."

It's likely that my grandmother and Burlingame had crossed paths prior to her becoming a guest at his asylum, though neither may have been aware of it. Long before he became the superintendent of the Institute of Living, Burlingame had worked at my grandmother's family business. When the Cheney Brothers Silk Manufacturing Company ran into labor problems in 1915, they hired him to assess the problem and come up with solutions. He worked full-time for the company for two years, becoming America's first "industrial psychiatrist." He would observe the daily toil at the factories, collecting worker complaints and attempting to generate remedies for them, tweaking rules and regulations to minimize friction between labor and management. As he made the rounds, he may have seen my tow-headed young grandmother pedaling her bicycle around the campus now and then.

But that's just speculation.

What is certain is that Burlingame had met my grandfather. Starting with the asylum's annual report of July 1, 1941, three years prior to my grandmother's arrival, Dr. William Beecher Scoville was listed as a member of the asylum's "consulting staff." The question of why a neurosurgeon would serve as a consultant to an asylum is a good one, and to answer it, it's necessary to leave my grandmother for a while, alone and afraid, and step back several years, and south several hundred miles, to a Halloween morning in Washington, D.C.

EIGHT

MELIUS ANCEPS REMEDIUM QUAM NULLUM

"O beautiful for spacious skies,
For amber waves of grain!"

The singing woman lay on her back on an operating table. It was October 31, 1939, and she was thirty-four years old. She worked in sales. She was having an operation because recently, while listening to CBS radio news broadcasts, she had realized that the announcers were talking about her. They didn't mention her by name, but she could tell. And she knew that other people could tell, too. Sometimes, even when the radio wasn't on, she still heard voices, and those voices said one thing to her, over and over: It was time for her to kill herself.

"For purple mountain majesties
Above the fruited plain!"

Her head rested on a small sandbag, and the table was at a slight incline, so her feet were somewhat below the level of her head. This was to minimize bleeding, although some blood was inevitable. Earlier they had shaved the hair from both sides of her head, as well as some from the front, so it didn't get in the way. They also administered injections of novocaine, numbing her temples completely, before cutting through the flesh and muscle on both sides and then pulling it back and away with a mastoid self-retaining retractor, exposing the bone. Once they found the spot they were looking for—approximately three centimeters back from the eye socket and six centimeters above the apex of the upper jaw—they drilled two one-centimeter burr holes into the sides of her skull, then removed the little plugs of bone. A razor then sliced neatly through the dura mater, and a jet of saline solution washed away the reddish mix of blood and cerebrospinal fluid beneath, exposing the pale, mottled surface of the woman's brain with its ripples like molten copper.

"America! America!
God shed his grace on thee!"

A neurologist stood behind the patient's head, looking down on her. His perspective was a useful one, and from the beginning of the operation he guided the surgeon, advising him, for example, how far into the hole on the right side of the woman's head to insert the scalpel, which was long and thin and had a two-sided blade. The key was to stop just short of the midpoint, to avoid the arteries there. Once the scalpel had reached a certain depth, about eight centimeters in, the neurologist told the surgeon to stop and to begin the incision of the upper right quadrant. (To visualize the upper right quadrant, imagine a cross section of a forward-facing human brain and then imagine superimposing an X/Y axis over it.) As the surgeon slowly levered the handle of the scalpel down against the lower portion of the burr hole, the blade inside the woman's skull pivoted upward, slicing through the upper right quadrant of her frontal lobes and severing tens of millions of connections between those lobes and the structures deeper in her brain.

"And crown thy good with brotherhood
From sea to shining sea!"

After that first cut, the neurologist told the surgeon to pivot the blade in the opposite direction so that it sliced slowly down through the lower right quadrant. Then the surgeon extracted the blade, walked around to the patient's left side, and inserted it into the burr hole there. The neurologist again told him when to halt the insertion. Symmetry was crucial. The surgeon then pivoted the scalpel up, and the blade sliced down through the lower left quadrant of the woman's frontal lobes, severing millions more connections.

"Oh beautiful for pilgrim feet,
Whose stern impassioned stress!"

Throughout the operation, the neurologist, peering down at the patient from above and behind her, invisible to her, his voice disembodied, occasionally asked her questions. She was only under a local anesthetic and was completely conscious. Her answers to the questions helped the neurologist monitor the progress of the operation. During the first two cuts, on her right side, her answers were accurate and lucid, just as the neurologist's experiences with fifty-one previous operations of this sort had led him to expect.

"Do you know my name?"

"Dr. Walter Freeman."

"Do you know where you are?"

"Washington, D.C."

"Where in Washington, D.C.?"

"George Washington University."

"Very good."

After the first cut on the left side, the woman's voice changed, becoming flatter, deader, slower. It lost its previous liveliness, was much more of a monotone than her normal speaking voice. This was also as expected. But the woman was still conscious and able to converse. When she began spontaneously telling Freeman about her love for music, and for singing, he asked her to sing him her favorite song. She picked "America the Beautiful."

"A thoroughfare of freedom beat
Across the wilderness!"

As the woman began to sing, Freeman took excited notes. This was something new, something unexpected. Not the singing, but the quality of it. While the woman's speaking voice had lapsed into the expected dullness after the cutting of the third quadrant, her singing voice seemed untouched. She was singing, Dr. Freeman wrote, "without quavering, showing excellent articulation, as well as true pitch and some expression."

> *"America! America!*
> *God mend thine every flaw!"*

As she completed the song's second stanza, Freeman told the surgeon to proceed with the final cut. The scalpel sliced through the upper left quadrant of the woman's frontal lobes, severing the final portion of the targeted connections. At that point the singing stopped, and the woman fell silent. Freeman looked down into her eyes for two minutes, waiting to see if she'd say anything. As expected, she did not. The effects of the fourth cut almost always included the cessation of spontaneous speech. Eventually he began to ask her questions again.

"What is my name?"

"I don't know."

He took off his hat and surgical mask and leaned over her so she could see him. He had a large, expressive face, with deep-set eyes and a meticulously trimmed goatee.

"Look at me, don't you recognize my head?"

"Sure."

The patient, lying there with her partially shaved head, waved up at Freeman's bald pate, then cracked a joke.

"Did they do that to you, too?"

Freeman made another note. He was always pleased when he found evidence of a preserved sense of humor. Recently, during another operation, just prior to the cutting of the fourth quadrant, he had asked a fifty-three-year-old patient what was going through his mind. The man thought for a minute, then said, "A knife." Now Freeman smiled down at his current patient and continued the questioning.

"Who am I?"

"William Randolph Hearst."

Freeman nodded, satisfied.

In the late nineteenth century, Dr. Gottlieb Burckhardt, a Swiss psychiatrist, performed the first modern neurosurgical attacks on mental illness. Burckhardt ran a private asylum in the town of Prefargier, in a rugged and mountainous region known for its watchmaking firms. Over the course of decades interacting with the mentally ill, he'd developed a theory about the neurological roots of madness and how the physical destruction of brain tissue might alleviate its symptoms. In 1888, he decided to put his theory to a test. He had no experience or training as a neurosurgeon, but he obtained a set of neurosurgical tools and set to work. One of the first patients he selected for these experiments was a "forever disturbed, unapproachable, noisy, fighting, spitting, all but straight-jacketed, untidy, smearing" fifty-one-year-old "particularly vicious woman," who'd been institutionalized for sixteen years. Over a yearlong period, and five different operations, Burckhardt opened up the left side of the woman's skull and removed a total of eighteen grams of her brain. The removals were done in a somewhat scattershot way, not focusing on any particular part of her neuroanatomy but instead targeting a broad sampling, including her postcentral, third frontal, parietal, and temporal cortices. As a result of the operations, Burckhardt reported that his patient had become "more tractable." Her previous intelligence, he added, "did not return," but he noted that by her final surgery, "Mrs. B. has changed from a dangerous and excited demented person to a quiet demented one."

By the end of 1889, Burckhardt had operated on five more patients. Two of them died, two became epileptic, and one committed suicide.

Still, he considered his experiments a success. "Doctors are different by nature," he wrote. "One stands fast on the old principle 'primum non nocere' ('first, do no harm'); the other states: 'melius anceps remedium quam nullum' ('it is better to do something than nothing'). I belong naturally to the latter category. . . . Every new surgical approach must first seek its special indications and contraindications and methods, and every path that leads to new victories is lined with the crosses of the dead. I do not believe that we should allow this to hold us back from approaching the goal, the curing of our patients by surgical methods. The purely medical side of our calling must and shall lead us nolens volens (whether we like it or not) along this pathway."

When Burckhardt published his account of these experiments, in 1892, the medical community reacted with revulsion. As one French psychiatrist put it in a pointed response to Burckhardt's better-to-do-something-than-nothing argument, "an absence of treatment was better than a bad treatment."

The path that Burckhardt began to trailblaze would remain blocked for the next four decades, until the 1930s, when a Portuguese neuroanatomist named Egas Moniz cut his own way forward. Unlike Burckhardt, Moniz focused his attacks on a specific part of the brain: the frontal lobes. His ideas had been gestating for years, but his decision to act probably crystallized in 1935, during a symposium at the Second International Congress of Neurology, in London.

The chairman of Yale's department of physiology, John Fulton, had been on the stage at the symposium describing the results of a series of experiments he and a colleague, Carlyle Jacobsen, had recently performed on two chimpanzees. The chimps were named Becky and Lucy, and Fulton and Jacobsen had run a number of basic cognitive tests on them, finding that they were, like most chimps, ornery. They quickly grew frustrated when they were unable to perform the tests correctly and would throw dangerous temper tantrums or just sulk in the corners of their cages. They would also grow impatient with the repeated test-taking itself, becoming less and less willing to participate, exhibiting what Fulton called "experimental neurosis." Then, Fulton told the audience, he and Jacobsen had surgically destroyed Becky's and Lucy's frontal lobes and readministered the same tests as earlier. They observed a profound change. "While the animal repeatedly failed and made a far greater number of errors than it had previously," Fulton said, "it was impossible to evoke even a suggestion of experimental neurosis." That is, the operation had disabled the chimps cognitively but had also made them more placid and less neurotic.

A question-and-answer period followed Fulton's presentation.

Moniz raised his hand.

"Do you think," he asked, "that this operation could be applied to humans, with the same results?"

Three months later, Moniz sought to answer the question on his own.

On November 12, 1935, he oversaw the performance of the world's first leucotomy. *Leuco* is from the Greek for "white matter," which is how the brain's connective neural tissue is commonly described. *Tome* is from the Greek for "to cut." A neurosurgeon colleague of Moniz's, Almeida Lima, drilled two holes in the forehead of a patient suffering from a deep depression, then inserted a long, narrow metal tube through the opening to a depth of approximately eight centimeters. The tube—which Moniz dubbed a leucotome—contained a coil of narrow-gauge wire, and when the tube reached a sufficient depth Lima pressed a plunger at the end of the instrument, causing a small loop of the wire to extrude into the white matter of the frontal lobes. He then spun the instrument 360 degrees, disconnecting a small core of tissue from the rest of the brain. Then he withdrew the wire, pulled the leucotome out about a centimeter, and pushed the plunger again, repeating the process. In all, Lima cut four cores out of each side of the patient's frontal lobes.

The results of the operation were "sufficiently encouraging," and over the next few months, Moniz oversaw nineteen more leucotomies on a variety of patients suffering from mental disorders before publishing his preliminary results in March 1936 in a French medical journal under the title "The Possibilities of Surgery for the Treatment of Certain Psychoses." Not long after the paper's publication, Walter Freeman stumbled across it. Reading that paper changed Freeman's life, as well as the lives of thousands of his future patients.

Freeman came from a family of physicians. His father had been a urologist, a decent and modest if somewhat reticent man, but Freeman seemed to inherit his ambition, along with a penchant for showmanship, from his grandfather William Keen, a famous surgeon who had served for many years as president of the American Medical Association. In 1888, Keen became one of the first surgeons to successfully remove a brain tumor and was known for putting on public demonstrations of his work, with as many as a hundred students and physicians crowding into the seats of his medical amphitheater as he performed his delicate procedures.

Freeman's first job as a neurologist was at an insane asylum in Washington, D.C., where he quickly saw how hopeless traditional therapies were in the face of the demons tormenting his patients. He was eager to employ any novel treatments as they came along, and experimented with the full battery of shock treatments—from insulin to Metrazol to electric—but he was never satisfied, partly because these treatments all seemed so blunt and imprecise, their mechanisms unclear. For a period of years he became obsessed with the idea of finding an actual physical cause for insanity, and he spent hours in the hospital morgue, autopsying the brains of his deceased patients, seeking out the neurological root of their madness. He always came up empty.

And then he came across Egas Moniz's monograph describing Moniz's first experiments with leucotomy. Immediately Freeman knew what he wanted to do, and just a few months later, on September 16, 1936, in collaboration with his neurosurgeon partner, James Watts, Freeman oversaw the first leucotomy on U.S. soil, on an anxious, insomniac housewife from Topeka, Kansas.

In 1938, Moniz was forced into a temporary professional hiatus after one of his former patients confronted him in his home and shot him through the spine. By that time, Freeman and Watts had already overtaken him as the most prolific practitioners of psychiatric surgery. Freeman had also given the new field a name—psychosurgery—and developed what he considered to be an improvement on Moniz's technique, replacing the wire-based leucotome with a scalpel and having Watts make his approach through the sides of the skull rather than taking Moniz's somewhat messier route through the forehead. Freeman described this as his "precision method" and also tweaked the name of the operation, replacing the Greek root *leuco* with the Greek root *lobo,* meaning "of the lobes."

Lobotomy.

That Halloween morning, as Freeman stared down into his fifty-second lobotomy patient's eyes and heard her mistake him for the publishing magnate William Randolph Hearst, his career was ascendant. He was a young, charismatic doctor wielding an exciting new treatment for ancient maladies. Soon he would be more famous than his grandfather ever was. His growing celebrity would be helped along by a number of fawning profiles in major publications (a few of which, coincidentally, were owned by Mr. Hearst). The first *New York Times* piece about Freeman, which ran on June 7, 1937, carried the headline "Surgery Used on the Soul-Sick" and gushed about his "new surgical technique, known as 'psycho-surgery,' which, it is claimed, cuts away sick parts of the human personality, and transforms wild animals into gentle creatures in the course of a few hours."

By 1939, the age of the lobotomy had arrived, and Freeman was its most fervent evangelist. He had been performing the operations at a frenetic pace, traveling around the country giving demonstrations to scores of curious physicians, who in many cases soon began eagerly trying it on their own patients. He was also preparing for the publication of his first book. The book—*Psychosurgery*—was as much a call to arms as it was a medical text, and it centered on a simple yet revolutionary thesis:

"In the past," Freeman wrote, "it's been considered that if a person does not think clearly and correctly, it is because he doesn't have 'brains enough.' It is our intention to show that under certain circumstances, an individual can think more clearly and more productively with less brain in actual operation."

He was glad she didn't recognize him.

If Patient 52 had remained lucid and continued to answer his questions correctly, if her voice had not suddenly been drained of emotion, if her eyes had remained sharp and inquisitive, if she had continued to sing that song, he would have told his colleague to keep cutting deeper, farther. He would have told him to cut until she became confused, until her thoughts became muddy and her personality ebbed away. This is why he kept his patients conscious during the operation: He wanted to make sure that their brains were receiving sufficient damage. He described this operative strategy as his "disorientation yardstick"; he explained that "impairment of memory, confusion, and disorientation usually come on within a few seconds to a few minutes after the fourth quadrant is sectioned" and that "when this disorientation occurs on the operating table we are satisfied that an effective operation has been performed. If the patient is still alert, oriented, and responsive, it is our custom to extend the incision into the upper and lower quadrants for fear that the relief obtained by the operation may be insufficient." In general, he adhered to a simple strategy: "The best method, of course, is just to cut until the patient becomes confused."

The trick, though, was to cut enough of the patient's brain to cause a state of confusion but not so much that the patient died or was permanently incapacitated. In successful operations, patients would become immediately disoriented, and perhaps incontinent, but would then over a period of weeks or months recover a measure of lucidity, recognizing those around them and remembering their pasts before the operation. They wouldn't be the same, though. That was the whole point, after all. To do with a few swipes of a blade what years on an analyst's couch, or in an asylum's cell, failed to accomplish.

Freeman expended a great deal of energy trying to gather evidence for these beneficial changes. He kept meticulous records and was diligent about staying in touch with all of his patients, monitoring their progress. He looked for signs of improvements everywhere. Sometimes he saw evidence where others didn't. For example, he always photographed his patients before their operations, then at some interval afterward. He developed the negatives himself and spent time peering into their pre- and postoperative eyes, reading them like tea leaves. In papers and presentations, he liked to point out how the eyes of most of his female patients looked notably more fearful and anxious preoperatively. (He failed to attribute this to the fact that for the preoperative pictures he almost always photographed the women while they were naked, while for the postoperative ones he almost always photographed them while they were fully clothed.)

He kept tabs on everything. Regarding postoperative eating habits, he found that there was "a high correlation between improvement and gain in weight" and noted approvingly that one woman, Patient 53, more than doubled her weight in the months after operation, from 85 to 210 pounds. Regarding postoperative personal grooming, he admitted that even his prying eyes were not "sufficiently acquainted with the mysteries of the boudoir to know just what happens following operation in regard to cosmetics, creams, lotions, rouge, lipstick, perfume, and the rest." But he was confident that most of his patients, after their lobotomies, "have again been able to pay some attention to their personal appearance," while preoperatively they "did not resort much to this socially acceptable feminine activity."

This idea of social acceptability was key to what Freeman hoped the lobotomy would achieve. In his view, the world was full of social misfits. Some were plain to see: the hopeless cases caterwauling in the back wards of asylums, the disheveled vagrants wandering the streets muttering to themselves. The majority, though, were less obvious. And although his initial focus was on the extreme cases, he had started to perform lobotomies on people suffering in much more subtle ways: the housewife who displayed "affective incontinence" and descended into crying jags every afternoon, the premature spinster "gradually drifting into seclusiveness," the obsessive-compulsive who washed his hands "so excessively that the skin was dry, rough, and cracking," boys and girls who had a tendency to misbehave, throw excessive tantrums, or display an excessive preoccupation with masturbation. They spanned all ages: Freeman's youngest patient was seven years old, his oldest seventy-two. All of these misfits, old and young alike, leading lives of quiet or not so quiet desperation, hungry for relief.

Freeman didn't pretend to understand what, exactly, his lobotomies were doing to his patients. He knew that the swipes of the scalpel were severing many of the connections between the frontal lobes and the brain's deeper structures. And he knew that the frontal lobes were important. (The evolution of Homo sapiens from lower orders of anthropoids can be distilled down to the fact that our simian ancestors have much smaller frontal lobes than we do. The frontal lobes, then, must be crucial to humanity itself.) One of Freeman's contemporaries, the anthropologist Frederick Tilney, expressed what was then the prevailing view when he described the frontal lobes as "the accumulator of experience, the director of behavior, and the instigator of progress."

But although their importance was unquestioned, their precise function was still a mystery. There were many different theories. The one Freeman was most drawn to imagined the frontal lobes as the rough physical analogue to Freud's concept of the superego. The emotional, animalistic impulses of the id originated in the deepest structures of the brain and radiated outward through the intermediate structures, which gave rise to the self (the ego) before reaching the frontal lobes, where emotions were processed and interpreted and reflected upon and controlled. In a functioning brain, the frontal lobes acted as a regulatory body. A feeling of profound sadness might rise up from the lower structures, and the frontal lobes would allow this sadness to be experienced fully, for a period of time, before tamping it down and cutting it off. In a dysfunctional brain, however, the frontal lobes might lock that feeling of sadness (or of fearfulness, paranoia, shyness, et cetera) into an unending cycle, or downward spiral, creating a neurosis.

Physicians might attempt to treat a neurosis with psychotherapy. Or with a change in environment. Or by placing patients in a copper box and heating them up until they developed a fever of 106 degrees. Freeman, who was nothing if not open-minded about the potential effectiveness of radical treatments, even suggested that a bullet to the brain, as long as one survived it, might have a salutary effect. "There can be no doubt that the first shock of the shooting produces a profound effect psychologically," he wrote. "From the purely physical side, that is, from the trauma, the pain, the shock, the fever, and possibly the surgical intervention, there is some resemblance to shock methods which are apparently quite effective in treating depressions. These same effects might be expected in relation to any severe trauma whether self-inflicted or not."

But Freeman believed that the lobotomy was a better, more direct, more scientific approach. By cutting a hole in a person's head, by inserting a scalpel and physically breaking up unhealthy "constellations of neuron patterns" while at the same time slicing through many of the pathways between the brain's emotional centers and the frontal lobes, he believed he could obliterate neuroses and prevent future ones.

Still, although he was a passionate believer in the lobotomy's potential, Freeman didn't think that potential had been fully realized yet. His patients, after their lobotomies, might no longer be the misfits they were before, but neither were they entirely normal. Whereas before they were prey to the whims of irrational emotions, postoperatively they often lacked the basic emotional responses we expect to see in human beings. One might be unable to cry at all, even when her mother died. Another might lose all interest in eating and would have to be prompted, bite by bite, through the course of a meal. A third might simply sit in a corner, mute, rocking back and forth for days and weeks and years on end, speaking only when spoken to. "Following operation," Freeman wrote in 1942, "there would seem to be a certain emotional bleaching of the individual's concept of himself. How much of [the brain] must be sectioned in order to relieve disabling mental symptoms, and how much must be preserved to enable the individual to function adequately in society, has not yet been definitely established."

Just as Freeman had refined Moniz's leucotomy into his own customized procedure, he believed future refinements surely lay ahead. He thought new approaches to the operation were necessary and that the ideal lobotomy had yet to be devised. Scores of doctors around the world, intoxicated by psychosurgery's promise, by the prospect of discovering surgical solutions to some of mankind's most intractable problems, had taken up Freeman's call to action. They were opening the skulls of misfits everywhere and rummaging around inside, attempting to find that one simple cut that might make them well.

None would perform as many lobotomies as Freeman, who was as prolific as he was passionate.

My grandfather, however, would come in a close second.

NINE

THE BROKEN

On the fifth floor of the Francis A. Countway Library, on the campus of Harvard Medical School in downtown Boston, there are a number of glass-fronted cabinets and glass-topped display cases. Their contents constitute the principal holdings of the Warren Anatomical Museum, which was founded in 1847 by a Boston physician who hoped it would stimulate curiosity and a spirit of inquiry among young medical students. There's a placard on one of the display cases with a Latin phrase that sums up the collection's animating principle: MORTUI VIVOS DOCENT.

The dead teach the living.

The cases are filled with the dead, or remnants of them. A gnarled skeleton of a woman with a severely contorted spine stands beside a photograph of her during life, naked, her face turned away from the camera. A row of four fetal human skeletons, ten, fourteen, eighteen, and twenty-two weeks old respectively, are posed in standing positions, as though they had learned to walk. One entire display case is dedicated to an assortment of kidney stones of all shapes and sizes and colors. In another, a plaster cast of the seven-fingered hand of a nineteenth-century Boston machinist grips a plaster cast of a rock.

All but one of these relics are from anonymous individuals. The exception is so famous that even just a glimpse of his skull might bring his name to your lips.

On September 13, 1848, a twenty-five-year-old construction foreman named Phineas Gage was leaning over a hole he had drilled in a shelf of rock, using a six-foot-long and two-inch-diameter iron tamping rod to jam a charge of gunpowder deep inside. He was in the wilderness of western Vermont, helping to clear a path for the construction of the Rutland and Burlington Railroad. He was by all accounts a diligent and conscientious man, so it was uncharacteristic of him to have forgotten to place a spark-inhibiting layer of sawdust over the gunpowder.

The blast jettisoned the tamping rod out of the hole in the rock like a missile out of a silo. The upper end of the rod, which was tapered to a dull point, penetrated Gage's skull just under his left cheekbone, then continued upward, moving at a diagonal slant through his frontal lobes before exiting through a hole in the upper right portion of the top of his skull. The rod flew a great distance but was eventually found, and a witness noted that it was "covered with blood and greasy to the touch." Gage was loaded into an oxcart and rushed to the nearest town. He remained conscious during the entire trip, then walked up a long flight of stairs to a hotel room. When the doctor arrived, Gage calmly showed him the holes in his head and said he hoped he was "not much hurt."

For the next two decades, Phineas Gage lived with an odd sort of fame. He attempted to go back to work at the railroad, but his co-workers found that the affable man they'd known had become a surly drunk who flew into unpredictable rages or inappropriate hysterics. A doctor's report indicated that he would indulge "at times in the grossest profanity" and that the balance "between his intellectual faculties and his animal propensities seems to have been destroyed." The railroad fired him, and P. T. Barnum later hired him to join his traveling circus, where he would sit with the now polished and engraved tamping rod across his knees, a gawker's delight. Eventually Gage tired of the freak show and on a sudden impulse decided to move down to South America, where he tried to start up a streetcar company in the port city of Valparaiso, Chile. A number of scientific papers were written about his case, which helped establish the general notion that the frontal lobes play a part in impulse control. One doctor noted that Gage was constantly on the move and that he "always found something that did not suit him in every place he tried."

Gage died in San Francisco in 1860, nearly twelve years after his accident. Seven years after that, his body was exhumed and his skull was shipped east, where it came to a permanent rest here in this museum, one shelf above the rod that pierced it.

I hadn't come to the library to see Phineas Gage. I hadn't even known he was there. I'd come to make copies of some old letters from the library's rare books and manuscripts archives and stumbled on the museum by chance.

The letters were between my grandfather and two Harvard scientists—the endocrinologist Fuller Albright and the neurologist Stanley Cobb—and they gave a glimpse of my grandfather's early ambitions, as well as some of his motivating impulses. He had written most of the letters while he was a neurology intern at Bellevue Hospital in New York City, exploring his next steps. He wanted, he wrote in his first letter to Albright, "advice and possible help in getting some first-hand experience in clinical endocrinology, especially as related to a neurological-psychiatric practice." Endocrinology is the study of the endocrine system, which regulates the body's hormones. He enclosed a copy of his only publication at the time, a solipsistic case study from the *Journal of the American Medical Association* chronicling his own bout with an at-first-mysterious illness. "A physician," the paper reads, "aged 28, in the summer of 1934, after drinking raw cow's milk and eating goat cheese in Norway, developed periodic exacerbations of malaise, easy fatigability, and generalized muscle and joint pains." He describes how the aching and exhaustion had laid him low for six months, causing him to put his young career on hold and leading to multiple diagnoses of neurasthenia, a catchall psychiatric term that at the time was often applied to describe people who were simply unable, mentally, to cope with high levels of stress. The paper chronicled his attempts to find an alternate explanation and ended with his discovery that he tested positive for brucella, an undulant-fever-causing bacteria carried by cows and goats in, among other places, Norway. The paper's single illustration was a black-and-white photograph of my grandfather's pallid forearm displaying a large abscess that a diagnostic skin test had provoked. He clearly relished proving his physicians wrong and finding a simple, easily treated, biological cause for what they had attributed to a vague and hard-to-target mental condition. "This paper," he wrote in the closing comment, "suggests one more substitute for that diagnostic wastebasket 'neurasthenia.'"

In the letter to Albright, my grandfather outlined the career path he'd followed so far: "I attended Yale College, B.A. 1928, Univ. of Penn medical school 1932, two years of general medicine and surgery at the Hartford Hospital and the Presbyterian medical center NYC; one year in psychiatry at the Cornell Medical center; and one year in neurology under Dr. Foster Kennedy at Bellevue Hospital." Kennedy, incidentally, was an aggressive proponent of eugenics, who in 1942, while he was president of the American Neurological Association, raised eyebrows by arguing that people who suffered from mental retardation should be killed, declaring that "the place for euthanasia, I believe, is for the completely hopeless defective; nature's mistake; something we hustle out of sight, which should not have been seen at all."

Cobb and Albright offered my grandfather a yearlong dual fellowship, splitting time between Cobb's clinical practice and Albright's laboratory. My grandfather accepted with enthusiasm and agreed to move to Boston as soon as his contract with Bellevue expired. He was excited at the prospect of tackling important research, though he was worried that his enthusiasm might outstrip the constraints of time. "The problems of urine-assay of sex hormones in homosexuals or menopausal or pregnancy psychoses seems too vast and a bit impractical to list in an application for a year's fellowship, don't you think?" he wrote. In the meantime, he wrote, he was open to suggestions for any experimental endocrinological work he might conduct before he left

Bellevue. He wondered whether there might be "a simple problem I might work at" while at the hospital, with its "wealth of material but poor laboratory facilities."

I had to read it twice before I understood what he meant by "material."

The dual fellowship with Cobb and Albright did not end up being particularly successful. After reviewing some of the research reports my grandfather produced in Albright's lab, Albright wrote to him that "I have been over your manuscripts. The problem with them is this: they contain a lot of different problems but not quite enough observations on any one problem to absolutely clinch it. I feel that there is very little in them which would materially help the progress of medical science." My grandfather didn't object to the harsh critique: "Thank you for your kind and clear letter re my various articles and data. I quite agree that they constitute a hodge-podge of information of no great value," he wrote.

But his stint with Cobb and Albright was fruitful in another way, since it was during this time that he discovered his passion for neurosurgery. Albright's laboratory was located at Massachusetts General Hospital, which had a peerless neurosurgery department, and in his free time my grandfather took the opportunity to observe some of the best neurosurgeons in the world at work. He was captivated by what he saw and decided to apply for a neurosurgical residency there, which began the following year, in 1938. He proved a quick study. After a whirlwind of additional residencies at the Boston City Hospital and Johns Hopkins, he founded his own department of neurosurgery at Hartford Hospital in 1939.

He kept in touch with Cobb and Albright, even as their careers diverged. He'd write to them on his Hartford Hospital stationery, telling them about the milestones in his life. "Emmie just had a 7 1/2 pound boy, with a magnificent Jewish nose," he wrote to Albright upon the birth of my uncle Peter, his third child, in late 1939. Albright wrote back, offering his "congratulations on the new prophet which has arrived in your house." He also let his former bosses know about any new material he came across that might be of interest to them. That same year, for example, he informed Albright about a twenty-eight-year-old man who presented a "beautiful picture of true pituitary hypofunction. . . . It is impossible to guess his age. He has a soft skin, a high voice, no facial nor body hair, long legs, asthenic habitus, low blood pressure, no libido, etc., etc." The following year, he wrote to inform him of a similar case, "a twenty-one-year-old dwarf who looks and acts as if he were in his early teens. . . . His genitalia is about one-half adult size, his pubic hair is abundant but silky and in feminine distribution. . . . Development and skeleton is symmetrical except for all structures being smaller than normal. Height approximately 3 feet 10 inches or so."

At the end of that letter he asked his old boss a question. "Are you interested in having him on your experimental ward for study?" He noted that the dwarf was "cooperative and passive" and told Albright to write back "if you want us to send him up."

The broken illuminate the unbroken.

An underdeveloped dwarf with misfiring adrenal glands might shine a light on the functional purpose of these glands. An impulsive man with rod-obliterated frontal lobes might provide clues to what intact frontal lobes do.

The history of modern brain science has been particularly reliant on broken brains, and almost every significant step forward in our understanding of cerebral localization—that is, in discovering what functions rely on which parts of the brain—has relied on breakthroughs provided by the study of individuals who lacked some portion of their gray matter.

This had not always been the case. Until the nineteenth century, most scientists viewed the brain as an undifferentiated mass. They recognized its importance, understood that it was the seat of emotion and intellect and consciousness, that it mediated our senses, that it more or less *was* us, but the reigning theory of brain function held that the brain was a perfect democracy, where every part was equal in potential and capability to every other part. From this view, injury to a particular part of the brain would simply cause a generalized diminishment of function, rather than any specific deficit. The movement away from this view faced a lot of resistance. One reason for this was the rise of phrenology, a pseudo science that became a worldwide fad in the mid-1800s and held that people's personalities and intellects could be minutely described simply by running your fingers over their heads and reading the contours of their skulls, which reflected the dimensions of the brains they encased. Phrenologists believed in cerebral localization, but the problem was, their arguments and theories about that localization had more in common with astrology than astronomy. The eventual debunking and stigmatization of phrenology made real scientists wary of accepting the reality of cerebral localization until inescapable evidence for it began to emerge in the form of brain-damaged individuals.

Phineas Gage was pivotal.

Then, in 1861, the year after Gage's death, a French neurosurgeon named Pierre Broca wrote a paper describing a new patient who was, in many ways, more scientifically significant than Gage. The patient's name was withheld, but he came to be known in the literature as Monsieur Tan, owing to the fact that he could not speak with any coherence and was able to say only the word *tan* over and over again. He had maintained his other faculties, however, and was able to understand everything he heard, and write legibly and intelligently. Upon the patient's death, Broca performed an autopsy and discovered that Monsieur Tan had a small and sharply circumscribed lesion in a small part of the left hemisphere of his inferior frontal lobe. He surmised, correctly it turns out, that this region of the brain was crucial for speaking. Today all basic human anatomy courses identify that spot corresponding to the damage in Monsieur Tan's brain as Broca's area, the center for speech articulation. (Monsieur Tan's brain, incidentally, went on to find a home in another museum of anatomical curiosities, this one located in Paris.)

Seventeen years later, in 1878, Carl Wernicke, a German neurologist, described a patient with damage to his posterior left temporal lobe, a man who spoke fluently but nonsensically, unable to form a logical sentence or understand the sentences of others. If Broca's area was responsible for speech articulation, then Wernicke's area, as it came to be known, must be responsible for language comprehension.

And so it went.

The broken illuminated the unbroken.

Among brain scientists, this approach, teasing out the functions of different areas of the brain by studying individuals who lacked those areas, became known as the lesion method, and by the middle of the twentieth century, it had become predominant. The notion that different areas of the brain corresponded to different functions was no longer controversial; it had become universally accepted dogma. Bit by bit, area by area, scientists were plotting out a functional map of the human brain.

But that map still contained immense expanses of uncharted territory.

For example, scientists might have succeeded in pinpointing a dime-size portion of the brain's superior temporal plane as necessary for the in-the-moment perception of sounds, and labeled it the primary auditory cortex, but they had no clue what part of the brain was responsible for our ability to recall and recognize a specific sound at a later date. Or, for that matter, how it could do the same for past sights or tastes or touches or smells or the multisensory stews that constitute human experience.

The brain was slowly giving up some of its secrets, its ancient functions coming to light, but memory, which more than anything else defines us, remained a dark mystery.

In the basement of the library, right before I took the elevator up to the fifth floor and stumbled on Phineas Gage, I read one of the earliest letters from my grandfather to Fuller Albright. In it he described some of his long-range goals.

"I have done no animal or experimental work since being a medical student in physiology and helping a little in the experimental surgery department," he wrote, "but my chief hobby is mechanics, so the technical side of research appeals to me strongly."

He wrote the letter in 1936. Within a decade, he would find a way to unite his passion for tinkering and his interest in experimental surgery.

And, in the back wards of asylums around the country, he would discover a nearly limitless source of material.

TEN

ROOM 2200

He drove the twenty miles from Hartford to Middletown in his new Buick convertible, top up to shield him from the approaching New England winter, following the highway alongside the Connecticut River due south and pulling off just before the river veered southeast toward the ocean. Middletown was once the largest and most prosperous city in Connecticut, a bustling port where traders made fortunes provisioning the Caribbean colonies with goods and slaves. Once the business of slavery declined, the residents found new purpose, building factories to assemble Colt firearms and Royal typewriters, and founding Wesleyan University, a liberal arts college on the west side of the city. Wesleyan was the city's most well-known institution, but on the opposite side of town was a second institution, built around the same time and almost double its size. That's where my grandfather was heading. It was November 14, 1946. He could see the redbrick and wrought-iron gate of the asylum on a hill in the distance long before he got there.

As he passed through the gate, my grandfather might have noted some similarities between Connecticut State Hospital and his wife's asylum, the Institute of Living. Both made good first impressions, and although the grounds at the asylum in Middletown weren't designed by Frederick Law Olmsted, they too were beautiful. The largest building, Shew Hall, towered at the head of a sloping circular drive and looked reminiscent, in architecture and scale, of Paris's Élysée Palace, the residence of the French president. This was the asylum's administration building, and from his office on the top floor the superintendent, Dr. Edgar C. Yerbury, could gaze west over all of Middletown, while the vast grounds of his institution sloped gently down toward the river to the east.

The Connecticut State Hospital was the largest public asylum in the world. It occupied 906 acres, 406 of which were taken up by the asylum's farm, with its piggery, chicken coops, and verdant rolling pastures full of cows. Although the patients worked the farm for free, and the profits could be considerable—the previous year more than fifty thousand dollars of asylum-sourced milk was sold to the local community—the asylum as a whole was struggling. The problem was this: There were too many inmates. Every year the asylum was tasked with feeding and sheltering and protecting—from each other and from themselves—growing numbers of mentally ill men and women, but its resources had not kept pace. This pattern had been repeated all over the country in the 1940s as state asylums filled to overflowing. The cause of this insanity epidemic was a mystery. Some argued that it had to do with all the disturbed soldiers coming home from World War II. Others believed that the lingering anxieties and uncertainties provoked by the Great Depression had finally come home to roost, or that Americans were simply less willing and able to care for their disturbed relatives at home, as was once the norm, and had become accustomed to the idea of turning them over to the state.

Regardless of the reasons, state asylums nationwide were in crisis. And while conditions had been deteriorating for years, the general public had become aware of this deterioration only recently. In May 1946, *Life* published an exposé, "Bedlam 1946," that hit newsstands with the force of an explosion. The text of the article painted a vivid picture of inmates being fed a "starvation diet" in "hundred-year-old firetraps in wards so crowded that the floors cannot be seen between the rickety cots." But the text was nothing compared to the photographs. A man named Jerry Cooke had spent weeks in a state asylum in Ohio and emerged with a horrific portfolio of images straight out of a Hieronymus Bosch painting. On one page, a group of naked men huddled together against a wall, some hiding their faces in shame while others gazed hollow-eyed into the lens. On another page, an old woman, also naked, sat neglected and withered on a dilapidated wooden bench. To the American public, just emerging from World War II, the photos were both shocking and shockingly familiar. State mental hospitals, the country's most popular magazine declared, had become "little more than concentration camps on the Belsen pattern."

The article provoked not just outrage but action. Congressional hearings were held in Washington, and within two months of the magazine hitting newsstands, President Truman signed the National Mental Health Act, which provided federal funds for psychiatric research. There was broad consensus that the current conditions were untenable and that something had to be done to ameliorate them, although opinions varied widely about the best strategies for doing so. Was it a question of throwing more money at the asylums, increasing their carrying capacities? Or should more outpatient facilities be built, allowing potential patients to remain in their homes?

Or perhaps there was a simpler, quicker solution.

Perhaps there was a solution that took aim, with surgical precision, at insanity itself.

My grandfather parked his Buick in the visitors lot in front of Shew Hall, and no doubt Superintendent Yerbury came down to greet him. The day's visit was a momentous one and had been almost a half year in the making. In June, just weeks after "Bedlam 1946" was published, Connecticut's Joint Committee of State Mental Hospitals held a meeting in Hartford. The committee was an agency composed of the governing bodies of the state's three public asylums: Connecticut State Hospital, Norwich State Hospital, and Fairfield State Hospital. While the *Life* exposé had not dealt specifically with these particular asylums, the article's clear implication was that the horrors it depicted applied to state asylums nationwide, and the joint committee knew that it was in no position to rebut that view. The Connecticut state asylums, the committee would admit, were "seriously overcrowded, which, with the bad housing facilities, made it extremely difficult to render the best of service to our mentally ill patients." Any measures that might reduce the numbers of inmates had to be examined, and it was clear that the time for action had arrived. At the meeting, Yerbury suggested that one solution might be for the three state asylums to adopt a "coordinated program of neurosurgery." His peers agreed, and two months later, in August, representatives of all three state asylums met again to begin working out the details. That second meeting took place at the Institute of Living.

Although the Institute of Living, as a private asylum catering to the wealthy, didn't face the same problems of overcrowding as its public counterparts, the latter had always looked to the former for guidance on the implementation of novel psychiatric treatments. The Institute of Living's forward-looking superintendent, Charles Burlingame, presided over the meeting. He had already introduced the lobotomy on a limited scale at his asylum several years before, and my grandfather, as the Institute of Living's chief consulting neurosurgeon, had performed the majority of them. In fact, my grandfather and several other neurosurgeons also attended that August meeting of the Joint Committee of State Mental Hospitals and "signified their willingness to be added to the state hospitals as consultants." A third meeting was held a month later, on September 18, 1946, and this time several researchers from Yale attended. A general way forward was agreed upon: My grandfather and the other neurosurgeons would begin operating on patients at all Connecticut asylums, public and private, at least twice a week. Asylum personnel, overseen by researchers from Yale, would meanwhile keep tabs on the patients' postoperative progress.

A vote was taken, and passed, and the Connecticut Cooperative Lobotomy Study was born. The next day, the *Hartford Courant* heralded the event, declaring that "Connecticut thus becomes the first state to undertake a scientific, controlled study and practice of the brain operation known as prefrontal lobotomy." The article also noted that the widespread promulgation of the lobotomy in Connecticut was "expected to ease the load on mental hospitals, both operational and financial."

And then, on that brisk November day, there at the mammoth Connecticut State Hospital with its hundreds of cows and pigs and chickens and its more than three thousand human inmates, the project was finally getting under way.

Most of the workers at the asylum referred to the operating room by its number: 2200. Because 2200 was not as well equipped as the OR at Hartford Hospital, my grandfather had brought along all the tools he needed. He also brought a crowd. A large group of the asylum's nursing and medical personnel, as well as its superintendent and administration officials, squeezed inside to watch the first lobotomy ever performed there.

The patient was a thirty-one-year-old man who had been at the asylum since he was fifteen. During his time there, he had often shown "periods of excitement and disturbed states." He also, despite coming from a poverty-stricken family, continued to express his belief that he was in fact a nobleman. Orderlies led the man into 2200 and strapped him down to the operating table. My grandfather laid out his tools, scrubbed up, pulled on his mask, and put on his loupes, the special glasses that magnified everything he looked at. The patient received several injections of a local anesthetic in the flesh of his temples, and then the operation began. The musty smell of bone dust filled the crowded room.

The clinical director of the asylum was a psychiatrist named Dr. Benjamin Simon, and as my grandfather began to slice the nerve tracts connecting the frontal lobes with the posterior parts of the brain, Simon leaned in and began to interview the patient so everyone could hear.

"Are you feeling any pain?"

"No, Doctor."

"What do you think of all this? What are they doing to you?"

"Well, he is doing some kind of operation on my head."

"What do you think it will do for you?"

"I want to be able to go back and work."

"Why can't you do that now?"

"I don't know. There is something the matter with my mind. . . . Oh, I felt something then!"

"What did you feel? Was it a pain?"

"No, it didn't hurt. It did something to the pressure."

"What do you mean by that?"

"It stopped the tension."

Simon stopped the questions then and let my grandfather finish the procedure. At the end, the patient, appearing "relaxed and somewhat sleepy," was wheeled off to a recovery room and another inmate took his place. After the second operation, my grandfather gathered his tools and returned to his Buick, and the crowd dispersed.

Like the Institute of Living's *Personews,* the Connecticut State Hospital had its own in-house newsletter that was distributed to the staff every two weeks. The one there was called *The Scribe,* and its next edition, which had a seasonally appropriate cover illustration of Christmas candles, would note "the 'Grand Central Station' atmosphere that 2200 assumed on last Tuesday, November 14th to be exact. 'Lobotomy' has become a sacred word all over the place, but to the people who were in the O.R. I doubt if even 'Webster' could find a proper definition for the word. It is a pity that a talking movie was not made of the day's events. Haw!"

No movie was made, but the events that took place in 2200 that afternoon were recounted a few months later, on March 31, 1947, during the next meeting of the Joint Committee of State Mental Hospitals. Benjamin Simon gave a vivid account of the day and of his odd operating-room interview, while Superintendent Yerbury declared that the patient in question was "now on the road to recovery." After hearing their testimony, the committee agreed that their experimental "program for study and furtherance of such neurosurgery in the state's mental hospitals" appeared to be bearing fruit and voted to continue funding the Connecticut Cooperative Lobotomy Study.

Three days after the meeting, my grandfather again loaded his surgical tools into his Buick and made what was becoming a familiar drive back to the Connecticut State Hospital. In the four months since his first visit, he'd gone back at least four times and made a similar number of visits to the two other state asylums. The lobotomy project was well under way by then, and its pace was picking up. My grandfather—always so good with his hands—was becoming increasingly skilled at performing the procedure. More fluid, more confident, faster. On that particular spring afternoon, he would end up lobotomizing three inmates in room 2200, not just two.

ELEVEN

SUNSET HILL

Clinical Notes

August 9, 1944: Mrs. Scoville continues in Group I activities. She does, however, seem a little superficial emotionally and although she apparently has a little insight, shows that she does not have a fully true understanding. She realizes she should not take on her family duties as yet or the care of her children; that she should have a period of adjustment outside before returning to her home. She wants to go on a vacation up in the Adirondacks with her folks. It does not seem the right place at this time but the family is very insistent that she have this opportunity to see what she can do under the conditions. The mountain resort is known as Keene Valley.

Sunset Hill. That's what we call it. The old house in Keene Valley, New York, about twenty miles southeast of Lake Placid, is still in our family, though its ownership has divided and subdivided as the property trickled down through the generations. Growing up, I spent several weeks there every summer, and now my daughter and I do the same. In most of my memories of my grandmother, she's there, at that house. Like all seven of her grandchildren, I called her Bambam, because that was how my older brother mangled the word *grandma* when he was a toddler, and the nickname stuck. Often, in these memories, Bambam is sitting on the front porch, reading or just looking out over the hill and the mountains beyond it. She can see Spread

Eagle, Noonmark, Cascade, and Porter, left to right. On a clear day she can see all the way to Mount Marcy, which at 5,343 feet is the tallest mountain in New York State. Marcy is a dwarf by Himalayan or Andean standards, but the Adirondacks were once just as imposing. Eons of glaciers and oceans of rain have ground them down, worn them away to stumps. Everest is simply younger.

There used to be a Ping-Pong table on that porch, and when I was a kid my grandmother and I would play. Summer by summer it became harder for her to see the ball. She could beat me at first and then she could hold her own and then I could beat her. Eventually, when I was maybe thirteen, she tried and tried and couldn't return a single serve. We stopped halfway through. "I guess I just can't do this anymore," she said.

During her final decade she didn't go out on the porch as much. She'd stay in the living room, lying on her back on the couch. A piano she used to sometimes play went slowly out of tune in the corner of the room, since no one else in the family had picked up the instrument. She'd listen to books on tape. Mysteries, mostly. She liked Agatha Christie. Sometimes I'd read *The New York Times* to her, and she'd correct my pronunciation. *Coll-eee-gial,* not *coll-eh-gial; Mue-nick,* not *Mue-nitch.* And then it became too difficult for her to follow the story lines of the books or the articles, so she'd just listen to classical music. Bach or Mozart or Chopin. It became harder and harder to tell if she was awake.

She died when she was 101 years old. Of course she was diminished. Time happens. She was not the woman I once knew, and there is nothing strange or unusual about that.

The question I have is whether the woman I once knew was anything like the woman she really was.

A few days after my grandmother clambered onto the hood of a neighbor's car and placed a noose around her neck, my grandfather wrote a letter to his parents explaining the situation. He asked them to keep the matter quiet, to not tell his wife's family. He also expressed his disbelief.

"Her breakdown came totally unexpectedly for me and I cannot yet realize it or accept it. I love her so—more than anyone in the world—which makes me believe that I can coax her back to us." Later he describes one of the most puzzling aspects of her breakdown. "She is," he wrote, "the most outgoing person I have ever known—entirely the opposite of schizophrenia." He had underlined the word *outgoing* with two fierce strokes of his pen.

I found that letter recently, and the line shocked me. My grandmother was one of the most withdrawn people I've ever met. Whatever the opposite of outgoing is, that's what she was. She spoke when spoken to. If you engaged her in a conversation she was articulate and intelligent, and her memory, at least when she was younger, was sharp. But she was passive. She could sit quietly for hours in a crowded room, never asking a single question or initiating a conversation.

That was the woman I knew. That was the woman my sisters and brother knew. That was the woman my mother and her brothers knew.

My grandfather's letter indicated she had once been different.

After reading it, I thought back to the old stories I'd heard about her youth and realized that there had always been a disconnect between those stories and her later self. This was a woman who'd partied with raucous gun-toting flapper friends drinking bathtub gin in the 1920s, who'd headed off to live in Vienna on her own after college, who'd dated and then ditched the younger brother of Katharine Hepburn. In old pictures in family albums, she's often out on some adventure, horseback riding on a friend's ranch out West, rock climbing in the Adirondacks, skiing in New Mexico. It's always hard to imagine old men or women as they were when they were young, but in my grandmother's case it was impossible to connect the woman I knew with the woman in those stories and pictures.

When I found that letter, I didn't know exactly how many courses of electric shock my grandmother had received. I still don't know. I don't know how often they locked her in that copper coffin and cooked her to 105 degrees, or how many times they strapped her into a tub full of cold water. I don't know whether they ever injected her with insulin and sent her into an artificial diabetic coma or knocked her out with a dose of primal-terror-inducing Metrazol. The records I have from her four-month-long residence at the asylum are incomplete. There are missing pages, monthlong holes in her history. Also, those documents deal only with her first stay at the asylum. She was reinstitutionalized several times, at the Institute of Living and elsewhere, and the records of what happened to her during those subsequent stays have been lost or destroyed.

Clinical Notes

> August 15, 1944: Mrs. Scoville was this morning discharged to the custody of her mother Mrs. Learned. They were to go immediately today to Keene Valley in the Adirondack Mountains, she would remain with members of the family for some time. The family was quite insistent that they go now rather than later, or that any other different arrangements be made now. They realize that she, perhaps, was not entirely well yet but was anxious to give her this opportunity; agreed that she should not return to her husband and family for some months until she was completely able to handle herself emotionally or until she had adjusted well. Mrs. Scoville was quite pleased to be leaving the Institute at this time.

A few months after that summer at Sunset Hill, my grandmother moved back in with my grandfather and their children at their temporary residence in Hopkinton, Massachusetts. My mother was six years old. In humans, long-term memories begin to stick sometime around that age. Prior to that, give or take a few years, we all suffer from what memory researchers describe as "infantile amnesia." The causes of that amnesia are still in debate: Some attribute it to the physical immaturity of young brains, while others argue that the language faculties of infants are simply too undeveloped and that we require words to remember events. In any case, some of my mother's earliest memories are from that house in Hopkinton. She remembers my grandmother spending hours in the backyard, silently tending to a victory garden, and she remembers that when the war ended later that year the community set fire to a derelict house in celebration. Everyone in town gathered to watch the old home burn.

TWELVE

EXPERIMENT SUCCESSFUL, BUT THE PATIENT DIED

Two weeks after the liberation of the Dachau concentration camp, three men sat in the ruins discussing atrocities. Two of the men were American soldiers—an interrogator and an interpreter—and the other was a liberated prisoner, an Austrian patent lawyer named Anton Pacholegg, who'd earned the wrath of the Nazi regime by "having dealings with the Jewish people." It was May 13, 1945. Pacholegg was shipped to Dachau at the end of 1942 and had been there ever since. At first, his duties at the camp were rudimentary: He swept the alleyways, helped pull a street roller, and worked in the gravel pit. In 1944, however, his professional background caught the attention of the camp administrators, and he was transferred to an office in a building known as the First Experimental Station of the Luftwaffe. During his tenure there, Pacholegg told the Americans, the station changed its name a couple of times, first being shortened to just Experimental Station and then, in March 1945, receiving its final name: Experimental Station: Experimenting on Living Humans for the Benefit of Mankind. Pacholegg's job at the station was to compile reports on the experiments conducted there.

The experiments were designed to provide information useful to the German war effort. For example, German aviators often had to eject from high altitudes, and when they did, the low-pressure environment could wreak havoc on their bodies, leading to ruptured lungs, burst blood vessels, and various other side effects that came with having all the oxygen in their bodies suddenly expand to many times its normal volume. Since the dawn of aviation, when it was observed that balloonists who ascended past a certain height found themselves in acute physical distress, scientists had conducted experiments to understand high-altitude physiology. That research often involved placing rats and other small mammals into pressure chambers and seeing what happened when the pressure was decreased, simulating high-altitude environments. Still, it was unclear how useful animal experimentation was to understanding the unique physiologies of human beings. At Dachau, as Pacholegg would explain there in the ruins of the camp, the Nazis devised a strategy to forgo animal experimentation altogether.

Interrogator: What was your function at this experimental station?
Pacholegg: I was a clerk.
Interrogator: In light of your being at this investigation, what would you say of interest to this proceeding as to what you know of this experimental station?
Pacholegg: First I want to talk about experiments about air pressure in connection with the Luftwaffe. The Luftwaffe delivered here at the concentration camp at Dachau a cabinet constructed of wood and metal measuring one meter square and two meters high. It was possible in this cabinet to either increase or decrease the air pressure. You could observe through a little window the reactions of the subjects inside the chamber. The purpose of these experiments inside the cabinet was to test human energy and the subject's capacity and ability to take large amounts of pure oxygen and then to test his reaction to a gradual decrease of oxygen—almost approaching infinity. . . . It was simply a method of testing a person's ability to withstand extreme air pressure. Some experiments would have no visual physical effect on a person but would only be indicated by meter recordings. There were extremes, however, in those experiments. I have personally seen through the observation window of the chamber when a prisoner would stand [in] a vacuum until his lungs ruptured. Some experi-

ments gave men such pressure in their heads that they would go mad and pull out their hair in an effort to relieve the pressure. They would tear their heads and faces with their fingers and nails in an attempt to maim themselves in their madness. They would beat the walls with their hands and head and scream in an effort to relieve pressure in their eardrums. Those cases of extremes of vacuums generally ended in the death of the subject. An extreme experiment was so certain to result in death that in many instances the chamber was used for routine execution purposes rather than an experiment. . . . The experiments were generally classified into two groups, one known as the living experiments and the other simply as the X experiment, which was a way of saying execution experiment.

The American soldiers probed deeper, pushing Pacholegg for more details, and he went on to describe a number of the other experiments conducted at the station. To test treatments for shrapnel injuries and other common frontline wounds, camp personnel would lesion the limbs of prisoners and fill them with bits of metal and wood. Sometimes they waited until gangrene had set in, and sometimes they accelerated the process by injecting them with gangrenous tissue samples. Typhus was another chronic problem in the battlefield, so hundreds of prisoners were infected with it and then administered a variety of unproven vaccines. Similar experiments were conducted related to the treatment of malaria, and tens of thousands of malarial mosquitoes were shipped to Dachau for use on the prisoners. During other experiments, prisoners were shot in the thigh with bullets coated with various poisons, and precise recordings were taken of the time that lapsed between penetration and death.

Then there were the so-called freezing experiments. Throughout the war, the German Luftwaffe lost numerous aircraft over the North Atlantic. Many of the aircrews successfully parachuted from their planes before impact, only to freeze to death in the cold ocean waters. Some were pulled from the ocean while still alive but died anyway due to hypothermia. At Dachau, the Nazis decided to conduct research into the limits of cold endurance, as well as the most efficient and effective ways to rewarm hypothermic human beings. These particular experiments were conducted mostly in a part of the station known as Block Number Five. There the Nazis had installed a large wooden basin, two meters long and two meters high. The basin was filled with water, and ice was added until its temperature dropped to 37.4 degrees, a few degrees above freezing. Prisoners were immersed either naked or while wearing the standard flight suits of the German air force. Their temperatures were monitored through rectal thermometers, and they were typically kept in the basins until their bodies were chilled to 77 degrees, although they almost always lost consciousness at 89 degrees. Eventually they were removed, and attempts were made to resuscitate them. Some were wrapped in blankets, while others were subjected to more aggressive rewarming tactics. As Pacholegg told his interrogators, "another experiment conducted with these half-frozen, unconscious people was to take a man and throw him in boiling water of varying temperatures and take readings on his physical reactions from extreme cold to extreme heat. The victims came out looking like lobsters. Some lived, but most of them died. Scientifically I cannot understand how they lived."

At the end of the interview, the army interrogators asked Pacholegg if there was anything else he wished to add about his time chronicling the activities at the Experimental Station. There was, he told them.

"I remember," he said, "[that] any report I made out almost always ended with the remark 'Experiment successful, but the patient died.'"

The nine-page transcript of Anton Pacholegg's interrogation eventually became document number 2428 in the first trial conducted by the United States Nuremberg Military Tribunals after the war. The Doctors Trial, as it became known, had twenty-three defendants, all of them Nazis, most of them doctors, and the opening line of the prosecution's opening statement was blunt and clear: "The defendants in this case are charged with murders, tortures, and other atrocities committed in the name of medical science." Indeed, "medical science" in Germany had been so corrupted, according to the chief prosecutor, that it demanded the coining of new words to describe it: "This case and these defendants have created this gruesome question for the lexicographer. For the moment, we will christen this macabre science 'thanatology,' the science of producing death."

The evidence against them was overwhelming. Apart from the damning eyewitness testimony from men like Pacholegg, the Nazis had meticulously documented their activities, producing a vast paper trail. In 1941, for example, a Luftwaffe physician named Sigmund Rascher, who would oversee the Experimental Station at Dachau, wrote a letter to Heinrich Himmler, the head of all medical services within the Third Reich. In the letter, Rascher lamented the fact that within the air force "no tests with human material had yet been possible for us, as such experiments are very dangerous and nobody volunteers for them." Rascher also noted that the tests "theretofore made with monkeys had not been satisfactory" and inquired whether concentration camp inmates might be provided for him to use instead. An assistant of Himmler's immediately wrote back, informing Rascher that "prisoners will, of course, gladly be made available."

Two years later, on February 17, 1943, Rascher sent an update to Himmler, this one including a short report on a new venture. The report, titled "Experiments for Rewarming of Intensely Chilled Human Beings by Animal Warmth," chronicled the use of Gypsy women shipped in from the Ravensbrück concentration camp in northern Germany to warm some of the frozen male prisoners during the hypothermia experiments. "In eight cases the experimental subjects were then placed between two naked women in a spacious bed. The women were supposed to nestle as closely as possible to the chilled person. Then all three persons were covered with blankets." The report included graphs depicting the relative rates of rewarming when one or two women were used, and noted that in certain rare cases the frozen men recovered sufficiently to perform sexual intercourse. In the cover letter to this report, Rascher mentioned that he was beginning to experiment with freezing the prisoners by simply "leaving them outdoors naked from 9–14 hours" in midwinter instead of using the ice water method, but that he believed such experiments would be better conducted elsewhere. "Auschwitz is in every way more suitable for such a large serial experiment than Dachau because it is colder there and the greater extent of open country within the camp would make the experiments less conspicuous (the experimental subjects yell when they freeze severely)."

The prosecution stressed that despite the clear monstrousness of the experiments, the monstrousness of the Nazis who conducted the experiments might be harder to recognize. In many cases, they did not conform to our usual understanding of what a monster is and isn't. "These defendants did not kill in hot blood, nor for personal enrichment," the chief prosecutor said. "Some of them may be sadists, who killed and tortured for sport, but they are not all perverts. They are not ignorant men. Most of them are trained physicians, and some of them are distinguished scientists. Yet these defendants, all of whom were fully able to comprehend the nature of their acts, and most of whom were exceptionally qualified to form a moral and professional judgment in this respect, are responsible for wholesale murder and unspeakably cruel tortures."

How was this possible?

The answer, according to the prosecution, was that the guiding principles of the Nazi state had caused a "moral degradation" of the German people and that moral degradation led to the physical degradation of other human beings. The crimes of the Nazi doctors, the prosecution argued, "were the inevitable result of the sinister doctrines which they espoused."

The trial lasted for almost a year. By the time it was nearing its end, the prosecution had effectively demonstrated that the Nazi "investigators had free and unrestricted access to human beings to be experimented upon" and had treated them like disposable "human guinea pigs."

. . .

The research conducted by the Nazis at Dachau and other concentration camps was perhaps history's most brutal and sustained example of inhumane human experimentation, but it wasn't the first. The broken have always illuminated the unbroken, and throughout history that breaking was often intentional. Around 300 B.C.E., in Alexandria, Egypt, two doctors named Herophilus and Erasistratus pioneered the craft of human dissection, and although most of their subjects were dead, there is evidence that some were not. Chronicling the work of those two doctors, the Ancient Greek historian Celsus described how, since "pains, and also various kinds of diseases, arise in the more internal parts, they hold that no one can apply remedies for these who is ignorant about the parts themselves; hence it becomes necessary to lay open the bodies of the dead and to scrutinize their viscera and intestines. They hold that Herophilus and Erasistratus did this in the best way by far, when they laid open men whilst alive—criminals received out prison from the kings—and whilst these were still breathing, observed parts which beforehand nature had concealed, their position, color, shape, size, arrangement, hardness, softness, smoothness, relation, processes, and depressions of each, and whether any part is inserted into or is received into another." A couple hundred years later, during the first century B.C.E., the Egyptian pharaoh Cleopatra supposedly ordered her own series of experimental vivisections on humans. At the time, there was a debate about whether male fetuses developed more slowly in the womb than female ones. In an attempt to settle the question, Cleopatra is said to have had a number of her own handmaidens forcibly impregnated, then dissected at various stages of their pregnancies while still alive.

Although vivisection was a rare extreme, the history of medical research is filled with unsettling experiments involving human beings. For example, in 1796, after noticing that workers on dairy farms almost never contracted smallpox, the British physician Edward Jenner decided to test a theory that this was because they had previously been exposed to the relatively benign disease known as cowpox. He made a series of small incisions in the arm of his gardener's son, eight-year-old James Phipps, then introduced the pus from a local milkmaid's cowpox blisters under Phipps's skin. During the following week, Phipps developed the mild fever, aches, and pains characteristic of cowpox, then recovered fully. Six weeks later, Jenner lanced his arm again and this time administered him smallpox, at the time the most deadly disease known to man. Phipps did not develop any symptoms, so Jenner exposed him again and again, twenty times in all, to no effect. Finally, Jenner concluded that he had discovered a smallpox vaccine. His discovery would change the world, leading not just to the eradication of smallpox but to the creation of modern immunology and the subsequent development of vaccines for hundreds of other diseases. Today it's possible to make a persuasive argument that Edward Jenner saved more human lives than any single person in history. Taking this into account, perhaps it's easy to argue that jeopardizing the life of an eight-year-old boy was acceptable.

In other cases, however, the experiments that led to medical breakthroughs were more troubling, and the calculus becomes murkier.

In 1845, a South Carolinian physician named J. Marion Sims undertook a four-year-long program of experimental surgeries on fourteen black women, all of whom were slaves and most of whom he had purchased and installed on his property as live-in test subjects. His surgeries were aimed at developing a treatment for vesicovaginal fistula, a potentially fatal complication of childbirth that was common at the time, and he operated on some of his slaves as many as thirty times each. Anesthesia was still in its infancy, and he didn't use any. After much trial and error, and many deaths from infection, Sims developed an effective surgical approach. Only then did he begin operating on white women. Sims went on to become president of the American Medical Association and is widely considered the father of modern gynecology. To this day, visitors to Central Park in New York City can see the larger-than-life bronze statue of him standing right across the street from the New York Academy of Medicine.

And in 1932, the U.S. Public Health Service launched the Tuskegee Syphilis Experiment, a long-term study that over the next four decades monitored the effects of syphilis on a group of black Alabaman men who were never told they'd been infected. Syphilis is fatal when left unchecked, but easily treated: The researchers could have saved many lives with a few prescriptions of penicillin, but they chose not to, preferring to observe the disease rather than cure it.

For most of human history, our attitudes toward human experimentation were strictly utilitarian. If the scientific benefits were great enough, then almost any cost was justified. In an 1895 article called "The Relative Value of Life and Learning," a prominent University of Chicago chemist named E. E. Slosson summed up this attitude when he wrote that "a human life is nothing compared with a new fact in science." He scoffed at those who held that "the aim of science is the cure of disease, the saving of human life," and argued that "quite the contrary, the aim of science is the advancement of human knowledge at any sacrifice of human life."

But the horrific experimentation laid bare during the Doctors Trial at Nuremberg demonstrated the "moral degradation" that such a mindset could lead to. And to those who cared to look, Nuremberg also cast a harsh light on the ethics of scientific research being conducted elsewhere. Indeed, the chief defense of the Nazi scientists was to argue that what they did was, at a fundamental level, what scientists had always done, and that while their experiments may have been uniquely brutal, human experimentation of one sort or another was ubiquitous. It was hard not to concede that they had a point.

On August 20, 1947, the tribunal delivered its verdict. To the surprise of no one, all twenty-three defendants were sentenced to die by hanging. However, in a concession that the Nazi experiments were perhaps only different in kind, not character, from medical research conducted elsewhere, the tribunal's verdict also included a new declaration of the fundamental principles that they believed should govern research on humans from that day forward. The rules became known as the Nuremberg Code, and although the code did not itself hold the force of law, it was a vastly influential template and inspired a spate of new laws related to the conduct of medical experiments worldwide.

This is the Nuremberg Code:

1. Required is the voluntary, well-informed, understanding consent of the human subject in a full legal capacity.
2. The experiment should aim at positive results for society that cannot be procured in some other way.
3. It should be based on previous knowledge (like an expectation derived from animal experiments) that justifies the experiment.
4. The experiment should be set up in a way that avoids unnecessary physical and mental suffering and injuries.
5. It should not be conducted when there is any reason to believe that it implies a risk of death or disabling injury.
6. The risks of the experiment should be in proportion to (that is, not exceed) the expected humanitarian benefits.
7. Preparations and facilities must be provided that adequately protect the subjects against the experiment's risks.
8. The staff who conduct or take part in the experiment must be fully trained and scientifically qualified.
9. The human subjects must be free to immediately quit the experiment at any point when they feel physically or mentally unable to go on.
10. Likewise, the medical staff must stop the experiment at any point when they observe that continuation would be dangerous.

While the Nuremberg trials were still in process, Charles Burlingame invited a man named Dr. Nolan Lewis to give a talk to his staff at the Institute of Living. Lewis was the "psychiatric adviser" to the international war crimes tribunal, and the talk he gave, which took place on January 15, 1947, was titled "Impressions of the Psychological Factors in Nazi Ideology." Lewis, in his analysis of the Third Reich, took the long view.

"In order to understand any ideology or social development one must return to the operation of primary laws or elementary principles," he said. "Anywhere along this great cosmic chain of evolutionary events . . . pathological events can and do happen. We detect and study pathological patterns in cells, tissues, organs, and in individuals, at chemical, physical, and psychological levels, and in higher cultural and political fields the so-called social pathological phenomenon appears often in actively destructive ways, throwing some parts of normal evolution either into a cul-de-sac or into an actual regression process, which retards or actually destroys some portion of civilization, the amount affected depending upon the size and virulence of the pathological tendency. The Nazi ideology was one of these pathological streaks."

Lewis explained to the audience that the social pathology of Nazism led to a remarkable transformation in the German people, leaving them with "a complete absence of human compassion as we understand it." He described how they were thus able to commit their crimes "indifferently, without undue emotional reactions," and that they were so morally debased that "the mud and other filth of the concentration camps affected the Nazis much more than human suffering." He then told the asylum staff that any attempts to understand the Nazis by seeking slivers of common ground were doomed to fail.

"We must stop thinking," he said, "that these Nazis are anything like us in their attitudes, thinking, or feeling."

After Lewis's talk, the employees of the Institute of Living went back to work, tending to the asylum's guests. In some ways, things had changed since the end of the war. There were no longer mandatory periodic blackouts, and the budget had expanded, allowing for the hiring of new staff and the launch of new construction projects. In other ways, however, things were the same as they'd always been. The asylum remained a place of constant activity, and Burlingame continued to encourage a multipronged approach to the treatment of his guests, prescribing the widespread application of heat, water, and electricity, not to mention the cold steel of my grandfather's surgical tools. He also continued to embrace new experimental treatments whenever they arose.

For example, at around the time of Lewis's visit, Burlingame hired a new staff psychiatrist named M. Marin-Foucher, who had recently developed a novel therapeutic technique. It involved a coffin-shaped cabinet much like the one in the Pyretotherapy Room, but in this case it was designed to have an opposite effect. Guests would be made to lie in the cabinet, strapped down between layers of blankets that contained rubber tubing. A freezing solution would then be pumped continuously through the tubing while a thermometer placed in the guests' rectums monitored their temperature. Once it dipped below 93 degrees, guests tended to remain unconscious until they were removed from the cabinet, between forty-eight to seventy-two hours later.

Marin-Foucher considered the treatment to be promising though inconclusive, and he eventually published his results in the asylum's in-house scientific journal. His paper was titled "Hypothermia: A New Treatment for Mental Illness," and like most academicians, he was careful to credit the relevant work by prior researchers. In this case, his scientific antecedents were clear: In the first paragraph of his report, he noted that he had been inspired by "the studies of the Germans on hypothermia in World War II."

THIRTEEN

UNLIMITED ACCESS

Dr. John Fulton sat in his book-stuffed office at Yale, speaking into his Dictaphone. He'd purchased the device, a primitive voice recorder that allowed you to temporarily preserve audio on wax cylinders, in 1927, and it quickly became a treasured possession. It allowed him to maintain an impressive correspondence with hundreds of friends and colleagues all over the world, ranging from Robert Oppenheimer to Thornton Wilder to Alfred A. Knopf. Most evenings, accompanied by a good bottle of Madeira wine, he would spend three to four hours dictating letters. The next morning one of his secretaries would transcribe them, then bring them to him for his signature. On this particular evening, August 24, 1948, Fulton was composing a letter to Max Zehnder, a young Swiss physiologist. Zehnder had visited Fulton's primate lab two years prior and had expressed interest in pursuing research there. His special area of expertise was the examination of "brain vessels under various pathological conditions," such as exposure to toxic chemicals. Fulton had not been able to accommodate Zehnder then, but a new opportunity presented itself.

"My dear Zehnder," he began. "I have a proposal which I think may appeal to you. It is this: The neurosurgical group at the Hartford Hospital, with the backing of Dr. Charles Burlingame, Director of the large psychiatric nursing home (the Institute of Living) at Hartford, is prepared to give you a staff appointment for a year which would guarantee your room and board during 1948–1949 with the understanding that you would be free to study their cases of frontal lobotomy."

Fulton sweetened the pot by offering Zehnder a simultaneous staff appointment at Fulton's own laboratory at Yale that, while unpaid, would look good on any young scientist's résumé. Mostly, though, he stressed the remarkable nature of the research opportunity at the asylum. "Dr. Scoville, a well trained neurosurgeon, has developed techniques for undercutting the orbital surface for removing the anterior cingulate and for undercutting Brodmann's areas 9 and 10. Dr. Burlingame has given Scoville unlimited access to his psychiatric material. They plan to carry out these procedures on several hundred cases during the year and they are eager for collaboration on the physiological side."

The next morning one of Fulton's secretaries transcribed the letter, and before presenting it to Fulton for his signature she made a mimeographed copy; that copy eventually migrated into the possession of the Yale archives, which stored it alongside Fulton's other Z-surnamed correspondents in a gray acid-free box; that box was placed in front of me on a long wooden table in a hushed, high-ceilinged reading room one afternoon six decades after the letter was written.

Institutions have memories.

Some institutions, such as Yale, do a good job of keeping those memories alive, preserving their documents and allowing the particulars of their pasts to be called up and reexamined days or years or decades later. This doesn't mean that their pasts are easy to understand or that their institutional memories are laid out in any sort of coherent narrative. They're not. But that's part of what makes archival research stimulating. It's an active process, and to get anything out of it, you have to put a lot in. The "finding aids" that help you navigate an archive are only rough guides, and you never know what you're going to unearth until you start looking. Sometimes you can get a little lost.

I once spent most of an afternoon reading through yellowing newspaper clippings that Fulton had taped into a scrapbook, most of them chronicling recent scientific "discoveries," such as the following Associated Press report from April 24, 1935:

"The old idea of [the] impassive, unemotional oriental was upheld by experiments described by Dr. G. M. Stratton of the University of California. He found white native Americans much more likely to react to their emotions than Chinese or Japanese. Emotions were tested by having a large, heavy hammer strike a hard blow within a few inches of the hand of men of each racial type. The Americans, generally speaking, had much less control over themselves when the hammer struck. They exhibited keen tendency to jerk their hands away, their blood pressure rose, their breathing speeded up, their pulse raced faster."

But sometimes those chance encounters reveal unexpected connections, and if you follow these connections from one to the other, they'll start to form a framework. In that way, institutional memories are just as subtly and complexly interconnected as human ones.

Max Zehnder accepted Fulton's offer, but he didn't last long at the Institute of Living. On January 6, 1948, about four months after taking up his position at the asylum, he wrote a letter of resignation to my grandfather, which he cc'd to John Fulton. Zehnder's English was imperfect, but it's clear that tensions between the Institute of Living superintendent, Charles Burlingame, and the Connecticut State Hospital superintendent, Benjamin Simon, had flared into some sort of turf war and that Max Zehnder had become either a pawn or a prisoner, depending on your perspective.

"Dear Dr. Scoville," Zehnder began, "I take the opportunity to thank you for all your kindness extended to me during my stay at Hartford. You are well informed about the situation and some of the difficulties encountered and as a foreigner I do not wish to interfere in these questions as I am unaware of the background in which I had no interest to penetrate." Among the difficulties Zehnder mentioned was that Superintendent Simon "declared me that I never could give any publication in connection with the Institute of Living in Hartford," after which Superintendent Burlingame "declared me that I should not leave the Institute even not for going over to the Hartford Hospital unless Dr. Scoville would ask for permission."

Most of Zehnder's complaints, however, had to do with the research he had come to America to conduct. Some of the problems he'd encountered appeared to result from simple miscommunication. For example, one of the first things Zehnder did after arriving was to take a full survey of all the patients who'd received lobotomies, making "abstracts of all their histories writing by long hand and without help." Then, after toiling for nearly a month, he discovered that "this whole survey was already made by the Connecticut Committee of Lobotomies," making his own survey valuable only as "an excellent Exercise in English, but not as scientific work."

More troubling to Zehnder was what he found when he reviewed the research being conducted in the Connecticut asylums. As far as he could tell, the lobotomy experiments lacked one of the cornerstones of the scientific method: proper controls. That is, the neurosurgeons and the neuroscientists who were cutting into patients' brains had been trying to glean the effects, and potential benefits, of the operations by studying only those patients and failing to conduct crucial comparative studies on similar patients whose brains they'd left intact. Zehnder stressed that "all observations before this time lack of any objective control necessary for real scientific work" and that for "exact investigation of effects" future lobotomy researchers would "need a simultaneous registrating control." As it was, Zehnder complained, "with the results of my work and the means employed I could not even reach the required standard of a scientifically serious publication."

After reading Zehnder's letter, I wanted to learn more about the apparently dysfunctional relationship between the two asylum administrators, Simon and Burlingame. I knew that Burlingame and

Fulton were close—in one of the letters Fulton wrote to my grandfather, he referred to Burlingame as "Burlie"—and I wondered whether letters between them might shed light on the matter. So on my next visit to the Yale archives, I requested to see box number 28, which contained Fulton's letters to and from B-surnamed individuals.

His correspondence with Burlingame turned out to deal mainly with scheduling the guest lectures that Fulton occasionally gave at the Institute of Living and shed no light on the tensions between the two old asylums. Frustrated at hitting a dead end, I rifled through some of the other folders in the box, which is how I stumbled upon Fulton's correspondence with a man named Paul Bucy.

Dr. Paul Bucy was a neurosurgeon and neuropathologist at the University of Chicago, but as a postdoc in the late 1920s he had spent six months in Fulton's lab at Yale. The two stayed in touch, and as their relationship deepened, their letters acquired a sort of breezy informality and warmth that was lacking in a lot of Fulton's other correspondence, which tended toward terseness. Bucy, unlike many in the scientific community, seemed unintimidated by Fulton and was willing to talk straight to him. For example, in 1948, Bucy sent Fulton a letter advising him that *The Precentral Motor Cortex,* a textbook Bucy had edited that included chapters by Fulton and a variety of other contributors, was being reissued. The letter was a boilerplate request for any revisions that Fulton might want to make. As a form letter, it was addressed "Dear Doctor," and at the bottom there was a stamp instead of a real signature. Fulton immediately wrote back:

> I have your bedbug letter signed with a rubber stamp about the new edition of your monograph on *The Precentral Motor Cortex.* Dear heavens, have you so many contributors that you are unable to write a personal letter and sign it yourself? You are really too young to go in for this sort of impersonal superficiality. If I weren't so fond of you I wouldn't have answered the letter.

Bucy responded with a pointedly handwritten note:

> My dear John, It must be nice to feel that you have so many friends that you can afford to throw gratuitous insults at the few loyal ones who remain. Believe me, I remain, sincerely yours, Paul.

It was not surprising that they'd become close. Their scientific interests were in line: Both of them were devotees of the lesion method and believed the best way to illuminate brain function was by destroying portions of the brain. Like most people in Fulton's lab, Bucy participated in the neurological lesioning of scores of primates while there—"I don't know that anyone ever accomplished quite as much as you did in six months," Fulton once wrote to him, "or who used so many monkeys"—and once Bucy left the lab he continued to mine the same vein, sectioning monkey brains in the service of science. At the University of Chicago, he teamed up with a similarly inclined German neurologist named Heinrich Klüver, and the two embarked on a fruitful campaign of vivisection, culminating in what they would declare, at the 1937 meeting of the American Physiological Society, to be "the most striking behavior changes ever produced in animals."

Their breakthrough had come about almost completely by accident. Like many brain scientists then and since, Klüver was fascinated by mind-altering substances. This fascination wasn't only clinical: Klüver often used himself as a test subject, ingesting massive doses of various drugs and then taking copious notes. Mescaline was his hallucinogen of choice, and he recruited Bucy to help him investigate its precise neurological effects. Since Klüver had noticed that humans and primates tended to make compulsive chewing motions while tripping on mescaline, he came up with a simple, if brutal, experimental approach: He would inject a number of macaque monkeys with the drug, then have Bucy remove various parts of their nervous systems. If the removal of a particular area caused the chewing motions to cease, he would take that as evidence that it was that particular area that mescaline acted upon.

Bucy first targeted the trigeminal nerves. This had no effect. Then he went after the facial nerves. The tripping monkeys continued their compulsive chewing. Lesioning both their trigeminal nerves and facial nerves didn't do anything, either. Finally, acting on the vague hypothesis that mescaline-induced chewing motions might be etiologically related to the spasmodic mouth movements made by patients suffering from temporal-lobe epilepsy, Bucy opened the skull of a monkey named Aurora and removed most of her temporal lobes bilaterally, including her hippocampus, uncus, and amygdala.

Once again, this did not stop Aurora's spasmodic chewing motions.

What it did do, however, was remarkable.

Aurora instantly acquired what Bucy and Klüver deemed "psychic blindness." The macaque, they wrote, "seems unable to recognize objects by the sense of sight. The hungry animal, if confronted with a variety of objects, will, for example, indiscriminately pick up a comb, a bakelite knob, a sunflower seed, a screw, a stick, a piece of apple, a live snake, a piece of banana, and a live rat. Each object is transferred to the mouth and then discarded if not edible." Aurora also developed a markedly more placid emotional affect, coupled with a loss of fearfulness and "increased sexual activity involving forms of heterosexual, homosexual and autosexual behavior." These results indicated that the temporal lobes, about which very little was known, seemed to be involved with the emotions and the sex drive. They also appeared to have an effect on memory, at least if a lack of memory was what caused Aurora's "psychic blindness," her inability to recognize previously known objects. Still, it was difficult to know how to interpret these findings precisely. After all, Aurora couldn't talk.

While Bucy was conducting his mescaline research with Klüver, he received a letter from Fulton complimenting him on his latest paper, which chronicled Bucy's experimental lesioning of the carotid sinus nerve in human beings: "The observations are exceedingly interesting, and they illustrate, as so many of your papers do, the unusual opportunities which any alert-minded neurosurgeon has to do significant physiological work, if he only has the gumption." Then

Fulton gave his former student a piece of advice, suggesting that if he wanted "to start something new, " he should look into the lobotomy, which Walter Freeman and James Watts had just begun performing in the United States. His advice had the phrasing of a hot stock tip: "I have an idea that in a year or so this will constitute one of the major phases of neurosurgery, and I think anyone in the field would do well to get in on the ground floor."

As it turned out, Fulton was simultaneously offering the same tip to many other people, often using almost exactly the same words. In a letter he wrote to another former student the same week, he stated that he thought the lobotomy was "going to be a highly significant procedure" and that he "would like to see someone here in New Haven get in on the ground floor and do a group of such cases." Then he added, "I feel certain that the procedure is justified."

It is hard to overestimate the effect that the endorsement of John Fulton had on the spread of the lobotomy. He was the most famous physiologist of his era, and his words instantly imbued what was initially a fringe-dwelling procedure with a new respectability. Fulton's proselytizing might have owed something to the fact that his own work with chimpanzees had inspired the procedure and that the lobotomy's ascendance would inevitably raise his own stature, but it would be unfair to call his endorsement entirely self-serving. Fulton had an ego, but he was also a passionate scientist. He hungered to understand the brain, and everything he had done up to that point in his career, the hundreds of experiments involving thousands of apes, monkeys, rats, and mice, was in service to that fundamental pursuit. Now, suddenly, if it took hold, psychosurgery offered an entirely new way to pursue that line of research, one that would involve a very different sort of animal.

Bucy, like most of his peers, responded enthusiastically to Fulton's suggestion. "I have been very much interested in what you have to say regarding the operations in psychotic states," he wrote a few days after receiving Fulton's letter. The fact that Freeman and Watts had begun performing lobotomies in the United States was news to Bucy, and he told Fulton that "it is not at all unlikely that we shall get going on something similar before long." Bucy also made clear that he would approach the practice of psychosurgery in a manner that would please his mentor: "We will go at it, of course, primarily from the aspect of studying these people from a psychological and physiological standpoint and of finding out in what respect their condition has been altered by the operation. It seems to me that most that is to be gained here is from the standpoint of study, certainly at the present time."

As his correspondence with Fulton continued, however, Bucy sounded several notes of caution.

To begin with, despite certain advantages humans had over other experimental test subjects, they had certain drawbacks, too. For instance, there was the fact that lobotomy patients were, well, patients.

"In animal experimentation," Bucy wrote, "one presumably always starts with a normal organism. At least that is the basis of the greater part of our animal experimentation. In the human cases, at the present time at least, such is never true." Bucy noted that there were historical exceptions, such as "in ancient Egypt where criminals were submitted to vivisection," but that in the present day, "so far as the patient is concerned, any experiments which we may perform are incidental to the care and treatment of disease. It is only for that reason that he submits himself to the experimentation."

Then there was the matter of the neurosurgeons themselves. Bucy was grateful to Fulton for inspiring a new generation to be interested in more than "mere mechanics," but he also worried that some of these surgeons were overreaching. He fretted about ambitious but oblivious men who "have had no real experience with investigation. They have no idea what the real scientific method or approach is. They know nothing about controls."

Finally, its experimental utility aside, the question of how to objectively evaluate the costs and benefits of the lobotomy as a treatment remained unanswered. In 1948, a respected Swedish neurosurgeon named Gosta Rylander took a subjective approach, publishing a critical paper based on extensive interviews with the families of patients he'd lobotomized. The interviews described, in chilling terms, the subtle but fundamental changes his patients had undergone. As one mother of a young woman put it, "She is my daughter but yet a different person. She is with me in body but her soul is in some way lost." Rylander's paper generated lots of attention within the psychosurgical community and inspired another blunt, straight-talking letter from Bucy to Fulton. "Rylander's statements are so obviously correct," Bucy wrote, "that no intelligent person who was not blinded by his own enthusiasm could doubt them. Frontal lobotomy is a worthwhile procedure in certain situations. But the procedure exacts a price."

Fulton replied somewhat defensively. "As you say the improvement may come at a price," he wrote, but "that is exactly what we are attempting to analyze through our Connecticut Lobotomy Committee. Bill Scoville has had a series of very striking responses to undercutting areas 9 and 10 and he feels that this less radical procedure does away with obsessions without causing intellectual impairment."

Bucy's correspondence with Fulton provides a raw glimpse of the strange tensions in the neuroscientific community in the middle of the twentieth century. On the one hand, Fulton's initial laboratory research and subsequent advocacy had sparked the rise of psychosurgery, which allowed for an entirely new type of neurological exploration. Asylums were suddenly giving neurosurgeons and physiologists "unlimited access" to their patients, and people like Fulton and my grandfather seized on the opportunities for research that this access allowed. But even within the ranks of these experimentalists were men like Paul Bucy and Max Zehnder, who worried that things were moving too fast and going too far.

If Fulton shared any of Zehnder's misgivings about the scientific usefulness of the research being conducted at the asylums, he didn't express them. Instead, in a letter to my grandfather after Zehnder's resignation, he took a potshot at Zehnder. "I think Dr. Zehnder is sincerely interested in studying lobotomy patients from a scientific standpoint," he wrote. "I also think he has ability, but he illustrates the futility of brains without technique." He ended the letter on an optimistic note: "I believe that our lobotomy work has an important future, and I tremendously appreciate your interest and cooperation in our lobotomy venture."

Fulton had reason to be optimistic. The press continued to describe lobotomies in effusive terms, running stories with headlines such as "Wizardry of Surgery Restores Sanity to Fifty Raving Maniacs" and "No Worse Than Removing Tooth." The Veterans Administration was promoting the procedure to shell-shocked World War II veterans and sponsoring lobotomy research at a variety of institutions, Yale included. Of course, even the staunchest defenders of psychosurgery realized there was something inherently ghoulish about it. The invitations to a Christmas party at Fulton's lab in 1948 featured a cartoonish illustration of Fulton holding a brain without frontal lobes in one hand and a carving knife in the other, and the evening's entertainment consisted of a bunch of singing graduate students who'd dubbed themselves the Lobotomy Quartet.

In the fifteen years since John Fulton's laboratory work had inspired Egas Moniz to invent the leucotomy, the landscape of brain research had changed in profound ways. While the therapeutic value of the lobotomy remained murky, its scientific potential was clear: Human beings were no longer off-limits as test subjects in brain-lesioning experiments. This was a fundamental shift. Broken men like Phineas Gage and Monsieur Tan may have always illuminated the unbroken, but in the past they had always become broken by accident. No longer. By the middle of the twentieth century, the breaking of human brains was intentional, premeditated, clinical.

Even Paul Bucy, despite his notes of caution, realized that this presented researchers with unprecedented opportunities. In one of his letters to Fulton, written in 1938, shortly after he had begun performing lobotomies himself, he couldn't help but gush over the excitement he felt transitioning from laboratories full of macaques to asylums full of Homo sapiens.

"Man," he wrote, "is certainly no poorer as an experimental animal merely because he can talk."

FOURTEEN

ECPHORY

Excerpt from a May 25, 1992, interview between Patient H.M. and a researcher at the Massachusetts Institute of Technology

Researcher: How long have you had trouble remembering things?
H.M.: That I don't know myself. I can't tell you because I don't remember.
Researcher: Well, do you think it's days, or weeks, months, years?
H.M.: Well, see, I can't put it exactly on a day, week, or month or year basis.
Researcher: But do you think it's been more than a year that you've had this problem?
H.M.: Well, I think it's about that. About a year. Or more. Because I believe I had an . . . this is just a thought that I'm having myself, well, I possibly have had an operation or something.
Researcher: Uh-huh. Tell me about that.
H.M.: And, uh, I remember, I don't remember just where it was done, in, uh . . .
Researcher: Do you remember your doctor's name?
H.M.: No, I don't.
Researcher: Does the name Dr. Scoville sound familiar?
H.M.: Yes! That does.
Researcher: Tell me about Dr. Scoville.

H.M.: Well, he, he would, he did some traveling around. He did, well, medical research on people. All kinds of people. In Europe, too! And the wealthy. And out on the movie stars, too.
Researcher: That's right. Did you ever meet him?
H.M.: Yes, I think I did. Several times . . .
Researcher: Was it in the hospital?
H.M.: No, the first time I met him was in his office. Before I went to a hospital.

It's unclear when that first meeting took place. I've gone through all of my grandfather's papers, at least the ones I've had access to, but records of that consultation do not seem to exist. I've read speculation that it probably happened sometime in 1943, but that doesn't make sense, because in 1943 my grandfather was still stationed at the military hospital in Walla Walla, Washington. More likely the consultation took place soon after the war, in 1945 or 1946.

Henry would have been in his late teens or, at most, twenty-one. A kid still. Maybe it was shortly after he graduated, three years late, from East Hartford High. Under his yearbook photo, Henry had placed a quote from Shakespeare's *Julius Caesar:* "There are no tricks in plain and simple faith." He'd had to watch from the crowd as his classmates walked across the stage, since he was ordered to stay in his seat to prevent the possibility of a ceremony-disrupting seizure. Maybe he'd recently started his first job after high school, salvaging scrap metal at the junkyard. Or maybe he'd already moved on to his second job, at Ace Electric Motors, spending his days winding coils of copper. At home, he spent his time listening to Roy Rogers on the radio and watching television. His hobby was collecting guns, and he had several—from rifles to an old flintlock pistol—which he took out into the fields behind his house for target practice. He wanted to build a model railroad but never did. He did build model airplanes, model cars. He couldn't drive. He had few friends.

Henry's parents and Henry's primary care physician, Harvey Goddard, had done what they could to manage his epilepsy, but it was clear that things were only getting worse, the seizures increasing in intensity and frequency. Out of ideas, Goddard suggested they consult my grandfather, who despite not being an epilepsy specialist had training as a neurologist that gave him more insight than a general practitioner.

Henry's mother or father would have accompanied Henry to the consultation. There, in my grandfather's office, they would have exchanged handshakes and hellos. Maybe the Molaisons dressed up for the occasion. My grandfather was surely his usual immaculate self, his thick, graying hair parted to the right, slick with olive oil, his handsome blue eyes direct and inquisitive. Maybe Gustave the electrician and my grandfather the mechanic made small talk and found some common ground in their tinkering. Or maybe not. The Molaisons were a poor family, with humble jobs. My grandfather was a prominent brain surgeon. There was a clear divide between them, a chasm of wealth, position, education, power.

Eventually my grandfather would have gotten down to business, methodically taking Henry's history, trying to find any clues that might explain his illness. He would have asked if there were any other cases of epilepsy in the Molaison family and learned that Henry's father had two cousins and a niece who all had some form of the disease, which hinted at a genetic component. He would have also asked if Henry had ever suffered any severe head injuries. Had he ever been knocked unconscious or received any violent blows that might have caused damage to his brain?

Henry would have told him about that distant day in July, about the bicycle he didn't see, about the fall, the blow to his head. Perhaps he leaned in to show the faint scar on his forehead.

If there was a record of this consultation it would consist of nothing more than my grandfather's clinical observations. A medical chart is no place for introspection, and he wouldn't have mentioned any personal memories Henry's story might have triggered. Even if Henry's bicycle accident reminded my grandfather of the first great loss in his own life, he wouldn't have written it down.

. . .

To understand that loss, it's necessary to step back about eight decades, to a courtroom in Brooklyn on February 24, 1875, when what was then the most scandalous trial in American history was nearing its end. The lead defense attorney, part of a formidable team that included a future secretary of state, paced the boards in front of the jury box and warned the twelve men sitting there that if they didn't acquit, they would be remembered as "actors in one of the greatest tragedies which has ever occupied the stage of human life." The lead prosecutor, meanwhile, had already told the jurors that "upon the result of your verdict, to a very large extent, will depend the integrity of the Christian religion."

The defendant was a preacher named Henry Ward Beecher. From the pulpit of Plymouth Church in Brooklyn Heights, Beecher had built a national following by railing eloquently against slavery, economic injustice, and inequality. He grabbed headlines with stunts like auctioning off a young slave at his church to raise enough money to buy her freedom, but it was his blunt, powerful oratory that earned him admirers like Oliver Wendell Holmes and Ulysses S. Grant. In 1863, after Beecher gave a series of speeches in England that helped persuade an ambivalent British public to side with the North in the Civil War, Abraham Lincoln said of Beecher that "there was not upon record, in ancient or modern biography, so productive a mind" and told his cabinet that if the North prevailed, "there would be but one man—Beecher—to raise the flag at Fort Sumter, for without Beecher in England there might have been no flag to raise." (Lincoln held Beecher's sister, *Uncle Tom's Cabin* author Harriet Beecher Stowe, in similar esteem, once supposedly describing her as "the little woman who started this big war.") By the beginning of the 1870s, Beecher was among the most famous and respected men in the country.

Then he was accused of adultery.

The accuser was a former friend, Theodore Tilton, who claimed that Beecher had seduced his wife, Elizabeth, and carried on an affair with her that lasted for years. Beecher denied everything and claimed that Tilton was just a desperate blackmailer. Tilton brought a civil suit in 1876, and the resulting trial was a sensation: The courtroom overflowed with journalists, witnesses, and curious onlookers, and *The New York Times* declared that "we can recall no one event since the death of Lincoln that has so moved the people as this question whether Henry Ward Beecher is the basest of men." There were many fifteen-cent tabloids devoted exclusively to coverage of the trial. "Pictorial History of the Beecher-Tilton Scandal. Its Origin, Progress and Trial. Illustrated with Fifty Engravings from Accurate Sketches," trumpeted the first page of one. Most of those sketches were of the courtroom proceedings—caricatures of witnesses becoming increasingly disheveled after hours of cross-examination—though some were re-creations of supposed events from the affair. One full-page drawing showed a woman in a billowing dress looking furtively over her shoulder as she knocked on the door of a Brooklyn brownstone, and bore the caption: "Mrs. Elizabeth R. Tilton Calling at the House of Her Pastor, The Rev. Henry Ward Beecher, on the 10th Day of October, 1868, When, as Alleged by Her Husband, She First Fell From Grace."

There was little hard evidence. The principal exhibits introduced during the trial were a series of letters Beecher had written to Elizabeth, her husband, and various mutual acquaintances. None of the letters were explicit confessions, but many hinted at some terrible secret Beecher was holding. One such letter, to a confidant named Frank Moulton, came to be known as the Ragged Edge Letter, and in it Beecher described his inner turmoil. "Nothing can possibly be so bad as the horror of great darkness in which I spend much of my time," he wrote. "I look upon death as sweeter faced than any friend I have in the world. Life would be pleasant if I could see that rebuilt which is shattered. But to live on the sharp and ragged edge of despair, and yet put on the appearance of serenity and happiness, cannot be endured much longer. . . . To keep serene as if I was not alarmed or disturbed, to be cheerful at home and among friends when I was suffering the torments of the damned; to pass sleepless nights often, and yet to come up fresh and full for Sunday—all this may be talked about, but the real thing cannot be understood from the outside, nor its wearing and grinding on the nervous system." The prosecutors described Beecher's agonizing introspection as the product of a guilt-stricken conscience, while the defense argued that the letters were simply the outpourings of an innocent but overly sensitive man. The letters were damning to some, exculpatory to others. It was a matter of interpretation, one that would be left up to the jury to decide.

Toward the end of the day, Beecher's lead counsel implored the jurors to consider the historical ramifications of their decision: "Gentlemen," he said, "by the judgment which you here pronounce you will yourselves be judged at the tribunal of the ages. What you do here will never die. When these scenes shall have passed away; when he who presides over this trial shall rest in the silent chambers of the dead; when the seats you occupy shall be filled by your children, or your children's children, strangers from distant climes will come to view the place from which was given back to the world, freed from cloud or passing shadow, the name of Henry Ward Beecher."

The jurors chose to acquit.

Beecher's name was never quite freed from passing shadow, however. Some suspicions are hard to shake, and the once unimpeachable reputation of the great preacher was forever tainted by doubt. Still, the Beecher clan remained fiercely proud of its lineage. My grandfather's father was Beecher's grandson, and he named his third son Henry Ward Beecher Scoville.

Henry was born in 1909, two years after my grandfather. The brothers had an idyllic childhood, close to each other and close to nature: Growing up, they often went on camping trips in the Connecticut woods or near their summer home on the coast of Maine. Their father would teach them the names of all the trees and the birds and his beloved snakes, and he gave the boys themselves goofy nicknames to mark these adventures—the eldest, Gurdon, became First

Lieutenant Trottie, my grandfather became Second Lieutenant Honey Bee, and young Henry was simply Henny Penny. Honey Bee and Henny Penny had a particularly tight bond. They were brothers and best friends.

My grandfather went away to college in August 1924, and a few months later terrible news came from home. There'd been an accident. Henry had been riding a bicycle, was hit by another vehicle. A fall, a blow to the head.

His brother Henry was gone.

Henry was there.

There in my grandfather's office, during their initial consultation. When Henry told him of his own accident—the bicycle, the fall, the blow to the head—was my grandfather reminded of his lost brother? Our brains do tend to make connections like that, crossing years in a heartbeat. Memory scientists have a word to describe those sorts of moments, when something in the immediate present triggers the recollection of something in the distant past: *ecphory.*

Here's what is known for sure: He prescribed Henry an antiepilepsy medication called Dilantin. Dilantin works by dampening the electrical activity in the brain, thereby reducing the frequency and severity of seizures. It only treats the symptoms, though; it's not a cure. He also made an appointment for Henry to return to Hartford Hospital for a pneumoencephalogram, which was then the state of the art in brain-imaging technologies. This would allow my grandfather to see if there was, for example, a tumor pressing upon some portion of Henry's cortex, causing his seizures. Something tangible, something targetable.

Records do exist for that next visit. On September 4, 1946, after checking in to the hospital, Henry changed into a surgical smock and was led to a room on an upper floor. The room contained a chair resembling something astronauts might train in, full of belts and restraints, capable of rotation in almost any direction, including upside down. Henry sat in the chair, and a technician strapped him in, fixing his arms to the armrests and using a harness under his chin to immobilize his head against a backrest. There was an opening in the rear of the chair, and a needle was inserted into Henry's lower back, into the base of his spinal column. This so-called lumbar puncture allowed the technician to do two things: First he extracted all of Henry's cerebrospinal fluid, drawing it out like blood into a syringe. (Although CSF is an essential part of the body's nervous system, protecting and cushioning the brain and spinal cord, there is actually less of it than you might think: The average adult has only about 150 milliliters—10 tablespoons—at any given moment.) Second, after the technician completed suctioning out Henry's CSF, he pumped oxygen in to replace it. In order for the technician to ensure that the oxygen diffused completely, not just into the areas surrounding Henry's brain but through all his brain's ventricles—the cavernous spaces usually filled with CSF—the chair was then lifted and spun slowly into a variety of positions, allowing the oxygen to bubble up and fill all the crevices. This hurt. A splitting headache was an inevitable side effect of pneumoencephalography. One woman who went through the procedure gave a typical account: "It was the most painful experience of my life . . . it would make an excellent torture program for the military."

A pneumoencephalogram was essentially an X-ray of the head, though it differed from standard ones in important ways. Normal X-ray images are created by directing beams of radiation at a portion of the body and placing a photographic plate on the other side. Since the radiation passes through soft tissues but is easily obstructed by denser substances such as bones, the resulting images show bones in sharp relief while softer tissues appear as faint blurs. X-raying the brain, however, is a challenge, since the cerebrospinal fluid that the brain floats in is so similar in density to the brain itself that the contours of the organ become indistinct. A standard X-ray of the head gives a clear view of the skull but leaves the skull's contents amorphous and vague. The pneumoencephalogram, which was invented in 1919, was designed to overcome these limitations.

Once the technician was satisfied that oxygen had completely replaced Henry's cerebrospinal fluid, he righted the chair and took a series of X-rays. The resulting images, unclouded by CSF, were much clearer than standard X-rays would have been. It was possible to see the intricate folds of Henry's outer cortex and even a hint of some of the deeper structures within his brain. Henry was then led to a hospital bed, where he stayed for the next two days as his body replenished its vital fluids and his headache slowly faded.

When my grandfather reviewed the pneumoencephalograms, what he saw was good and bad. Good because there was no indication of a tumor or other brain abnormality that might be causing Henry's seizures. Bad because this meant that the root cause of the epilepsy remained a mystery. He scheduled a second pneumoencephalogram for Henry, in case the first had missed something. Then he sent Henry home with a discharge sheet indicating that he was to "continue on Dilantin indefinitely."

"These are the days of the New Psychology and the New Biographers. The first substitutes complexes for principles and inhibitions for morality. The latter are convinced that all great reputations of the past are illusions, to be shattered as rapidly as possible."

That was my great-grandfather, the lawyer/naturalist/novelist Samuel Scoville, Jr., writing in *The Philadelphia Inquirer* in 1927 in defense of his grandfather Henry Ward Beecher. An unflattering and bestselling biography of Beecher had just been published, one that portrayed him as a habitual adulterer and troubled soul, and my great-grandfather mounted a sort of anti-publicity campaign, flooding the pages of *The New York Times, The Atlantic,* and other publications with letters attacking the book. His point was that its author had focused so intently on the negative elements of Beecher's story that he was blind to the positive. His point was that anyone trying to understand another man's life has a responsibility to look at the man in full.

I found those letters almost a century after they were written and winced when I read them. I had started researching my own grandfather's story and in the process was catching glimpses of darkness. The darkness fascinated me—all those asylums, all those lesions, all those broken men and women—but I also knew that the letters were right: It's easy to judge the dead, and they can't defend themselves. None of us are all light or all dark, and most are both at once. My grandfather was no exception.

Henry went home, and my grandfather continued his work.

Maybe he performed an operation later that day. Maybe he removed a meningioma, a cancerous tumor that clings to the brain stem like an ornery crab. Maybe he teased out a damaged disk from between compacted vertebrae or tended to a car crash victim, performing an emergency craniotomy to drain a potentially lethal hematoma. Maybe he used some of his many custom-made and patented tools: a Scoville retractor to hold an incision open, a Scoville clip to pinch a vein shut. Maybe he saved a life. He saved a lot of lives. That was his job, and he did it well. Or maybe he lost one. He was always candid and honest when he lost one. These were the days before malpractice suits made doctors reticent to admit their mistakes. A former colleague of my grandfather's remembers him once emerging from the operating room after a procedure had gone awry. He'd accidently sliced open a patient's middle cerebral artery, tried to sew it back together with silk sutures as the blood spurted, couldn't finish in time. A woman was waiting outside the OR, nervous, hoping for good news. My grandfather walked straight to her.

"Ma'am," he said, "I think I killed your husband."

He saved a lot more than he lost.

FIFTEEN

THE VACUUM AND THE ICE PICK

The Institute of Living, November 19, 1948

The bright lights of the operating room, positioned just so, illuminated Marie Copasso's exposed skull. Earlier, the scrub nurse, Florence Dudin, had shaved the hair from the top front portion of Copasso's head, making it easy for my grandfather to scalpel a half-moon incision and then roll the skin of her forehead down like a carpet. The white bone beneath, once the blood had been flushed away, gleamed.

From a small, elevated observation room above the OR, a crowd of people watched the proceedings. The operation on Copasso was a standing-room-only performance: There were at least two dozen people in the audience, most of them neurosurgeons. The two who had traveled the farthest were probably Eben Alexander, who had flown to Hartford from his home in North Carolina, and Kenneth George McKenzie, Canada's first neurosurgeon and the current president of the international Society of Neurological Surgeons, who'd arrived from Toronto. Like Alexander and McKenzie, the other surgeons here were among the biggest names in the field: Buckley, Fox, German, Maltby, Whitcomb. Apart from all the surgeons, there were a number of psychiatrists, neurologists, and research scientists present as well.

The keenest observer of them all was probably Walter Freeman, the man who'd introduced the lobotomy to America. This was because

Freeman, apart from my grandfather and Marie Copasso, had the most at stake. What was unfolding that day at the asylum was to be a surgical debate of sorts, and Freeman and my grandfather were pitted against each other. Freeman would have peered over the edge of the observation deck down into the theater and watched carefully as my grandfather lowered the drill onto Marie Copasso's skull.

The drill, or at least the drill bit, was something of a revelation. For the past few years, my grandfather had been renting one of the garage bays at the Mobil station on the corner of Washington and Jefferson in downtown Hartford, and most Sundays he would take one of his sports cars there and tinker with it. Recently he'd found a tool at the garage, a drill with a one-and-a-half-inch-diameter circular bit called a trephine, strong enough to penetrate autobody steel. It occurred to him that such a trephine might be perfect for cutting holes in skulls. He took one and sent it off with some instructions to a medical equipment manufacturer called Codman & Shurtleff; it custom-machined several just for him, slightly modified so that they would penetrate only as deep as the bone and stop short of the meninges, the delicate membranes surrounding the brain. Now he positioned one of these trephines on the exposed skull above one of Copasso's closed eyes and pressed the trigger on the drill, causing a violent whirring and a faintly visible plume of moist calcium. The drill itself, apart from its new bit, was the exact same type he might use to cut a hole in the body of his Buick.

He removed a small plug of bone from Copasso's forehead, then repositioned the drill above her other eye and repeated the process. When done, he placed the drill to the side, took up a scalpel, and sliced through the dura, arachnoid, and pia mater, exposing the rippled surface of Copasso's frontal lobes.

The operating room was state-of-the-art, all white tiles and antiseptic sheen. It had opened the month before, in the Institute of Living's newly constructed Burlingame Building, which sat adjacent to the

Burlingame Research Building on the northern edge of the asylum's grounds. It was the world's first and only operating room designed exclusively for psychosurgery. Like my grandfather's custom trephine, his operating room had been outfitted to his precise specifications: There were six germicidal lamps, a pair of X-ray-viewing boxes, a dissection-and-coagulation machine, and an adjustable operating table that could be fixed in any position and extended or shortened at will, accommodating any patient, no matter their size.

The room was part of Superintendent Burlingame's full-court press on the lobotomy. The Institute of Living had been providing lobotomies to its guests for a decade, since 1938, and Burlingame's longstanding familiarity with the procedure gave him, he thought, unparalleled practical knowledge regarding how to care for psychosurgery patients. The basic benefit of the operation, in his view, was obvious: It obliterated a patient's preexisting personality, allowing enterprising psychiatrists such as himself an opportunity to rebuild a new and improved personality in its place. "Leucotomy," he wrote, " 'clears the decks' for the construction of a more adequate personality."

With an eye toward taking advantage of this malleability, Burlingame had built an independent psychosurgical unit at his asylum, a unit that began with my grandfather's operating suite and extended to an entire floor of classrooms and a segregated residence ward in the same building. Patients received special training meant to reacquaint them with the expectations of society, training that included "constant guidance and drill in matters of good personal hygiene," vocational instruction, and even sexual reorientation. A significant portion of the Institute of Living's guests had been sent there for homosexuality, which Burlingame considered a mental illness, and he believed that gay patients' sexual preferences, postoperatively, were as flexible as their personalities. At a psychiatric conference earlier in 1948, he'd boasted of reversing the orientation of at least two lobotomized guests. There was the "young man who had always had difficulty in adjusting to the opposite sex. For some months postoperatively, he progressed in all respects except this one. But gradually, and carefully, he was introduced to the social hour and it is a source of considerable satisfaction to us that he is now dancing and acquiring quite a polished manner with members of the fair sex." Then there was "the woman in whose illness the precipitating factor had been an unfortunate love affair. Preoperatively and postoperatively she was viciously antagonistic toward all men, but much the same thing has been accomplished with her as with the young man. In both instances, it is doubtful that resocialization would have been spontaneously accomplished merely by sending these patients home or allowing them to follow their own bent."

Burlingame's affection for psychosurgery had made him both a leader in the field and a somewhat controversial figure: A few months earlier, during the 1948 meeting of the American Psychiatric Association, the nominating committee had backed Burlingame as its preferred candidate for the association's presidency, but Burlingame's presidential aspirations were upended by an anti-psychosurgery, pro-psychotherapy candidate named George Stevenson, who defeated Burlingame in large part by attacking him for being too quick to turn to the scalpel when it came to the treatment of his institution's guests. Burlingame, however, hadn't let that setback shake his faith in psychosurgery.

Soon Burlingame would begin overseeing the creation of Marie Copasso's new personality. First, though, her old personality had to be destroyed, and that part of her treatment was not Burlingame's domain. So that day in November, looking down on that shiny new operating room, Burlingame was just one more person in the audience watching my grandfather work.

Without him having to ask for it, the scrub nurse passed my grandfather a long, thin tool called a flat brain spatula, reminiscent of a shoehorn, which he inserted carefully into the hole in the right side of Copasso's forehead. He levered up that hemisphere of her frontal lobes and peered inside. He was looking for the neural fibers connecting the lower, orbital portions of the frontal lobes to some of the deeper structures in the brain. Once he spotted his targets, he inserted another tool, a suction catheter—a very small, slender, electric-powered vacuum—and sucked the fibers out. Then he retracted the spatula and the catheter and moved to the other hole.

He worked quickly, with gravity. He was often prickly while operating, snapping at nurses or residents if they displayed even a hint of sloppiness. He prided himself on precision. And that's what this operation was all about: precision. The problem with Walter Freeman and James Watts's standard lobotomy, like Egas Moniz's original leucotomy, which it descended from, was that it was sloppy. They made a mess of the frontal lobes. Even Freeman described his procedure as "mutilative." My grandfather was now two years into his participation in the Connecticut Cooperative Lobotomy Study, and while he had become intimately familiar with the traditional approach that was the focus of the study, he was also somewhat disillusioned with it. Although he was now so expert and fluid a lobotomist that he sometimes performed as many as five in a single day, he'd come to believe that the standard lobotomy caused too much frontal-lobe mutilation, which in turn caused an insidious and irreversible blunting of the personality. Whatever good results came of the lobotomy, he believed, were due to the simple and specific severing of the fibers between the frontal lobes and other parts of the brain. So why not just focus on those fibers and leave the rest of the frontal lobes in peace?

That's what he was attempting with this operation. He called it a fractional lobotomy, or an orbital undercutting, since it focused on the lower, orbital portions of the frontal lobes. He'd been practicing this novel approach for several months now, at the Institute of Living and at Connecticut's three state asylums, alternating between the traditional lobotomy and this new one of his own design and comparing the results. The initial results, anecdotal though they were, generated excitement at the highest levels. John Fulton, whose work on frontal lobe ablations in chimpanzees had inspired Moniz to begin leucotomizing humans, had become particularly intrigued by this new technique. Fulton had been unable to make it to the Institute of Living, since he had a prior engagement at the Surgeon General's office in Washington, D.C., but a number of researchers from his laboratory were in attendance.

In all, the operation took about an hour. Afterward my grandfather replaced the bone plugs in Copasso's forehead and stitched her up. Orderlies wheeled her out to the recovery room. Soon she'd be brought to the fourth floor of the same building, to the special ward for post-operative lobotomy patients that included Superintendent Burlingame's "retraining classroom for psychosurgery guests." My grandfather washed up and joined his friends and colleagues in the audience.

Another woman took Copasso's place on the operating table.

It was Walter Freeman's turn.

Freeman, too, had been working to come up with a new and improved lobotomy. Like my grandfather's procedure, with his customized autobody trephine, Freeman's new approach also involved a customized tool drastically repurposed from its original use. Whereas my grandfather's epiphany occurred in a garage, though, Freeman's happened in a morgue.

In Freeman's view, the principal problem with the lobotomy, as he and James Watts had devised it, was not that it mutilated the frontal lobes by slicing through too much of them but that it was too complicated to perform. As a major surgical operation, it required not only cutting holes in a patient's skull but also all the attendant complications and costs in terms of hospitalization and extended recovery times. And, most frustrating to Freeman, it required a neurosurgeon. The standard lobotomy was Freeman's idea, and he was its most vocal proselytizer, but when it came down to it he was relegated to being an observer in the OR while his colleague the neurosurgeon Watts did the actual cutting.

Watts was not there that day.

In fact, Freeman and Watts had become somewhat estranged recently, ever since Watts had arrived unexpectedly one afternoon at the office he and Freeman shared and was horrified to discover Freeman performing, unassisted, the new operation he was now about to demonstrate for this audience at the asylum.

The basic idea behind the new approach was to gain direct access to the frontal lobes without having to penetrate the walls of the skull. To perfect his technique, Freeman practiced on cadavers at the George Washington University morgue until he'd located their point of least resistance: the orbital bones at the rear of the eye sockets. Whereas most of the skull is thick and armored, this portion of it was comparatively thin and delicate. In the morgue, he experimented with several existing surgical tools, searching for something that was sufficiently thin to reach the orbital bones without damaging the eye and sufficiently strong to penetrate the bones themselves. He failed repeatedly, as the tips of these tools tended to snap off against the bone, a result that would be catastrophic in a living patient.

Then Freeman had an idea. It occurred to him that a certain slender, strong tool might be perfectly suited for breaking through the orbital socket.

The woman on the operating table was named Rebecca Adams. She was secured to the table with leather straps, and the electrodes of an electroconvulsive therapy machine had been attached to her forehead. Somebody flipped a switch on the generator, applying a burst of current that caused Adams to convulse against the straps and rendered her instantly unconscious. The electrodes were removed.

Freeman picked up his ice pick.

It didn't take long.

He pulled up one of Rebecca Adams's eyelids and inserted the ice pick just above her eyeball, sliding it back until it encountered the resistance of her orbital bone. He used a small hammer to tap it through, then pushed it approximately three inches into her frontal lobes. Once it reached the correct depth, he swished the ice pick quickly back and forth, slightly less than parallel to the horizontal plane of her skull. Then he returned the ice pick to its original position and drew it out the same way it went in. He wiped it clean and turned to Adams's other eye. Much like the earlier, standard lobotomy that he had pioneered, this operation was by its nature imprecise, since he could not see the connections he was hoping to sever and simply hoped that the ice pick was cutting them as it pivoted through her brain. But precision was not his goal. The transorbital lobotomy, as he had dubbed the procedure, was designed to be quick and easy. He believed that the only thing holding psychosurgery back from mainstream acceptance was the fact that it still constituted a major and expensive medical procedure, requiring the services of a neurosurgeon and a number of attending medical staff. In contrast, the transorbital lobotomy could be performed almost anywhere, by almost anyone. Freeman believed he could train any reasonably competent psychiatrist how to perform an ice pick lobotomy in an afternoon. He envisioned a day when patients might have their mental illnesses plucked from them even more easily than they might have a troublesome tooth pulled.

The entire operation, from the application of the electrodes to the final removal of the ice pick, took no more than fifteen minutes. Soon Rebecca Adams began to stir. Dark bruises were forming under her eyes, but other than that she was, physically at least, unchanged.

As I look back at that long-gone day, squinting through a haze of dusty surgical logs and old letters, it's tempting to ask whether the men in charge of the asylum were as mad as its inmates. From the perspective of the present, it can look as though the surgeons, the psychiatrists, the administrators, the whole lot of them, were trampling blithely, arrogantly, even insanely through the delicate soil of other people's brains.

But you can't put ghosts on the couch. Besides, madness usually implies a break with reality, a loss of rationality. This was something else. Those men in that operating room, like the men observing from the operating theater, were, as far as we know, all reasonable, intelligent, rational human beings. Their motivations in most cases were straightforward.

There was Walter Freeman, the striving son of a renowned surgeon, making his own play for greatness. There was Charles Burlingame, the asylum superintendent famous for his embrace of innovative new treatments, doubling down on his convictions. There was the audience, that assembly of surgeons and psychiatrists and physiologists, some seeing therapeutic promise, some seeing scientific potential, some seeing a little of both.

And there was my grandfather.

My grandfather's motivations were maybe more complicated than most. To begin with, his passion for medical research had always been equal to his interest in the purely healing aspects of the art. There was a through line of experimental curiosity you could trace from his early fascination with the misfiring adrenal glands of dwarfs to his later attacks on the frontal lobes of asylum dwellers. By 1948, the border between his work as a doctor and his work as a scientist had become impossibly blurred. For example, in the weeks leading up to that day's surgical showdown, he'd been putting the finishing touches on a paper that would announce his new procedure to the world. The paper's title—"Selective Orbital Undercutting as a Means of Modifying and Studying Frontal Lobe Function in Man"—perfectly captured the odd way in which medical practice and medical research overlapped in my grandfather's mind. His more precise operative approach to the lobotomy had opened up an entirely fresh cerebral landscape for him to alter at will, and while he hoped this new approach would be beneficial to his patients, he knew it also offered tantalizing and novel opportunities to illuminate the functions of the brain by means of experimental methods that were once, not long ago, limited to animals.

Plenty of neurosurgeons of his era straddled the line between practice and research, but none embraced psychosurgery with quite my grandfather's enthusiasm. By 1948, he was already well on his way to becoming one of the most prolific lobotomists in history, second only to Walter Freeman. Although there were other subfields within neurosurgery to which my grandfather also applied his inventive zeal—the treatment of spinal trauma, the removal of tumors, the clamping of aneurysms—psychosurgery had become his dominant interest.

The genesis of his passion was something he never discussed, at least not in public. He revealed it explicitly only once that I'm aware of, and that was in a very private forum, in a note he wrote to his second son, my uncle Peter, in the 1950s. Peter and my grandfather had a difficult relationship, and the note was clearly written after some sort of argument: "My constant bullying of you is only a desire to make you strong so that you can be independent of me and other 'bosses.' I shall not bring up the past," my grandfather wrote. Their argument apparently had something to do with my grandmother, whom my grandfather referred to as "E," short for "Emily."

"E's illness was constitutional and grave, and not environmental," my grandfather wrote. "I don't wish to discuss it but you will have to believe me in that, for I have spent 20 years in studying and operating on mental illness in the hopes of contributing to a cure of it—for E's sake."

Before the day was done, two more women would be led into the operating theater. Walter Freeman performed another transorbital lobotomy, and my grandfather concluded the session with a final orbital undercutting. By two P.M., the last patient had been wheeled away and the men were ready to leave the asylum. They met up a short while later at the University Club in downtown Hartford for dinner and debate.

The ostensible reason for that day's Psychosurgical Conference, as the Institute of Living had dubbed the exhibition, was to review and assess the preliminary results of the Connecticut Cooperative Lobotomy Study. The numbers were impressive: In the previous two years, under the auspices of the study, 550 men and women had been lobotomized in Connecticut's public and private asylums. By some measures, the clinical results of these lobotomies were encouraging: Some patients left the asylums and returned home, while others at least became more sedate and easier for staff to manage. In a note in the 1948 annual report of Connecticut State Hospital, Superintendent Yerbury, after noting that his asylum remained overburdened and that conditions there were "unfair to the patients and their families and continue unduly an economic burden to the State," had thrown in the following dash of optimism: "On the other hand, there have been brighter sides to the picture during the biennium. Most notable has been the introduction of the procedure known as the prefrontal lobotomy." The report goes on to enumerate a number of interesting facts about the procedure's rapid adoption at his asylum, including the fact that during the previous year more inmates had received lobotomies than had received dentures.

The drawbacks of the traditional lobotomy, however, were by then becoming well known, and there was a hunger for a new direction. Psychosurgery was at a crossroads, and my grandfather and Walter Freeman had just demonstrated two starkly different ways forward. My grandfather's "fractional lobotomy" was technical, complicated, precise, requiring the deft hands of a skilled neurosurgeon. Freeman's operation was the opposite: Anyone could do it. Freeman himself, after all, wasn't a trained surgeon of any sort. In fact, the operations he'd just performed were technically illegal, since he'd never had surgical privileges in Connecticut, or any other state for that matter.

There is no record of how the debate at the club unfolded, though later writings by both my grandfather and Walter Freeman give a hint of what might have been the basic thrusts of their arguments. "In spite of its extreme simplicity," my grandfather would write, Freeman's transorbital lobotomy was "undesirable" because of "its complete lack of precision." He added pointedly that even if such a lobotomy were ever to be a viable option, "it should be performed by surgeons and not psychiatrists." Walter Freeman, for his part, and somewhat bizarrely considering the fundamentally blind and rough aspects of his own method, criticized my grandfather's use of a suction catheter in his orbital undercuttings as being less precise than using a scalpel, deriding it as being like "using a vacuum cleaner over a bathtub of spaghetti."

Unsurprisingly in this contest waged largely in front of neurosurgeons, my grandfather won the day. Freeman's approach, after all, was a direct attack on their livelihood, since it opened up the practice of psychosurgery to nonsurgeons. And this was putting aside all questions of the blind, mutilative guesswork that lobotomizing by ice pick necessitated. The speed and ease of the operation was a poor trade-off, and many people were as horrified on a visceral level by the transorbital lobotomy as Freeman's estranged partner, James Watts. John Fulton captured the prevailing opinion when he wrote, in a letter to Freeman about his new approach, "Why not use a shotgun? It would be quicker." After that day at the asylum, despite having introduced the lobotomy to America, Freeman found himself on the outside of the tight-knit neurosurgical clique.

In another sense, both approaches prevailed. Freeman might not have been able to win over the neurosurgeons, but his whole point was that he didn't need to. He and a platoon of eager psychiatrist acolytes would soon spark a rapid uptake of his ice pick method. Over the next decade, Freeman, touring asylums around the country in a camper van he dubbed his Lobotomobile, would perform over 2,400 transorbital lobotomies, sometimes as many as twenty-five in one day.

My grandfather's orbital undercutting procedure, on the other hand, quickly replaced the old Freeman-Watts method as the preferred lobotomy technique among neurosurgeons. It was also clear that this first type of selective lobotomy was just the beginning. Indeed, from that day forward, whenever my grandfather performed an orbital undercutting, hoisting up the frontal lobes to get at the connections beneath, he would catch a glimpse of not only the frontal tracts that were his target but also some of the deeper structures that lay beyond them, specifically the mysterious uncharted territory of the medial temporal lobes and their network of intriguing structures, the purposes of which remained unknown: the uncus, the amygdala, the entorhinal cortex.

The hippocampus.

One of the neurosurgeons who knew my grandfather best once said to him, "Bill, the only criticism I have of you is that you never do the same operation twice." He was known as a tinkerer, a restless explorer in the operating room, never satisfied with existing techniques or methods, even the ones he had invented.

Maybe it was inevitable that once he had spotted that new and mysterious landscape on the horizon, he would eventually attempt to reach it.

PART III

THE HUNT

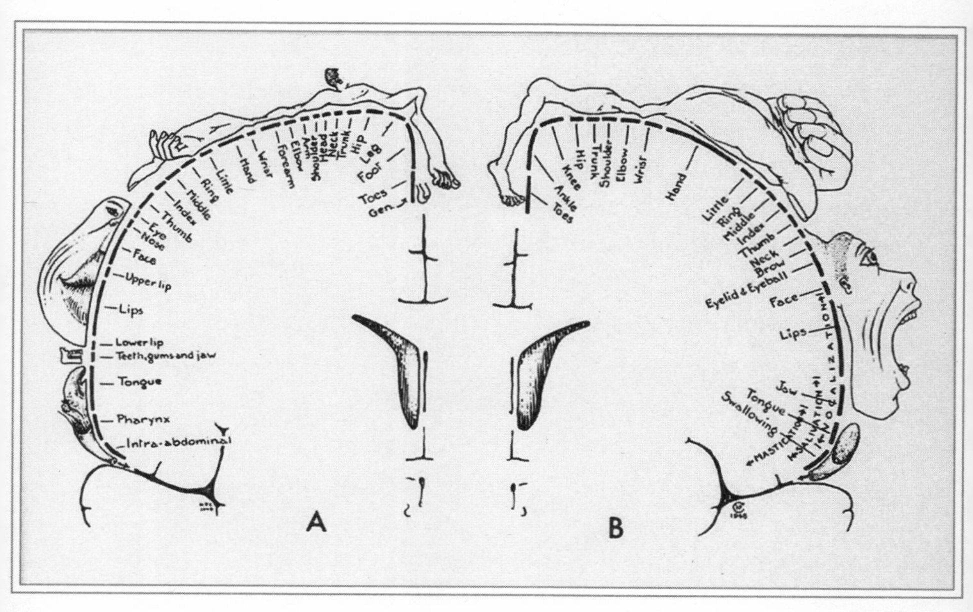

SIXTEEN

IT WAS BROUGHT INTO THE SEA

The psychologist told the man a story. She asked him to listen closely. She was reading from a booklet on her lap, and she pronounced each word carefully in her brisk British accent.

"Anna Thompson," she said, "of South Boston, employed as a scrub woman in an office building, reported at the City Hall Station that she had been held up on State Street the night before and robbed of fifteen dollars. She had four little children, the rent was due, and they had not eaten for two days. The officers, touched by the woman's story, made up a purse for her."

The man, who would later be known to the world as Patient P.B., listened attentively with a somewhat anxious expression on his face. It was October 23, 1951, and he was a forty-eight-year-old civil engineer. For most of the past thirteen years he had been suffering from recurring seizures. Unlike the grand mal, full-body seizures that the public generally associated with epilepsy, his were of the type known as psychomotor seizures. A person with psychomotor epilepsy experienced a temporary bout of automatism, perhaps performing a nonsensical motion or repeating a single word or simply standing still. For the duration of their seizures, usually less than a minute, these patients would be entirely absent. In the case of the civil engineer, his seizures usually began with a dawning sensation that the world around him had suddenly become absurd, that reality itself had become, in some ineffable way, unreal. Then he would lose awareness. He would stare straight ahead, moving his mouth as though he were chewing and fumbling the air with his hands. This would go on for several minutes. The man was an amateur meteorologist, kept a barometer outside his house, and was in the habit of making daily notations about the weather. If one of his attacks occurred while he was at home, he would often go to the porch, check the barometer during his period of automatism, and make an accurate note of what he found, though he never remembered having done so once the seizure subsided.

His first surgery had taken place five years previously. The surgeon removed a small portion of the brain region known as the medial temporal lobes. The operation did not have any apparent adverse effects, but it was not a success, either: Soon after he left the hospital, the seizures returned. So he came back here, to the Montreal Neurological Institute, for a second operation. This time, the same surgeon performed a more extensive removal, lesioning a larger portion of his brain. And now, three weeks later, the psychologist, whose name was Brenda Milner, was examining him to see if the second operation had caused any functional deficits.

The initial results were impressive. He didn't seem to demonstrate any of the usual postoperative aphasia that people who undergo brain surgery often do. That is to say, words came to him without effort. His intelligence quotient, measured on the standard Wechsler scale, had gone up slightly, from 119 to 120. This wasn't entirely unexpected: Epilepsy often left people somewhat foggy-headed, and the alleviation of symptoms following an operation often brought about a crispness of thought. In any case, both scores were indicative of a superior intelligence, markedly above the average. Then she moved on to testing the patient's memory. Like intelligence, memory was also usually unaffected by the operations performed here. After reading him the story about the mugging victim, Milner asked him to recount it to her, including every detail he could recall.

He paused for a moment, gathering his thoughts, then began.

"Anna X," he said, "reported that she had been robbed last night of several hundred dollars that she needed to pay for the sustenance of her children. They said that they would do what they could immediately to help her."

Milner made a note that "although the gist of the story has been retained," apparently "all the precise details of names, places, and quantities have been lost." Then she immediately read him another story, asking him again to listen carefully and repeat it back to her.

"The American liner *New York*," she read, "struck a mine near Liverpool Monday evening. In spite of a blinding snowstorm and darkness the sixty passengers, including eighteen women, were all rescued, though the boats were tossed about like corks in the heavy sea. They were brought into port the next day by a British steamer."

The patient recounted this second story back to her as follows:

"The American liner *New York* struck a mine near Liverpool in a blinding snowstorm. The people vainly attempted to be brought into port by boats, but great dismay was sustained."

Milner noted that his version of the story "begins accurately and well, but after the first sentence the thread is lost and the story is completely distorted."

Then she asked him to tell her what he remembered of the first story.

He couldn't remember a word of it. He told her that this was because he had been "concentrating too heavily on the second one" and added that trying to remember the first story "created terrible confusion and pandemonium."

They moved on to some additional tests, which took about five minutes to complete. Afterward, Milner asked him to tell her whatever he remembered of the two stories she had told him just a little while earlier.

He looked at her blankly.

He did not remember that she had read any stories to him at all.

Brenda Milner had herself once sailed through a sea full of mines in an American liner off the coast of the United Kingdom. In 1944, during World War II, she and her husband, Peter, left their homes in Britain, bound for North America. At night, out in the open water, any light would have made them a target for U-boats, so the captain ordered a complete blackout, with not even candles permitted. They churned westward for more than a week, zigzagging over moonlit seas, avoiding predictable routes, and when they reached Boston, with its incandescent skyline, Milner felt giddy, drunk on light itself, and flush with the sudden sensation of safety and possibility. In England, she and Peter, who was an electrical engineer, had spent the previous year working together at an isolated military radar research center, battling the ever-present worry that one or the other of them would be kidnapped, spirited across borders, and put to work for the Reich, as was rumored to have happened to many other promising young Allied scientists. Brilliance made you a target, too.

Milner was born in Manchester in 1918 and grew up in an old house surrounded by delphiniums. Her father was a music critic and piano instructor—her mother, twenty-six years younger, had been one of his students—and his somewhat ad hoc work schedule allowed him to spend time with his daughter, immersing her in Shakespeare and German and math. He homeschooled her until the age of seven, when he died of tuberculosis. Her widowed mother sent Milner to a girls' boarding school, where she excelled. She went on to attend Cambridge University and contemplated majoring in philosophy before settling on psychology, mainly because she decided there would be more job opportunities for psychologists than philosophers.

She graduated in 1939, but her examination results won her a prestigious fellowship, which allowed her to stay on at Cambridge for an additional two years. At first her research there explored sensory conflict—what happens when external cues clash with internal ones, such as when a visual illusion makes it look as though you're leaning to the left even if your proprioceptive senses tell you you're standing up straight. That project ended abruptly when the war broke out in September 1939, and Milner and a group of her Cambridge colleagues were put to work on a project designing psychological aptitude tests meant to help the military determine which recruits were better suited to being, say, bombardiers as opposed to fighter pilots. In 1941, she was reassigned again, sent to that isolated radar research center, which is where she met Peter, who in 1944 surprised her by telling her that A) he had just been assigned to an atomic-energy research facility in Montreal, and B) he wanted her to marry him and come along.

After an evening eating planked steak and ice cream in Boston and a night at the Copley Plaza hotel, the newlyweds headed north to Montreal. Languages were one of Milner's strengths, and while her husband spent his days working on the top-secret problem of transforming matter into energy, she took a nonfaculty job at a francophone university, lecturing undergraduates on the basics of experimental psychology. She found honing her French stimulating but missed research. After taking a seminar at nearby McGill University with Donald Hebb—a psychologist who had just written *The Organization of Behavior,* a soon-to-be-classic book that postulated that every aspect of human behavior could be understood on the basis of neural circuitry in the brain—Milner was inspired to apply for the doctoral program in psychology at McGill. Hebb became her adviser and mentor, and in June 1950 he told her about a unique research opportunity at the nearby Montreal Neurological Institute.

The Neuro, as the institute was commonly known, was the most prestigious center for neurosurgery in Canada. It was helmed by Canada's most renowned neurosurgeon, an imperious American transplant named Wilder Penfield. Penfield had pioneered a new surgical method for the treatment of epilepsy, one that involved targeting the temporal lobes of the brain. Because the general function of the temporal lobes was a mystery, Penfield decided that it would be good to have a psychologist study his patients before and after their operations, to see whether his surgical lesions were causing any notable changes beyond their purely therapeutic effects. He only had room for one researcher. Hebb told her that the job was hers if she wanted it.

Brenda Milner had spent her academic career up until then trying to glean a deeper understanding of the hidden processes taking place inside the black box of the human skull. As soon as she arrived at the Neuro, the black box was, for the first time, flung wide open to her.

In 1892, at a meeting of the New York Neurological Society, a prominent doctor named Joseph Price presented a paper called "The Surgical Treatment of Epilepsy," which was later published in *The Journal of Nervous and Mental Disease.* In his introduction, Price described the long, hard road that generations of doctors had walked in their attempts to treat this stubborn and ancient sickness. "Its history, from a therapeutical standpoint," he wrote, "is one that has taxed the efforts of supremest superstition and defied the resources of scientific medication. Its treatment has been one of trial and disappointment, for it still remains one of the greatest opprobria of medicine."

There were, however, grounds for optimism. The age of reason had dawned, and superstitions were dropping away. The principal causes of the disease, Price declared, were finally coming to light.

"Debauchery leads to it," he said. "Young widows are prone to it, and its origin outside of physical causes may be traced to amorous songs and certain stimulants, such as chocolate and coffee." Onanism, Price continued, was also associated with many cases of epilepsy, and in those cases there was a straightforward remedy. "In women, efforts in a surgical way have long been tried for its relief," he noted. "One table I have consulted gives as high as 73.7 per cent of cases cured of masturbation by clitoridectomy. This surely makes it not presumptive in its claims for recognition."

Not all epilepsies had their roots in lust. Price noted that some epileptic fits were characterized by guttural, convulsive noises originating in the throat. This presented surgeons with an obvious area of attack. "Tracheotomy," he said, "was urged by Marshall Hall and others on the ground that many convulsions began in and were limited to the larynx." Other surgical approaches that he listed with approval included male castration and female ovariectomies, the benefits of which, he noted, would accrue to the epileptics as well as to the societies in which they lived. "So far as unsexing an epileptic is concerned," he continued, "I do not understand how or why there is reason to feel compunction at such a suggestion. I can hardly question the protective value to society, not only of forbidding epileptics to marry, but of rendering them unable to procreate." The amputation of the left arm also reportedly had occasional success, and nonsurgical treatments included doses of belladonna, a plant-derived poison with its roots in Europe, and curare, a plant-derived poison with its roots in South America. Leeches, too, showed promise.

Price wrote about both medical and surgical treatments for epilepsy, though he tended to favor the surgical approach. "Operation has the best of the argument," he wrote. "Out of seventy-one cases treated medically, and out of a second series of seventy-one treated surgically, the statistics as exhibited in my collection, the advantage is all with the surgical treatment, as in these four all were at least benefited, while in the medical series a great proportion showed no effect at all, and some grew worse. The exact percentage someone may calculate who has a greater taste for such work than I."

Six decades later, when Brenda Milner arrived at the Montreal Neurological Institute, epilepsy treatments still consisted of an assortment of medications and operations, although the medications and the operations had changed. Doctors had by then zeroed in on the disease's true point of origin, which, it turned out, had never been in the nether regions, the throat, or the limbs. Instead, the modern medical establishment rediscovered that, as Hippocrates had prophesied 2,400 years earlier, "the brain is the cause of this affection."

While the fact that epilepsy is a brain disorder may seem obvious today, that knowledge was hard-won and was built, once again, upon the backs of broken men and women whose damaged brains illuminated the functions of our own. The 1861 postmortem examination of the brain of Monsieur Tan, the monosyllabic patient discovered by the French doctor Paul Broca, provided neurologists with the first clear evidence that a specific faculty—in this case, speech—was localized in a specific patch of cortex, and it wasn't long before doctors interested in the etiology of epilepsy correctly guessed that epileptic seizures—which often consistently provoked convulsions of speech or movement—might be caused by damage to the associated parts of the brain. A complementary development, also in the late 1800s, was the discovery that by electrically stimulating portions of exposed brains of living humans or animals, it was possible to create automatic, convulsive physical responses: Shock a particular portion of the surface of the brain and a patient's fingers might twitch. Shock another and his biceps might flex. Slowly doctors developed the concept that epilepsy might sometimes result from a similar effect, with portions of the brain discharging uncontrollably as though touched with an electrode. This convulsiveness might be caused, they surmised, by tumors or scar tissue or other damage to the brain. Careful observation of a patient's seizures could give surgeons a clue to where to look for this "epileptic focus": If a patient tended to convulse on the left side, for example, there might be reason to believe that there was an abnormality in the brain's right hemisphere. Surgeons would then simply open the right side of the patient's skull and look around for damage. If they found the problem, they would take it out.

Eventually, surgeons began using electrical stimulation *during* the operations, helping them identify the functional components of the brain before they made their irrevocable cuts, ensuring that they wouldn't, for example, inadvertently slice away the patient's ability to hear or taste or see. And while electrically stimulating the brain became an essential part of making brain surgery safer, it also became a key tool in exploring the organ's subtle geography. Neurosurgeons became neural cartographers, charting out not just the broad contours of the motor cortex, for example, but all the intricate differentiations within it, right down to which part of the brain controlled the fingers as opposed to the toes.

In 1950, when Brenda Milner arrived at the Montreal Neurological Institute, Wilder Penfield, the institute's founder and chief neurosurgeon, was the king of the mapmakers.

• • •

Like anyone interested in how the brain works, Milner had seen and studied Wilder Penfield's "homunculi" illustrations. These illustrations, published in Penfield's bestselling book, *The Cerebral Cortex of Man,* and reproduced in dozens of academic and nonacademic publications, were simple but powerful guides to neuroanatomical function. There were two diagrams, each showing a grotesque cartoon figure of a man draped over a portion of a human brain. In the first diagram, the man represented the motor cortex, which is a narrow strip of neural tissue about three-quarters of an inch wide that arches over the top of one side of the brain, dips slightly in the middle, then proceeds down over the next hemisphere. Since the motor cortex is more or less equally represented bilaterally, Penfield's diagram showed only one hemisphere. At the very top, his homunculus dipped its toes into the chasm between the hemispheres, the so-called central sulcus, because that is the portion of the cortex that would cause Penfield's patients' toes to twitch when he stimulated it with his electrode. Next, in predictable nursery-rhyme order, came the ankles, knees, hips, and trunk, right up to the homunculus's shoulder, which rests on the spot right where the brain's flattish top begins to slope noticeably downward. Then things got strange. First of all, instead of proceeding from the shoulders to the neck, as you might expect, Penfield's homunculus was for the moment headless, as the elbows and wrist came next, draped over the next centimeter or so of motor cortex. And then, blown out of all proportion, occupying more space than the entirety of every other part of the body that came before it: the hand. The hand was enormous. Just the heel of the palm took up more cerebral real estate than the entire leg, and its thumb would dwarf Little Jack Horner's. The tip of the thumbnail occupied a spot about halfway down the cortex, and about a millimeter below it the homunculus's head finally made its appearance. The brow was truncated, almost Neanderthal, because Penfield was only able to elicit brow twitches from a lilliputian patch of neurons. Its ears were tiny, almost invisible, for the same reason. Its mouth, however, was almost as gigantic as its hand. Penfield had discovered, through the application of electric shocks to the brains of his patients, that "the cortical representations of the act of eating" were spread widely across a large portion of the surface of the motor cortex, so that applying an electrode to one spot might cause, for example, "mastication with movement of tongue," while giving a jolt to another spot nearby might cause "mastication with vocalization."

Penfield's second homunculus complemented his first and represented the somatosensory as opposed to the motor cortex, using another cartoon to illustrate the portions of the brain that govern our sense of touch. The somatosensory cortex was roughly the same size as the motor cortex and lay just behind it. If you looked at a girl wearing two headbands, one behind the other, the somatosensory cortex would lie under the second one, slightly closer to the back of her head. Penfield's careful mapmaking had revealed that the cortical representation of the different body parts along the somatosensory cortex ran almost exactly parallel to the cortical representation of those same body parts along the motor cortex. So if he jolted a point on a patient's motor cortex that caused the patient's ankle to flex, then applied the electrode to a point about a half-inch behind the first one, the patient would usually feel as though someone were touching his ankle. There were a few exceptions. For example, Penfield's motor cortex homunculus lacked genitalia, while his somatosensory homunculus had a little penis nestled in the central sulcus, just below its toes.

The most important distinctions between Penfield's maps of the motor cortex and the somatosensory cortex, however, were the methods by which the maps were obtained. Penfield had been able to chart the motor cortex by simple observation: When he stimulated a point on the patient's brain, he would note what part of the patient's body moved, and make a record of it. The mapping of the somatosensory cortex, however, required the active participation of the patient. Patients would have to report to Penfield where, exactly, they felt as though their bodies were being touched. And this meant, of course, that the patients had to be conscious.

Penfield was not the first neurosurgeon to operate on conscious patients. In the late 1800s, a British surgeon named Victor Horsley found that patients could withstand brain surgery while awake so long as large amounts of cocaine were first injected into their dura. By Penfield's time, local anesthesia was almost the norm, injections of synthetically derived novocaine having supplanted the coca-plant-based alternatives. But although Penfield was not the first to keep his patients awake on the operating table or to stimulate their brains electrically, a unique combination of reportorial meticulousness and operative approach made his work groundbreaking. While other surgeons typically operated through small holes in the skull and could therefore only see a small portion of the surface of the brain, Penfield preferred to open a large "bone flap," roughly five by four inches, exposing a prodigious expanse of cortex that he could then cut or stimulate at will. This allowed him to make more progress in charting the mysterious territories inside our skulls than any who'd come before him.

As early as 1928, Penfield began to muse about the possibility of "an institute where neurologists could work with neurosurgeons and where basic scientists would join the common cause, bringing new approaches." He lobbied for it relentlessly until, in 1934, with the help of millions of dollars from the Rockefeller Foundation, the Montreal Neurological Institute finally opened its doors. There was nothing like it: a place where scientists and medical doctors were thrown together and jumbled about until the distinctions between them blurred. Neurosurgeons, neurologists, psychiatrists . . . each approached the study of the brain from a different angle, and at the Neuro they could combine their individual strengths, and compensate for their individual weaknesses, in pursuit of a common cause. And that cause was straightforward and far-reaching: The Neuro, Penfield hoped, "would open the way to brain physiology and psychology. And then, sometime perhaps, we would make a more effective approach to the mind of man."

When Brenda Milner walked through the front door of the Neuro in 1950, it had already established itself as one of the world's leading centers for the interdisciplinary study of the brain. Penfield's ambitions for the place were literally carved in stone: One of the first things Milner saw, climbing up the steps into the atrium, was a huge marble statue of the goddess Nature, who was coyly pulling open her robes, naked underneath, symbolically revealing herself to science. The statue, like the rest of the atrium, was commissioned personally by Penfield in collaboration with an architect, and as Milner looked around, her eyes would have caught on dozens of other subtle messages hidden around the room, a secret code readable only by those fluent in the language of the brain. A decorative design on the ceiling revealed itself, upon closer inspection, to be an artistic representation of the cells of the cerebellum, and in the center of the ceiling was the institute's logo, an emblem of a ram's head surrounded by odd symbols. Only a few of those entering here would remember that Aries, the astrological ram, ruled over the brain, and fewer still that the odd symbols were hieroglyphs lifted from the Edwin Smith Papyrus and considered to be the first written representations of the word *brain.*

Her degree adviser, Donald Hebb, had given Milner two directives prior to her arrival. One, to be as helpful as possible. Two, don't get in anyone's way. She would recall spending her first couple of weeks there "hugging the walls." Eventually, however, it dawned on her that to follow the first directive she would have to ignore the second, and so she began to step forward and speak up. This was not easy. Wilder Penfield turned fifty-nine in 1950, but he was if anything a more willful, and intimidating, presence than ever. Still big and muscular—he'd been a first-string tackle on the football team at Princeton University—Penfield dominated the Neuro in a very real sense. Every week, he convened a staff meeting in a conference room on the third floor to discuss the status of the institute's patients and research. Dozens of neurosurgeons, neurologists, and electroencephalographers would crowd in, though only the four or five people Penfield considered to be his stars were allowed to sit at the table with him; the rest were relegated to chairs around the periphery of the room. Whenever somebody in attendance said something that displeased him, Penfield would remove his glasses and deliver a hard stare. His subordinates learned to fear that stare in a visceral, Pavlovian way. But for all his occasional bullying, Penfield respected, without prejudice, brilliance. If you stood up to him, and could support yourself with solid facts and arguments, he would back down. And although Brenda Milner, with her sylph-like proportions and gentle, Cambridge-educated accent, was physically unintimidating, her mind quickly revealed razor-sharp claws. Penfield had been hesitant to invite a psychologist into his institute at all—he considered the field wishy-washy and unscientific, dominated by sex-obsessed Freudians—but it wasn't long after Milner's arrival that he asked her to join him at the conference table.

Once there, she would never leave.

For Milner, the institute felt like home. Her curiosity and inquisitiveness, the qualities that had animated her since childhood, found a place where they could flourish, where the puzzles she confronted were intoxicatingly difficult and the stakes bracingly high. This didn't mean the day-to-day work situation was idyllic: Penfield allowed Milner access to his patients and provided her with a small office, but for the first several years of her tenure he didn't pay her a dime and she was forced to "borrow" her notebooks and pencils and other office supplies from the University of Montreal. Even some of the basic psychological tests and manuals that formed the bedrock of her work she had to finagle from generous researchers at other institutions, to whom she pleaded poverty in heartfelt letters, since she couldn't afford to pay for them herself. In the end, what mattered to her was the work she was able to do. And the work she was able to do was everything she'd ever dreamed of.

Her basic mandate was to determine whether Penfield's epilepsy surgeries were having any adverse effects on his patients. She was particularly focused on his psychomotor epilepsy patients, the men and women whose seizures were often characterized by moments of inattention, odd behavior, and short periods of amnesia. By employing his stimulating electrode, Penfield had demonstrated that these specific types of epileptic attacks could be reproduced on the operating table by shocking a part of the brain known as the medial temporal lobes. In addition, the institute's new electroencephalography equipment, which allowed researchers to monitor the brain's electrical spikes and discharges in real time noninvasively, without opening the cranium, also indicated that psychomotor seizures often originated in this portion of the brain. So in 1949, Penfield began implementing what to him was the obvious surgical approach to the treatment of these cases: If he could determine that patients' psychomotor seizures originated in a particular hemisphere, he would open up that side of their skulls and perform a "unilateral partial temporal lobectomy," removing that hemisphere of their medial temporal lobes. The hemisphere he left behind, Penfield hoped, would pick up the slack for its departed twin. Still, removing brain regions whose function he didn't understand, even unilaterally, and even if doing so appeared to have a therapeutic benefit, bothered Penfield. That's why he had asked Donald Hebb to send a psychologist to the Neuro. Brenda Milner's mission was to determine whether these removals Penfield was making were having an effect on his patients. And, if so, what that effect was.

In its most common definition, *temporal* means "of or relating to time." In the case of the temporal lobes, however, the word is just an anatomical guidepost and refers to the fact that the temporal lobes lie directly behind the parts of the head referred to as the temples. "Medial" means "in the middle," so the medial temporal lobes are simply the centermost parts of the region.

Penfield's operations focused on a specific portion of the medial temporal lobes, an area usually called the limbic lobe. Here again, the French physiologist Paul Broca, the man who introduced the world to Monsieur Tan, played a critical role: The French word *limbique* means "egg-shaped," and the limbic lobe, when viewed from the side, appeared roughly oval in shape, so that's what Broca named it. Within the limbic lobe were several other structures whose names also had clear and simple visual inspirations. The amygdala, dense and compact, came from the Latin word for "almond." The uncus, which ended in a sharp curve, derived from the Latin for "hook." And in the center of this centermost portion of the brain was the hippocampus, the largest structure of the group. It had a broad, thin-snouted top and a thin, curling tail. Hippocampus is Greek for "seahorse."

The functions of the limbic lobe were an almost unmitigated mystery. For years, a popular view based on studying a few anosmic patients—people unable to detect odors—held that the sole function of the limbic lobes was to enable our sense of smell. (An alternative name for the limbic lobes, in fact, was the olfactory cortex.) In the decade before Milner's arrival at the Neuro, however, that view had begun to change. Critically, the studies conducted by Paul Bucy and Heinrich Klüver at the University of Chicago, where chimpanzees tripping on mescaline had their temporal lobes removed bilaterally, seemed to indicate that more important functions than smell were at stake in the region.

Wilder Penfield, for one, had a feeling that the limbic lobe, and in particular its central and largest organ, the hippocampus, must serve an important purpose. This was based in large part on aesthetic reasons: He found it hard to believe that such a delicately formed, and in his view beautiful, structure—a structure that he'd noticed seemed to be directly connected by neural tracts to most of the other major structures in the brain—would be useless. The hippocampus didn't look like some sort of vestigial, peripheral throwback, a cerebral appendix. Nestled there in the center of our heads, enigmatic but interconnected, it looked crucial.

Brenda Milner reviewed the existing literature on the temporal lobes. Two things were clear to her.

One, the purpose of the hippocampus and its limbic brethren remained a mystery.

Two, she was in as good a position as anyone in the world to solve it.

She began working with Penfield's patients.

She would spend time with them before their operations, getting to know them, conducting long interviews, and writing up detailed impressions of everything she could think of, from their style of dress to their sense of humor. She would complement these subjective notes with extensive batteries of tests, measuring their intelligence, their memories, their problem-solving abilities, their temperaments.

On the day of the operation, she would often watch Penfield in action. There was no particular reason she had to be in the OR; she simply found it fascinating. She would squeeze down a narrow hallway that led to the spectators' gallery of operating room 2, Penfield's preferred arena, on the second floor. The gallery was cramped, with three rows of wooden benches fronting a large glass window overlooking the so-called operating theater. The gallery was almost always crowded with residents and visiting doctors coming to see the master at work. Another fixture there was the electrophysiologist Herbert Jasper, whose EEG (electroencephalograph) machine was installed front and center in the gallery, receiving electrical information from the sensors Penfield placed on the surface of his patients' brains. Jasper interpreted the signals and relayed those interpretations back to Penfield out loud to help guide the knife. In addition, there was always a secretary present to transcribe Penfield's observations as well as any comments made by the patient. And that was the most fascinating part for Milner: watching and listening to the patients as they responded to the direct manipulation of their brains.

Penfield quickly and deftly used his electrode, and the reactions it provoked, to orient himself so that he always had a sense of where the motor and somatosensory cortices were, ensuring that he wouldn't inadvertently cause paralysis or permanent numbness. In a similar way, he mapped out and identified all the other critical landmarks, like the auditory cortex, so that he wouldn't leave deafness or other deficits in his wake. Then he dug in, finding the area that intuition and electricity told him was responsible for his patient's disease and destroying it. The spectators in the gallery usually applauded at the operation's conclusion, before Penfield cleaned up and the nurses wheeled the patient off to the recovery room.

A few days later, once the swelling had subsided, Milner would meet with the patient again, pull out her notebook, and resume her testing.

For months, she came up empty.

She met and tested dozens of patients, watched them go under the knife, then retested them afterward. Almost all were helped by the operations: During the preoperative testing, Milner grew accustomed to the sessions being interrupted by seizures, which would manifest themselves as verbal tics or physical movements or momentary, blank-eyed absences. After the operations, these seizures came less frequently or not at all. As for whatever other changes Penfield's operations were causing—the changes she had come to the Neuro to ferret out—they appeared to be either too small to perceive or not testable by the methods she was employing. Or maybe they didn't exist at all? Was that possible? Could the removal of such an extensive portion of central structures in the brain truly have a negligible effect?

One of Milner's problems was that Penfield's operations were unilateral. In other words, he was removing only half of his patients' limbic lobes. The opposite hemisphere was no doubt compensating for most of the loss, which made finding deficits more challenging. There was no way around this. Penfield wasn't going to bilaterally lesion brain structures whose functions were unknown. To do so, he believed, would be a terrible thing, analogous to playing Russian roulette with his patients. Maybe they would only lose something minor, such as their sense of smell. But could he justify the risk?

What if they lost something more important?

Milner understood this. The humans she was studying weren't lab rats or chimpanzees, to be sacrificed at will. They were, above all, patients, and the mandate of the Montreal Neurological Institute was to make them better. Their role as research subjects was strictly secondary, and Milner believed that the lure of making scientific discoveries should never trump their safety.

Nevertheless, it was frustrating. Patient after patient, test after test, day after day, very little discernible change. This didn't mean that the patients weren't each interesting in their own right, or that there weren't minor breakthroughs along the way. They were, and there were. Everyone had their own story, and she took satisfaction in getting to know them all. There was the twenty-three-year-old bank clerk from Ann Arbor, Michigan, for example, whose physician had sent him to Montreal after nine years of chronic psychomotor epilepsy. The origin of his disease was unclear, though it may have had something to do with a case of the measles he'd contracted at the age of ten. His seizure pattern was odd: Suddenly, at unpredictable times, he would stop whatever he was doing and then say the word *yes*. Then he would turn his head slowly to the left and contort his face in that direction as well, while staring into space. "This would be followed," Brenda Milner noted, "by symmetrical licking movements."

Milner ran her gamut of tests on the bank clerk preoperatively, watched the operation, and then reran the tests afterward. Two days after the operation, the patient became aphasic, unable to speak, and the lower portion of one side of his face sagged, as though he'd been stricken with Bell's palsy. Both the aphasia and the facial paralysis were, as Milner recorded, "transient," likely just the result of the intracranial swelling that occurs after any brain operation. Those negative symptoms soon disappeared, and three days after the operation he seemed to recover most of his faculties. One of the tests she ran that day involved asking him to reproduce, from memory, a series of four geometric drawings that she'd shown him six days earlier, three days prior to the operation. He reproduced three of the four drawings perfectly, an accomplishment that would have been impressive for anyone, even a person who had not just experienced a procedure in which "the anterior portion of the left temporal lobe was removed, going back 5 cm along the lateral surface and 6.5 cm along the base."

Then, more than a year into Milner's study of Penfield's patients, she sat down with Patient P.B., the civil engineer who'd received the same operation. She told him those stories about a mugging victim named Anna Thompson and a boat trip across perilous seas, and found that the stories slipped almost immediately out of his mind. She struggled to understand it. Why had the operation had an effect on him that it hadn't on anyone else? Why had it seemingly left him amnesic, while it left the memory systems of other patients, who'd undergone almost identical procedures, intact?

Soon afterward, she sat down with another patient, a twenty-eight-year-old man, Patient F.C., who worked as a glove cutter in a factory. He'd suffered from epilepsy since a young age, and after several years of treatment with antiseizure medications, he decided to undergo Penfield's operation at the Neuro. The operation took place on October 21, 1952, and by the time Milner reexamined him weeks later, the swelling had subsided, although there was a large reddish crescent-shaped scar slightly visible on the right side of his cranium, under his lengthening hair. Milner told him the first story, the one about the mugging victim in South Boston, and asked him to repeat it back to her.

"Anna Thompson," he said tentatively, looking at her for support. "She robbed fifteen dollars. She had four children."

When working with patients, Milner tried to keep her face composed and not betray anything she was thinking, so as not to give them clues as to how they were doing or cues that might influence their answers. She nodded, made some notes, then told him the next story, reading it out, as always, clearly and carefully.

"The American liner *New York,*" she said, "struck a mine near Liverpool Monday evening. In spite of a blinding snowstorm and darkness the sixty passengers, including eighteen women, were all rescued, though the boats were tossed about like corks in the heavy sea. They were brought into port the next day by a British steamer."

As soon as she finished, she asked him to repeat it back to her.

"They were brought into port by a steamer," he said, echoing the last line he had heard. Then he stopped. He was unable to add any more detail.

Milner asked him to recall the first story.

"It seems to me you mentioned Anna Thompson," he said. "It was brought into the sea."

She made some additional notes, administered some additional tests. About an hour and a half later, before the patient left for the day, she asked him to recount anything he could remember about the two stories she had told him.

He looked at her blankly.

SEVENTEEN

PROUST ON THE OPERATING TABLE

A few years ago two friends of mine got married on Sifnos, a beautiful Greek island. I rented a small apartment right on the beach, and my daughter, Anwyn, and I flew and ferried there from our home in Whitehorse, in Canada's Yukon Territory. We stayed for a week. Anwyn was five years old, and Sifnos was the greatest playground she'd ever seen. Turquoise water, bright sun, breakfasts of fresh yogurt and honey; goats and cats roaming down whitewashed ancient alleys. Countless scenes of wonderful strangeness imprinted themselves indelibly upon her memory, like the potbellied man we saw some mornings down by the water, holding a cellphone to his ear with one hand while using a broomstick in the other to swish a huge, fresh-caught octopus back and forth along the ground, tenderizing it.

At night, sandy and sunburned and exhausted, we read stories from *D'Aulaires' Book of Greek Myths,* which I had bought for the trip. Anwyn loved the story of Hermes most, loved his mischievousness, his cleverness, his cheekiness. When Hermes's mother accused him of stealing a herd of cows from Apollo, Hermes feigned innocence, saying, "But I'm just an innocent baby." When I read that line to Anwyn I'd stretch the word *baby* out like taffy, amping up the whininess. The line quickly became one of Anwyn's own catchphrases, and she started lobbing it back at me anytime she could.

"But I'm just an innocent baaaaaaaaaaby!"

She loved the story of Artemis the Huntress, too, and Hercules, and blustering, trident-waving Poseidon. She loved that near the front of the book there was a map of Greece and its islands and that it wasn't hard to imagine these tales we were reading taking place somewhere that looked very much like Sifnos. One afternoon, exploring some steep, thorny hills high above a rocky beach, we found some old stone walls in front of a small, empty cave. She wondered about the history of the cave and whether any gods had ever visited it.

Toward the end of the week, we came to a scene in the book that was a pleasing mirror to the scene we were living. It was a Greek myth about a parent telling the Greek myths to her children.

"The nine muses were the daughters of Zeus and the Titaness Mnemosyne," it read. "Their mother's memory was as long as her beautiful hair, for she was the goddess of memory and knew all that had happened since the beginning of time. She gathered her nine daughters around her and told them wondrous tales. She told them about the creation of earth and the fall of the Titans, about the glorious Olympians and their rise to power, about Prometheus, who stole the heavenly fire, about the sun and the stars. . . . The nine muses listened to her with wide, sparkling eyes and turned her stories into poems and songs so they would never be forgotten."

Anwyn listened closely, looking at the colorful drawing of the nine muses and their beautiful, long-tressed mother. Then her eyes grew heavier and heavier until they finally closed.

Sometimes strange things happened in Wilder Penfield's operating room.

Once, while he was in the preliminary stages of operating on an epileptic patient with the initials S.B., Penfield determined that S.B.'s epileptogenic focus was somewhere in his right medial temporal lobe, then opened up that side of his skull and began seeking out the culprit. To avoid inadvertently destroying something he shouldn't, he first probed with his electrode to locate the borders of the auditory, motor, speech, and visual cortices, getting a feel for his patient's neural landscape. As Penfield probed near a large vein that rose upward from a portion of the surface of the brain known as the fissure of Sylvius, S.B. began to speak.

"There was a piano there and somebody playing! I could hear the song."

Penfield paused. He removed the electrode, waited a beat, then placed it in exactly the same place. This time, the song came into focus. "Yes," S.B. said. " 'Oh Marie, Oh Marie!' Someone is singing it." Again Penfield stimulated the point, and again S.B. heard the song. This time, S.B. explained that it was the theme song to a radio program he listened to.

Penfield removed the electrode and touched it to another spot nearby.

"Something brings back a memory," S.B. said. "I can see Seven-Up Bottling Company . . . Harrison Bakery."

Penfield was puzzled. He wondered if the patient was inventing these visions, making things up because he knew he was being stimulated. He decided to test this. He told the patient that he was about to place the electrode on his brain again. But, instead of placing it, he simply held it above the surface of the cortex, not touching it.

"Nothing," S.B. said.

A second, similar episode involved a woman with the initials D.F. That time, the point that elicited the response was within the fissure of Sylvius itself. In her case, the patient began hearing music, an orchestral arrangement of a popular song. Whenever Penfield stimulated this particular spot, D.F. reported that the same piece of music leapt into her mind. She even hummed along, accompanying the song for a full chorus and verse.

Whenever one of these odd, electrode-provoked delusions happened, a secretary in the gallery overlooking the OR made sure to record every detail. In one patient, Penfield's probing triggered a vision of a dog walking along a country road. Another time, a female patient began to hear a voice speaking faintly, indecipherably. Penfield moved the electrode slightly to a different part of her brain, and the voice came into focus: Somebody was calling out a single name, over and over: "Jimmie, Jimmie, Jimmie . . ." Jimmie was the name of the patient's husband. And once, when Penfield stimulated a point near the top of a twelve-year-old boy's right temporal lobe, the boy announced that he could hear a telephone conversation between his mother and his aunt. Penfield removed the electrode, and the conversation stopped. As soon as he reapplied it, the conversation resumed. "The same as before," the boy said. "My mother telling my aunt to come up and visit us tonight." Penfield asked the boy how he knew the conversation was taking place over the telephone, and the boy said that he could tell by the way the conversation sounded, and he knew his mother was on the line with his aunt because he recognized the tone of voice she used when speaking with her sister. Penfield tried again to determine if the patient was just making things up and concluded that the boy was an accurate witness. "Every effort was made to mislead him by stimulations without warnings and warnings without stimulation, but at no time could he be deceived. When in doubt, he asked thoughtfully to have the stimulation repeated before committing himself to a reply!"

To Penfield, it was clear that these unexpected responses he was getting by stimulating the medial temporal lobes differed fundamentally from the responses produced by stimulation of other parts of the brain. The response of his patients to the jolting of other areas might be marked by "a tingling feeling, an absence of feeling called numbness, a sense of movement; olfactory sensation, by a disagreeable odor; gustatory, by a strong taste," but all of those sensations had one thing in common: They were generalized, not tied to any particular moment in time. The responses he was triggering by stimulating the medial temporal lobes were, Penfield noted, "of an entirely different order. They are made up of the acquired experience of that particular individual. It is the difference between a simple sound and a conversation or a symphony. It is the difference between the sight of colored squares and the moving spectacle of friends who walk and talk and laugh with you. The one is a simple element of sensation. The other is a recollective hallucination."

A recollective hallucination.

In other words, a memory.

Penfield wrestled with the implications of what he'd observed. In his famous homunculus illustrations, he'd already mapped out the human brain's sensory and somatosensory cortices. Now he'd stumbled into an entirely new domain: the memory cortex.

But how was this possible?

What made a spark of electricity spark the past?

"The answers to those questions are of great psychological importance," Penfield wrote, before admitting that he was going to have to "venture from the firm ground of observation onto the dizzying scaffolding of hypothesis." His patients' experiences in the OR led him to believe that the human brain retained absolutely every experience that ever crossed its synapses. Every moment, every waking hour, even every dream. Anything that it had seen or heard or tasted or smelled or thought. Anything and everything, so long as during the moment of raw experience some attention had been paid to it. "Whenever a normal person is paying conscious attention to something," Penfield wrote, "he is simultaneously recording it in the temporal cortex of each hemisphere. Every conscious aspect of the experience seems to be included in these records." It was, as Penfield would later describe it to *Time,* as though there were a "tape recorder" in the brain, activated at the moment of birth and stopping only at death. Each event of a person's life was stored away as a distinct "neurone pathway." Even events that people might later have no ability to recall of their own volition, the ephemera of the everyday, were all carefully preserved. "It would appear," Penfield said, "that the memory record continues intact even after the subject's ability to recall it disappears."

Penfield further speculated that when people engaged in an act of normal remembering, without the aid of an external electrode, they were still doing exactly what Penfield was doing to his patients on the operating table. That is, a self-generated electrical jolt to the medial temporal lobe's "memory cortex" was triggering the playback of particular memories. "This would seem to be absurdly simple," Penfield wrote, "and yet the new evidence is inescapable."

• • •

Penfield first presented his theory of memory during his presidential address at the seventy-sixth annual meeting of the American Neurological Association on June 18, 1951. The meeting took place in the grand ballroom of the tallest building in New Jersey, the twenty-four-story Claridge Hotel, high above the boardwalk in Atlantic City. His presentation caused a stir. The first audience member to comment was Lawrence Kubie, a psychiatrist and psychoanalyst who had acquired a sort of reflected fame by gathering an impressive collection of famous clients, including Vladimir Horowitz and Tennessee Williams.

"I am profoundly grateful for this opportunity to discuss Dr. Penfield's paper," he said, describing how it had put him in a "state of ferment," as though he were "watching pieces of a jigsaw puzzle fit into place and a picture emerge." The address, he said, had "been as exciting a moment as I have spent in a scientific meeting in recent years. I can sense the shades of Harvey Cushing [the founder of modern neurosurgery] and Sigmund Freud shaking hands over this long-deferred meeting between psychoanalysis and modern neurology and neurosurgery, through the experimental work which Dr. Penfield has reported."

Kubie then said that he hoped Penfield might begin adding dream studies and free association exercises to his battery of preoperative routines and wondered how many of the electrically stimulated memories might have been in fact the sorts of repressed memories that analysts such as himself trafficked in. Casting "a hopeful glance into the future," Kubie imagined a day when even nonepileptics might have their brains "stimulated on the operating table" to "see whether the reliving of the past through electrical stimulation of the temporal cortex exercises any influence on preexisting neurotic symptoms and mechanisms, and on preexisting associative patterns and emotional storm centers"—a sort of open-brain psychoanalysis.

Penfield didn't comment on Kubie's suggestions, perhaps because he had always distrusted psychoanalysis and psychoanalysts. He was almost certainly pleased, however, by Kubie's memorable summation of his discoveries.

"This is Proust on the operating table," Kubie declared. "An electrical *recherche aux temps perdu.* Yet is it *perdu*?"

Others in the audience were less impressed.

A neurologist named Karl Lashley, who was at the time the world's leading expert on the science of memory, was also in attendance. For the past two decades, working first at the University of Minnesota and then at his own laboratory in Orange Park, Florida, Lashley had conducted a series of experiments on rats, attempting to determine where their memories were stored. To do this, he would teach the rats a task, say, the proper way to navigate a maze, and then cut out various parts of their brains. What he found, to his surprise, was that no particular lesion would make the rats less able to remember how to navigate the maze. Instead the rat's navigational skills became muddled in proportion to how much of their brain tissue he removed, regardless of where exactly he removed that tissue from. Also, he was able to teach the rats new tasks, regardless of what parts of their brains he removed. The conclusions he drew from this were that no particular part of the brain stored memories and, likewise, that no particular part of the brain was responsible for the task of storing memories. Instead, he theorized, when it came to memory, if you removed a specific part of the brain, the remaining parts would attempt to pick up the slack, taking over the tasks the lost part had once been responsible for. He called this theory equipotentiality, to indicate his belief that every part of the brain had equal potential. And equipotentiality, or replaceability, as others often referred to it, became the prevailing view of how memory worked. Looking for the specific sites responsible for memory creation and storage, in this view, was pointless, since those sites could be anywhere, changing from brain to brain.

In Atlantic City, Lashley's critique of Penfield began somewhat obliquely, with a comment about the use of metaphor. This criticism was most likely directed toward Penfield's reference to the brain as a tape recorder, though he didn't say so specifically. Instead, Lashley simply dismissed "the analogies of various machines and neural activity" and pointed to "a curious parallel in the histories of neurological theories and of paranoid delusional systems. In Mesmer's day the paranoiac was persecuted by malicious animal magnetism; his successors, by galvanic shocks, by the telegraph, by radio, and by radar, keeping their delusional systems up-to-date with the latest fashions in physics. Descartes was impressed by the hydraulic figures in the royal gardens and developed a hydraulic theory of the action of the brain. We have since had telephone theories, electrical field theories and, now, theories based on the computing machines and automatic rudders. I suggest that we are more likely to find out how the brain works by studying the brain itself and the phenomena of behavior than by indulging in far-fetched physical analogies."

As for the "memory mechanisms" that Penfield proposed, Lashley was unconvinced. "Dr. Penfield's observations on stimulating the temporal lobe raise many problems," Lashley said, "but I do not believe that they justify the conclusion that memories are stored specifically in that region." He admitted that he had "no clear alternative to offer in explanation of Dr. Penfield's data" but added that whatever that data proved or didn't prove, it would be hubristic to give it too much credence. The functions of the temporal region were still, he said, "completely obscure," noting that when he, Lashley, had destroyed the visual processing areas in animals, it did "not abolish visual memories" and that when he likewise destroyed the tactile areas, it did "not abolish tactile memories." Finally, Lashley wasn't even convinced that the so-called memories Penfield described provoking were memories at all. Despite the fact that "Dr. Penfield considers that he is stimulating specific memory pathways," Lashley said, we still had no idea "what cerebral processes arouse memories."

The transcript of the meeting does not record whether Penfield at this point removed his glasses. He did, however, respond.

"Dr. Lashley," he said, "pointed out, as I feared that he would, that in his opinion there are no specific memory traces. That is in keeping with his observations in the early days in Minneapolis, when he worked with rats. It is in keeping with his demonstration of the replaceability of areas of brain, functional areas of brain, one by the other. Yet if there are no recording patterns in the cortex, how is it that an electrical stimulus can cause the patient to reexperience an earlier experience?"

Then he added a jab of his own.

"I would point out," he said, "that the replaceability seems to be somewhat less as one rises in the evolutionary scale."

In other words, Lashley might be an expert on the minds of rats, but Penfield's expertise came about through his work with an entirely different category of animal.

At the end of the day, and the end of the conference, Penfield had presented some intriguing case studies, but he lacked the necessary evidence to support a real theory of how memory worked. Even if it was assumed that the brain contained a sort of tape recorder—or, for that matter, a telegraph, computer, or hydraulic pump—that allowed us to preserve our experiences, Penfield's operations hadn't given any idea as to its location. Just because stimulating a certain part of the brain cued up a particular recording didn't say much about where that recording originated or how it was made. To use another metaphor Lashley would disapprove of, when you tune a radio to a particular station and hear a particular song coming out of it, it doesn't tell you where the radio station is located physically, or the studio where the song was originally recorded.

Was there a seat of memory in the brain?

Karl Lashley would say no, but Wilder Penfield felt otherwise. The brains of rats might exhibit equipotentiality, each part equally important and capable when it came to memory, but Penfield's experiences had led him to a less democratic view. He was already famous for mapping out in great detail how various parts of the brain were dedicated to different actions and sensations. Why should the brain exhibit any less specialization when it came to something as fundamental as memory?

Penfield was a devout Christian. He had faith in the existence of a higher being, though of course he had no proof.

He also believed in the seat of memory.

He had no proof of that, either.

When Anwyn and I got back home from Sifnos, I did some research on Mnemosyne, the goddess of memory. I wanted to see if her story was any more fleshed out in books geared toward adults than it was in the version presented for kids in *D'Aulaires' Book of Greek Myths*. The answer, it turned out, was no, not really. Despite her obvious importance—Mnemosyne is credited with not just the creation of memory but the creation of language itself—her biography is always paper-thin, a collection of isolated fragments. She had long hair, she slept with Zeus, she gave birth to the Muses, she wore a golden robe. That's about all there is.

This doesn't mean people didn't write about her and reify her. They did.

Here was Homer, describing how Hermes viewed her:

"First among the gods he honored Mnemosyne, mother of the Muses."

Here was Hesiod, describing her love life:

"For nine nights did wise Zeus lie with her, entering her holy bed remote from the immortals. And when a year was passed and the seasons came round as the months waned, and many days were accomplished, she bore nine daughters, all of one mind, whose hearts are set upon song and their spirit free from care, a little way from the topmost peak of snowy Olympus."

Here was Pindar paying tribute in one of his odes, making the point that no glory endures if nobody remembers it:

"Even high strength, lacking song, goes down into the great darkness.

There are means to but one glass that mirrors deeds of splendor; by the shining waters of Mnemosyne is found recompense for strain in poetry that rings far."

But Mnemosyne's own story, her trials and tribulations, whatever they were, has been forgotten. The Greek pantheon of gods and goddesses is so vast that some characters are bound to be more fleshed out than others, but it struck me as unjust somehow that the story of the mother of all stories would get such short shrift.

Plato, in one of his Socratic dialogues, chronicled an interesting conversation that the Greek mathematician Theaetetus had with Socrates in 369 B.C.E. As far as I can tell, this conversation contains the earliest known attempt at a scientific explanation for how memory works. It straddles both worlds, though: Socrates's concept of memory doesn't entirely let go of Mnemosyne. Instead he steps tentatively into the realm of secular reason while keeping one foot in the old mythos. He also uses yet another metaphor for the inner workings of the mind that no doubt would have gotten under the skin of Karl Lashley.

> Socrates: Please assume, then, for the sake of argument, that there is in our souls a block of wax, in one case larger, in another smaller, in one case the wax is purer, in another more impure and harder, in some cases softer, and in some of proper quality.
>
> Theaetetus: I assume all that.
>
> Socrates: Let us, then, say that this is the gift of Mnemosyne, the mother of the Muses, and that whenever we wish to remember anything we see or hear or think of in our own minds, we hold this wax under the perceptions and thoughts and imprint them upon it, just as we make impressions from seal rings; and whatever is imprinted we remember and know as long as its image lasts, but whatever is rubbed out or cannot be imprinted we forget and do not know.

The world evolved in countless ways during the 2,320 years between the end of that conversation and the end of Wilder Penfield's presentation to the American Neurological Association.

Our understanding of memory, however, had advanced very little.

EIGHTEEN

FORTUNATE MISFORTUNES

My mother remembers that during the years immediately following the war, my grandmother would continue to go away sometimes, for days or weeks or months. The children never understood these absences. One day their mother would be home, the next she'd be gone, and a nanny would move into the house until my grandmother returned. My mother remembers that by 1950 the absences were fewer, and my grandmother seemed more stable, less agitated, less upset. She even became a girl-scout troop leader. She would host my mother—just entering her teens—and the other scouts, teaching them sewing and jewelry making, two skills that she might have honed during her time at the Institute of Living, in the little arts and crafts colony near Pomander Walk.

My grandfather pushed my mother's brothers, hard. He had their IQs tested, and then told them whose was higher. He yelled, he bullied. My mother remembers thinking that the reason he didn't put equivalent pressure on her—she was never a good student, and he never seemed to care—was that she was a girl, and he didn't believe that a girl's grades mattered much. She remembers being grateful for this. Her older brother, Barrett, responded well to the pressure. He excelled. Her younger brother, Peter, did not. Peter had a difficult time, growing up. He moved from school to school. He acted out.

My mother remembers that the best days, for the family as a whole, were the ski trips. When my grandmother was not away on one of her unexplained absences, and my grandfather was not at work; when nobody was fighting; when, early on a Sunday morning, they would squeeze into their ski clothes and their ski boots, then squeeze into a car and head off to a mountain: to Stowe, Bromley, Otis, Mohawk, or Mad River Glen. Often, on the way to the slopes, my grandfather would suddenly pull into the lot of some small church in some small town. My grandmother was not religious, and held her nonbelief as a point of pride until the day she died, but she would march into the church along with the rest of them. My mother remembers wondering what the men and women in the pews thought of her strange, complicated family as they entered those sacred spaces and clomped up the aisles in their ski boots.

On August 31, 1950, orderlies at Connecticut State Hospital led a woman with the initials D.M. from her ward to room 2200, where my grandfather was waiting. She was twenty-eight years old and had been at the asylum for ten years. She had been diagnosed as a "homosexual schizhophrenic, actively hallucinating." She lay down on the operating table. The records don't indicate whether she was sufficiently cooperative to undergo surgery under local anesthesia or whether a general anesthetic was required to subdue her. In either event, my grandfather proceeded to slice a wide arc across the top of her head, roll her forehead down, and use his custom trephine to drill his usual two holes in the front of her skull. After using a scalpel to slice an opening in her arachnoid mater membrane, he inserted his flat brain spatula into one of the holes and levered up her frontal lobes. He squinted through the magnifying lenses of his loupes, peering inside. He oriented himself, taking visual notes of the various cerebral landmarks as his eyes traced a path past the frontal lobes and toward the structures beyond. He spotted the "slight bulge" of the uncus, approximately three centimeters past the tips of the temporal lobe, opposite an area known as the dural ridge.

Although the day's operation would be a landmark one, my

grandfather's first, tentative steps into the mysterious landscape of the "hippocampal zone" had actually taken place more than a year earlier. That was when, in collaboration with an electrophysiologist colleague named W. T. Liberson, he administered electric shocks to the uncuses of eight "sufficiently cooperative" lobotomy patients. The uncus is a hook-shaped, dime-size tangle of neurons that is either the farthest-forward portion of the hippocampus or its own independent structure, depending on which neuroanatomist you ask. Research with monkeys had hinted that removing the uncus might pacify agitated primates, but regardless, like with the rest of the medial temporal lobes, the purpose of the uncus was unknown.

"Striking effects were exhibited," my grandfather later wrote of his electrical experiments. Specifically, "in all but one patient complete and prolonged apnea was recorded after stimulation." In other words, they stopped breathing. Although the apnea often "considerably outlasted the duration of the stimulation," all the patients eventually began breathing again, though one required artificial respiration to do so and another remained in a state of only "periodic respiration" for at least an hour. Many of the patients experienced seizures, and several fell into prolonged states of unconsciousness.

The function of the uncus, however, remained an open question, and my grandfather eventually decided that he'd need more than electricity to find the answers.

So on that late August day in 1950, instead of inserting one of his electrodes and giving D.M.'s uncus a jolt, my grandfather picked up his suction catheter and its attached "electrosurgical coagulating wire." He fed the tool into her head, moving it carefully under the spatula, trying not to touch or damage anything he shouldn't. Three centimeters past the tip of the temporal lobe, he reached the uncus. He activated the tool. The suction catheter came to life and began vacuuming out D.M.'s uncus while the wire cauterized any veins that the vacuuming caused to burst. If D.M. was in fact under local anesthetic, she would then notice that the musty smell of her bone dust had been joined by a richer, more pungent smell as portions of her neural tissue were burnt away.

She did not stop breathing. This may have been something of a surprise, given his previous experiments. D.M.'s ability to continue breathing as he destroyed her uncus was, as my grandfather later pointed out, "in marked contrast to the profound physiologic changes resulting from electrical stimulation." Once he was satisfied that he'd removed the entire uncus, a quantity of brain matter measuring approximately three centimeters long, two centimeters high, and one and a half centimeters wide, he removed the vacuum and proceeded to the other hole.

Immediately after her uncotomy, which is what my grandfather named the procedure, D.M. appeared to be more stuporous than patients who'd undergone his orbital undercutting lobotomy. The orderlies wheeled her away, and my grandfather changed out of his scrubs and drove back to Hartford. Initial reports on her condition weren't particularly encouraging, as she exhibited little psychiatric improvement, but they weren't particularly discouraging, either, since she didn't appear to have been compromised at any essential physiological level. She continued to breathe, for example. So ten weeks later, on November 16, 1950, my grandfather returned to room 2200 and performed four more uncotomies on four more women.

First was patient I.S., a forty-eight-year-old paranoid schizophrenic with a history of suicide attempts. During this operation, my grandfather's hand slipped and his electrocautery device accidentally caused "severe damage" to an untargeted part of I.S.'s midbrain. This damage provoked a "violent jerk on the operating table," and his patient immediately fell unconscious. He proceeded with the uncotomy and noted that I.S.'s limbs continued to spasm unpredictably throughout the procedure.

Then there was patient E.M., a twenty-seven-year-old schizophrenic who'd been hospitalized for four years. She'd recently shown "temporary improvement" after shock therapy but was still "lacking in initiative and activity" and displayed "impaired judgment." E.M. was cooperative, which allowed my grandfather to perform the surgery under local anesthesia. This time, his hand didn't slip.

The third patient, B.P., was twenty-five years old and had been hospitalized for two and a half years for, among other things, "religious delusions," "excessive masturbation," and "homosexual trends." She vomited while my grandfather suctioned away her uncus, but the operation went smoothly otherwise.

His final operation of the day was on M.D., an "occasionally mute" and "actively hallucinating" twenty-five-year-old woman. She also vomited.

In attempting to assess the subsequent psychiatric effects of these uncotomies, my grandfather borrowed the protocols of the Connecticut Cooperative Lobotomy Committee. For each patient, at some indeterminate time following the operation, he tallied the opinions of five people, each of whom was asked to rate the patient's improvement on a scale from negative one to plus four. These five people were the asylum's ward physician, charge nurse, supervisor, ward attendant, and my grandfather himself. Occasionally a relative of the patient would be allowed to contribute to the scoring as well. A score of negative one indicated that the patient had become worse, while a positive score would indicate varying degrees of improvement. The highest score, positive four, was reserved for patients who'd been able to leave the institution altogether.

What he found was that, in general, lesioning the uncus didn't appear to have much negative or beneficial effect. Instead, four of the five patients he'd operated on showed little to no change whatsoever in their conditions, and received scores of zero, zero, zero, and one. The exception was Patient I.S., the woman who had received extensive accidental damage to some of her deep midbrain structures when my grandfather's electrocautery slipped. During the eight hours immediately following operation, I.S. remained in a stupor, her legs and arms periodically spasming. Her spasticity cleared up after a week, though she remained "vegetative and withdrawn" for two weeks.

Then, after a month, she suddenly began to show marked improvement. After five months, she had improved to such a degree that she was able to leave the asylum and return home. This gave her a rating of four plus.

My grandfather continued his experiments, pushing deeper into uncharted territories of the human brain. On the morning of Thursday, December 14, 1950, he performed his first complete medial temporal lobotomy. This was "a far more extensive resection" than the uncotomy, which had served as a prelude to this more drastic procedure. The setup was similar, however: He drilled open his patient's skull using the same trephines, levered up her frontal lobes using the same flat brain spatula, and vacuumed and burned his patient's gray matter using the same suction catheter and custom electrocautery tip. Only this time, after destroying the uncus, he kept on going, suctioning out her amygdala and most of her hippocampus. Although the messy mechanics of burning and suctioning made it impossible to preserve what he'd removed for later histological inspection, he was able to weigh most of it and found that he'd removed twelve and a half grams of brain tissue out of each lobe, for a total of twenty-five grams. Twenty-five grams is approximately what two tablespoons of water weigh. The question he was trying to answer, however, was more qualitative than quantitative.

That is, what did those twenty-five grams of brain, with their millions of neurons and billions of synaptic connections, do?

He once explained the line of reasoning that had led him to target the medial temporal lobes, pointing out that anatomical studies seemed to indicate a "close functional relationship" between the medial temporal lobes and the frontal lobes. So, his reasoning went, why not follow those connections back from the frontal lobes to the medial temporal lobes to see if the latter were the root cause of madness? He brought up Paul Bucy's work on the mescaline-dosed macaques who appeared "tamer" after having their temporal lobes removed, and explained that those "previously reported alterations in the behavior of experimental animals following temporal lobe surgery were primarily responsible" for his decision to see what effects similar surgeries might produce in man. Or, more accurately, in woman. My grandfather, like most lobotomists, performed a disproportionate number of psychosurgeries on women. This discrepancy never received a satisfactory explanation, but it seems worth pointing out that the known clinical effects of lobotomy—including tractability, passivity, and docility—overlapped nicely with what many men of the time considered to be ideal feminine traits.

That same Thursday, as soon as my grandfather finished his first medial temporal lobotomy, he proceeded to perform three more.

The first was V.M., a "destructive, assaultive, noisy" twenty-eight-year-old woman who'd been "hospitalized since 1946, unimproved on shock therapy" and whose aggressive tendencies often required "packs and seclusion." ("Packs" was asylum shorthand for ice packs, a form of tranquilizing therapy in which patients were bound tightly in soaking wet, ice-cold bedsheets.) Immediately after the operation V.M. vomited, then became restless and hunched over. She "wished to be left alone," my grandfather wrote, before adding that the final result of the operation was to make her "more childish" and "more active with self-mutilation." She rated a zero to one plus.

There was E.S., a thirty-eight-year-old "mental defective with psychosis," who also had epilepsy and whom my grandfather described as follows: "Impulsive, assaultive, resistive, mutters, huddles in fetal position in a chair." She also vomited right after the operation and had seven major seizures during the first month of recovery. In the year that followed, however, those seizures steadily improved in "number and severity." Her behavior improved somewhat, and by the end of 1951 she no longer required seclusion all of the time and was "slightly less assaultive." Final rating: one plus.

There was G.M., fifty-eight years old, who despite "temporary improvement on shock" remained largely "deteriorated, untidy, assaultive, impulsive." After her medial temporal lobes were removed, she experienced an “emotional regression with baby talk” but was otherwise “cheerful” and “no longer assaultive.” She did, however, appear aged and was “still actively hallucinating.” One plus.

As with the uncotomies, the preliminary results of the medial temporal lobotomies were inconclusive. Before he could evaluate the operation’s therapeutic promise or begin to answer the larger question of what the medial temporal lobes did, there was more to be done. Like most experimentalists, my grandfather believed that the more research subjects you worked with, the better. An N of one does not count for much. Luckily for my grandfather, he suffered no shortage of material. The lobotomy continued to rise in popularity—just the year before, in December 1949, Egas Moniz had received the Nobel Prize in medicine for his invention—and asylum superintendents around the world were still giving neurosurgeons unlimited access to their patients. By the 1950s, a dizzying variety of approaches to the procedure had been developed, each targeting different parts of the brain: topectomy, gyrectomy, cingulotomy, capsulotomy. The Nobel committee had endowed psychosurgery with a patina of nobility, demonstrating that future breakthroughs in the field might pay great professional, therapeutic, and scientific dividends. For ambitious tinkerers like my grandfather, the lure was irresistible.

In the weeks following his first four medial temporal lobotomies, he performed ten more:

Patient B.B.
Patient C.G.
Patient A.G.
Patient A.R.
Patient G.D.
Patient R.B.
Patient D.B.
Patient M.S.
Patient A.D.
Patient A.Z.

The case of Patient A.Z. was interesting.

She'd been institutionalized for the past eight years at Connecticut State Hospital. She was thirty years old, and although prior to surgery she'd been "temporarily helped by extensive shock therapy," she was nevertheless classified as "tense," "assaultive," "tidy," and "impulsive." She was also, my grandfather said, "preoccupied with sex thoughts" and "sex threats," which he classified as paranoid delusions. He operated on her on November 19, 1950. She was cooperative and under local anesthetic during the operation. She remained conscious throughout until, while my grandfather was in the process of suctioning out portions of her right hippocampal cortex, he "inadvertently went through the arachnoid and injured by suction a portion of the right peduncle, geniculate or hypothalamic region." A.Z.'s immediate response was to fall into a deep coma. She remained in the coma for seventy-two hours, incontinent, spasming periodically. Then she slowly came to, remaining in something of a stupor for a week but was eventually able to walk without support and regain control of her bladder.

She emerged from the coma with what my grandfather described as "complete remission in her delusions, anxiety, and paranoid trends." This "immediate and marked" result, he said, had "delighted" her family.

The case of A.Z., like the earlier case of I.S.—in which surgical slipups led to unexpected benefits—reminded my grandfather of a story he was told while at one of the asylums, about a female patient who'd been tied down in a bath for twenty-four hours. The bath had a broken thermostat, and the water was far hotter than intended. This resulted in the patient developing "extreme hyperthermia," "beyond the limits of the thermometer." Presumably, according to my grandfather, such prolonged overheating would damage some of the same parts of the brain that he had accidentally lesioned in I.S. and A.Z. So it seemed to him significant that the hyperthermic woman also "underwent a remarkable remission of all psychotic trends."

He wondered whether "the unexpected benefit accruing from deep central damage," as exhibited in these three "fortunate misfortunes," indicated that "the primary mechanisms of mental disease" might lie in regions even deeper within the brain than his own aggressive procedures were targeting.

But those were questions for another time. For now, in contemplating the case of A.Z., my grandfather thought that one more thing might be a significant factor in her "excellent results." Not only did she seem to have experienced a dramatic remission, but she also exhibited a "retrograde amnesia for her entire psychosis (of three years' duration)."

She hadn't just recovered from her illness. She didn't even remember it.

My mother had no idea what my grandfather was doing in the asylums back in those days. He worked a lot and talked about it very little. His career, in her mind, remains vague and indistinct. What she remembers is his presence. She remembers the times when he was home.

He was a good father, that's what she remembers.

A busy one, of course. On a typical workday, he came home late after dinner, past eight or nine P.M. He would retreat to his study for an hour or two, where he'd sit surrounded by his collection of neurosurgical bric-a-brac—ancient blades, old books, a bleached and anonymous skull—and use a Dictaphone to keep up with his correspondence. My mother knew not to bother him while he was in his study. She did other things instead, getting ready for bed. She read, or listened to *The Shadow* or *The Lone Ranger* on the radio, or gossiped on a tin can telephone with a girl who lived across the street. Just before my mother went to sleep, though, my grandfather would always go to her room and say good night. Often, she remembers, he told her a story, one he made up on the spot. It was a serial, a continuing saga about three animals: a deer, a bear, and a talking monkey. They'd go on new adventures every night. She doesn't remember the details of these adventures.

I've asked her if she remembers the monkey's name. She doesn't. But the details don't matter. What matters is that these stories came from her father. What matters is that he took the time, late at night, exhausted from whatever he'd done that day, to sit with her and tell her stories. What matters is that even now, more than a half century later, those acts of storytelling glow warm and golden in her memory. Afterward, he'd tell her to say her prayers, then he'd leave, shutting the door softly behind him. She was never sure where he went, but she imagines him going back to his study to continue with his work, whatever that work was, while she drifted off to sleep, the latest escapades of the deer and the bear and the monkey tumbling in her head.

A person can be many different things to many different people.

He was good to her.

NINETEEN

HENRY GUSTAVE MOLAISON (1926–1953)

MIT Neuropsychology Laboratory, February 1986

H.M.: At one time that's what I wanted to be.

Researcher: Is it? What?

H.M.: A brain surgeon.

Researcher: A brain surgeon?

H.M.: Yeah. And I said no to myself. Before I had any kind of epilepsy.

Researcher: Did you? Why is that?

H.M.: Because I wore glasses. I said, suppose you are making an incision in someone, and you could get the blood on your glasses, or an attendant could be mopping your brow and go too low and throw your glass off.

Researcher: That would be bad, wouldn't it?

H.M.: Yeah, 'cause you'd make the wrong movement then.

Researcher: And then what might happen?

H.M.: And that person could be dead or paralyzed.

Researcher: Yes. So it's a good job you decided not to be a brain surgeon!

H.M.: Yeah. I thought mostly dead. But could be paralyzed in a way. You could be making the incision right, and then a little deviation. Might be a leg or an arm. Or maybe an eye, too. On one side, in fact.

Researcher: Do you remember when you had your operation?

H.M.: No, I don't.

Researcher: What do you think happened there?

H.M.: Well, I think I was, I'm having an argument with myself right away, the third or fourth person to have it. And I think they, well, possibly didn't make the right movement at the right time, themselves then. But they learned something that would help other people around the world, too.

Researcher: They never did it again.

H.M.: They never did it again, because by learning it. And a funny part is, I always thought of being a brain surgeon myself.

Researcher: Did you?

H.M.: Yeah, and I said no to myself.

Researcher: Why's that?

H.M.: Because I said, an attendant might mop your brow, and might knock your glasses over a little bit, and you make the wrong movement.

Researcher: What would happen then if you made the wrong movement?

H.M.: And that would affect all the other operations you had then.

Researcher: Would it? How?

H.M.: Because that person was paralyzed on one side. Or you made the wrong movement in a way, and you possibly couldn't hear on one side. Or one eye, tight. You would wonder to yourself then, and it would make you more nervous.

Researcher: Yes it would.

H.M.: Because every time you did, you'd try to be extra-careful, and it might be detrimental to that person. Perform the operation right on that time. Because you'd have that thought and that might slow you up then. As you make the movement. And you could have continued right on.

Researcher: Do you remember the surgeon who did your operation?

H.M.: No, I don't.

Researcher: I'll give you a hint. Sco . . .
H.M.: Scoville.

By 1953, it was obvious that the drugs hadn't helped. Henry was on massive doses of powerful, brain-dampening epilepsy medications—Dilantin, five times a day; Mesantoin, three times a day; phenobarbital, twice a day; Tridione, three times a day—and they hadn't helped, or at least they hadn't helped enough. Henry was still seizing several times a day, sometimes falling to the ground, sometimes just falling silent. Those second type of seizures, the petit mal ones, were often described as "absences." When he was in their grip he became, briefly, a human husk, his lungs working and his heart beating but his mind on pause. The truth was, though, that even when he wasn't seizing, Henry was never entirely present, in the sense that his epilepsy had caused him to withdraw from the richer life that might otherwise have been his.

He was a smart, strong twenty-seven-year-old man, but he existed within borders as circumscribed as a child's. Weekday mornings he'd catch a ride downtown to the Underwood factory, on Capitol Avenue, where he'd sit on the line and help assemble the typewriters, a blue-collar worker making white-collar tools. In the afternoon he'd catch a ride back home to his parents' house in East Hartford. He was unable to drive a car, of course, just as he was unable to go off to war, or to college, or to any of the other places that his old friends had gone. Instead he just stayed home, where his parents could take care of and watch over him. He would spend evenings listening to the radio. He liked the big bands—Benny Goodman, Duke Ellington—and he liked some of the new rock and roll—the jive music, as he called it—that tinned through the speakers. He liked dance music but he never danced. He read magazines, soaked up Hollywood gossip, learned about scandals and successes of the sort he knew he would never experience. On a good weekend he would take a rifle a short walk into some nearby woods, heft the stock to his shoulder, sight down the barrel at a target, pull the trigger, and feel the kick. On a bad weekend he wouldn't do much of anything at all.

In 1953, Henry's past was still clear to him. It was his future that was growing dark. If things continued as they were, if his seizures continued to increase in frequency and severity, it wasn't hard to imagine that he would soon become too big a burden for his aging parents. If he became unable to work, unable to contribute, Gustave and Elizabeth might have to let him go. They might have to send him someplace like the nearby Mansfield Training School, an institution founded in 1930 through the merger of two older institutions, the Connecticut Training School for the Feebleminded and the Connecticut Colony for Epileptics. There the strictures that bound his life would be cinched even tighter. Like many of the other residents, he might be put to work in the onsite factory, making bricks. Or he might just sit in one of the crowded wards, becoming more and more absent, continuing his slow slide toward an uncertain fate.

Unless.

In a large banquet hall at the Hollywood Beach Hotel, in Hollywood, Florida, on the afternoon of April 23, 1953, my grandfather stepped to a podium to give a speech to the Harvey Cushing Society, America's preeminent association of brain surgeons. It was the closing address of that year's neurophysiological symposium. A little earlier, John Fulton had given the symposium's opening remarks, during which he'd made a joke about Becky, the chimpanzee from his laboratory who had inspired Egas Moniz to begin lobotomizing humans seventeen years before. "Was this the face that lopped ten thousand lobes?" Fulton asked, referencing a photo of Becky's wrinkled features. Then Fulton made a now familiar entreaty, urging the many psychosurgeons in the crowd "to study their patients with the same thoroughness with which chimpanzees are studied, for you have a much finer opportunity to gain insight into some of the basic problems of frontal lobe function than we who are limited to gaining information from inarticulate beasts."

Paul MacLean, another Yale researcher, spoke after Fulton.

MacLean was considered the world's leading authority on the limbic region in animals, and his speech opened with a literary flourish: "Today, with the annual celebration of Shakespeare's birthday, it may be expected that the occasion will arouse renewed discussion among those interested in English literature as to whether Shakespeare or Bacon wrote the plays. This points up for contrast an equally bewildering problem that faces those whose major concern is with the functions of the brain. In the brain, the authors of function—the structures themselves—are easily identified. But what do these authors write? That is the question. This is no better illustrated than by our lack of knowledge regarding the functions of parts of the limbic system that will concern us here, particularly that sizable author known alternatively as the hippocampus or Ammon's horn."

MacLean then gave a comprehensive survey of how very little was known about the true functions of the hippocampus, before ending with a complaint and a Fulton-esque challenge: "Animal experimentation can contribute next to nothing about the 'subjective' functions of the hippocampal formation," he said. "To corrupt a statement by Wiener, psyche is information, not matter or energy. The animal cannot communicate how he feels. Here is the rub for the physiologist. Realizing that Aladdin's lamp is not for him, he obviously looks, as he has long been accustomed, to the neurosurgeon!"

The stage was set for my grandfather.

My grandfather looked up from the podium and out over the group of surgeons and scientists. Many giants of his field were there: Bill Sweet, Leo Davidoff, Gilbert Horrax. As for my grandfather, he was forty-seven years old, no longer the ambitious young striver he'd once been. He was now a peer, or more than a peer, of many of the people here, a teacher as much as a student. The neurosurgical residency programs at Yale and the University of Connecticut had recently merged, and he had become the co-director, training the next generation from his old alma mater, teaching them the brutal subtleties of his craft, instructing them in the use of the numerous techniques and tools that he'd invented and that many of the men in that ballroom had begun to use in their own practices. He had become, in the eyes of the neurosurgical community, something of a giant in his own right.

"For the past four years in Hartford," my grandfather began, "we have been embarked on a study of the limbic lobe in man."

Coming as it did immediately after MacLean had outlined both the enduring mysteries of the limbic lobe and the difficulty of solving those mysteries through animal research, my grandfather's announcement was bound to cause a stir. Hearing it, the attendees may have hoped that he was about to announce a breakthrough, a revelation of some sort.

If so, he dashed those hopes right away.

After describing the operations through which he'd conducted his study—the uncotomies, in which he lesioned just a part of the limbic region; the medial temporal lobotomies, in which he lesioned almost all of it—my grandfather gave a gloomy general assessment of his results. "I speak with all humility," he said, "of the small bits of passing data we have accumulated in carrying out these operations on some two hundred thirty patients."

He had reason to be humble. The cuts he'd made in the brains of hundreds of human beings truly hadn't contributed much useful knowledge. He dutifully recounted some of the more interesting phenomena he'd encountered over the course of his experiments, such as the fact that "vomiting and temporary loss of consciousness occurred commonly during manipulation of the uncal region, but following resection they disappeared," and mentioned that one of his psychotic patients had suffered postoperative memory problems, but he didn't pretend that these scattered tidbits did much to illuminate the larger questions of what the limbic lobe—or medial temporal lobe or hippocampal region or whatever you wanted to call it—actually *did.*

One explanation for my grandfather's unimpressive results had been best articulated more than a decade before, in one of those letters between Paul Bucy and John Fulton, the one in which Bucy complained that psychosurgery-based research was hobbled by the fact that one never "starts with a normal organism." This was certainly the case for my grandfather's limbic lobe investigation, in which his subjects had all been "long-standing, seriously deteriorated" asylum residents. If it was true that it was difficult to understand how the human mind worked by operating on animals, it was also true that attempting to understand how the normal human brain worked by lesioning the deeply abnormal ones belonging to hopeless psychotics was challenging at best, a fool's errand at worst.

And when it came to *treating* mental illness—which after all was what these procedures were designed to do—my grandfather's medial temporal lobotomies had proved equally useless. They'd produced, he told the audience, only "meager" psychiatric improvement, and "no marked physiologic or behavioral changes."

An observer listening to my grandfather describe the therapeutic and scientific failures of his limbic lobe studies that day might reasonably conclude that these studies had hit a dead end and should be abandoned. Instead my grandfather told the audience that although his project had so far failed to provide much useful information about the mechanics of the mind, he hoped that "continuing limbic lobe studies may bring us one blind step nearer to the location of these deeper mechanisms." And at the end of his talk he hinted that he'd already begun thinking of a way to expand his studies, a strategy that would also avoid the scientific pitfalls that were inevitable when you worked with asylum-sourced research subjects. He described how some of his psychotic patients had also been epileptic, and how his operations had seemed to provide them relief from their seizures. Now, he said, "an interesting query comes to mind—could *bilateral* resection of such known epileptogenic areas as the uncus raise the threshold for all fits, as do pharmaceutical anticonvulsants?" Or, to put it another way: What would happen if, rather than performing his limbic lobotomies only on the mentally ill, he began performing them on perfectly sane people who suffered only from epilepsy?

It was an open question, one awaiting an answer.

Or, at least, a patient.

• • •

In March 1953, the month before my grandfather traveled to Hollywood, Florida, for the Harvey Cushing Society meeting, he had another consultation with the Molaisons. Although a detailed record of this consultation doesn't exist, it's reasonable to make certain assumptions. He would have questioned them about the progress, or lack of progress, of Henry's treatment, and they would have told him about Henry's increasing difficulties. They would have made it clear that the drugs hadn't helped—or hadn't helped enough. Given the frequency of his petit mal seizures, it's possible that at some point during the consultation Henry experienced one of them, his mouth going slack, his head tilting to one side, his eyes open and blank, his fingers scratching listlessly, repetitively, mindlessly at his pant leg. If so, my grandfather would have watched closely, waiting for Henry to come to. By the end of the consultation, my grandfather would have been able to take full stock of the Molaisons' hopelessness.

Then he would have offered them hope.

The drugs hadn't helped, but something else might. Maybe he told them about Wilder Penfield's operations, the unilateral ones pioneered at the Montreal Neurological Institute. Maybe he told them about his own bilateral operations, the ones he'd honed at asylums around New England. Maybe he gave the Molaisons a quick primer in neuroanatomy, leaning in and tapping gently at the sides of Henry's head, just behind his temples, just above his ears, explaining that the source of Henry's affliction probably lay somewhere in his medial temporal lobes, a couple of inches beyond the tips of my grandfather's fingers. Maybe he told the Molaisons that he might, if they'd let him, be able to remove that affliction altogether.

The Molaisons—Henry, Gustave, Elizabeth—thought it over. They must have been frightened, as there is no medical prospect more frightening than brain surgery. They must also have been trusting, as my grandfather was an esteemed doctor in a position of authority, a professor at Yale, a man radiating competence. Whatever calculus the

Molaisons used, however they weighed the pros and cons, debating the opaque risks of future surgery against the clear desperation of the status quo, is unknown. They may have taken their time, arguing among themselves, interrogating my grandfather. Or they may have come to a decision quickly.

What is known is this: They said yes.

The surgery was scheduled for August 25, 1953.

The week before, on August 17, Henry returned to Hartford Hospital to receive an electroencephalograph. Unlike the excruciating pneumoencephalograms, which had required the draining of his cerebrospinal fluid, the electroencephalograph, or EEG, as it was known, was painless. Henry lay on his back on a gurney, and a number of electrodes were affixed to his scalp. The electrodes registered Henry's brain activity, picking up on the faint currents passing between his neurons. The operator of the device was able to see that activity in real time, conveyed in visual form in spikes and waves that a pen made across a roll of crosshatched paper like a seismograph. An unusual amount of spikes coming from one hemisphere of his medial temporal lobes would be evidence that Henry's epileptic focus lay there, which would be evidence that surgically lesioning just that particular hemisphere might bring Henry relief. At one point during the exam, Henry had one of his petit mal seizures, going absent right there on the table. Despite this, the EEG failed to reveal an epileptic focus.

A psychologist named Liselotte Fischer met with Henry on August 24 to administer a battery of psychological tests, a baseline against which the effects of the operation could later be measured. Henry, Fischer noted, "admits to being 'somewhat nervous' because of the impending operation, but expresses the hope that it will help him, or at least others, to have it performed. His attitude was cooperative and friendly throughout."

When Fischer gave Henry a pad and pen and asked him to draw a man and woman, he drew the man first: a hospital patient, in a hospital gown, with a "crosspatch" mark on his temple. Fischer interpreted this as a manifestation of Henry's "acute anxious involvement with the impending operation." Then, Henry began to draw the woman.

"She ain't going to be pretty," he said as he sketched out a figure with an oversize head and bulging breasts. Fischer eyed the drawing and wrote that "with its aggressive stance and domineering features it is in glaring opposition to the male figure, and invites the interpretation of 'aggressive, castrating mother figure.'"

Fischer gave Henry a Rorschach examination, showing him a series of inkblots and asking him to describe what he perceived. Looking at one splotch, Henry said he saw a deer without horns, which turned into a doe. Fischer saw that as further evidence of Henry's preoccupation with castration. Another inkblot spurred a description of "a lion who moves away from the subject, so that his tail is oversized and 'right in my lap,'" which Fischer interpreted as an indication of "sexual confusion" and a "homosexual trend." She also noted "some repetition of the concept of fleeing," and of "concepts of mutilation."

Finally, Fischer administered an IQ test. Henry scored 104, higher than average but lower than he would score postoperatively. This may or may not have had something to do with the fact that he'd been taken completely off his antiepilepsy medications in the weeks leading up to the surgery and had experienced as many as twelve petit mal seizures during the hours he spent with Fischer. She'd watch and take notes as he'd go absent for ten to fifteen seconds, swaying and breathing heavily, scratching at his arms, his clothes, his belt, before regaining his senses.

"I gotta come out of this again," he'd say.

Henry spent the night in the hospital. The following day, August 25, somebody on the nursing staff shaved his head, then brought him to the operating room.

Just as he had done so many times before in the asylums, my grandfather injected a local anesthetic into his patient's scalp, sliced an arc across the top of his head, and rolled the skin of his forehead down like a carpet. He then used his trephine drill to remove two silver-dollar-size plugs of bone, a scalpel to cut through the meninges that protected the cerebrum, and a flat brain spatula to lever up the frontal lobes, exposing the deeper structures beyond. He scanned the region visually, his eyes picking out the glistening pink outlines of the hippocampus, the amygdala, the uncus, the entorhinal cortex, trying to identify any obvious defects, any coarse or atrophied tissue, any scars or tumors or other defects that might be the source of Henry's epilepsy. He saw nothing. Before proceeding further, he did something he didn't normally do in the asylums: Under the direction of the electroencephalographer W. T. Liberson, he used a slender, forcepslike instrument to reach into the holes and apply tiny wire-trailing electrodes to a number of spots along the surface of Henry's medial temporal lobes. He and the rest of his surgical team then waited while Liberson monitored the EEG readouts, making one last attempt to find a discrete epileptic focus. Liberson peered at the wavy lines on the scroll, looking for a telltale pattern, one that could point my grandfather toward a specific target in a specific hemisphere. He told my grandfather that once again he'd come up empty and had failed to find a focus.

Though the anesthetic ensured that Henry felt no pain, he was conscious, and throughout all the slicing and peeling and drilling, a symphony of unsettling and unfamiliar sensations had trickled through his blunted nervous system. When my grandfather leaned over to extract the electrodes, Henry had a direct view of his upside-down face, or at least the parts of his face that weren't covered by his surgical mask and surgical bonnet and surgical loupes. Henry's pupils contracted against the blinding light of the headlamp.

Maybe, at that moment, Henry told himself that this whole experience would all be worth it, that the frightening things that were happening to him would finally free him of epilepsy's burden, would allow him at last to be fully present, fully alive, able to achieve his potential. Whatever Henry was thinking, though, it didn't really matter at that point. He had said yes, had agreed to be operated upon, and whatever happened next was out of his control.

The same could not be said about my grandfather, who now had an important decision to make.

There was no focus.

This meant, of course, that there was no target, no specific place in Henry's medial temporal lobes to attack, not even a hint of which hemisphere Henry's seizures originated in.

If another neurosurgeon had been in my grandfather's shoes that day, things might have turned out differently. Wilder Penfield, for example, would have conceded defeat. Penfield had clear rules of engagement in the operating room: If he couldn't determine a focus either visually or through EEG, he wouldn't make any lesions. In fact, even if the EEG hinted at the presence of an epileptic focus, but visual inspection of the brain revealed no abnormalities, Penfield made it a point to do nothing rather than make an excision that might do more harm than good. "The neurosurgeon," Penfield once wrote, "must balance the chance of freeing his patients from seizures against the risks and functional losses that may be associated with ablation." In that balancing act, Penfield always erred on the side of caution, and in Henry's case, with no target, he would have decided not to proceed with the operation. He would have stitched Henry up, kept him a few days for observation, and sent him home with an apology and a refill of his prescriptions. He would have told him that there didn't appear to be anything he could do for him surgically, at least not then, and not with the information they had.

My grandfather was not Wilder Penfield.

Standing there in his operating room, looking down at the wet expanse of Henry's skull, glimpsing his exposed brain through the two trephine holes, my grandfather could have admitted defeat, could have ended the operation. This would have been the safest move. There was no chance Henry would be improved by following that course of inaction, but there was also no chance he would be hurt by it.

Alternately, he could have chosen to take one, and only one, of the paths ahead. He could have operated on Henry's left hemisphere, or Henry's right hemisphere, then withdrawn, patched him up, and seen what happened. He had no target, no specific evidence of an epileptic focus in either hemisphere, but maybe he would get lucky. This would be the surgical equivalent of a coin toss: If one hemisphere of Henry's medial temporal lobes was the hidden source of his epilepsy, then that approach would have a 50 percent chance of eliminating it. It would be much riskier, of course, than doing nothing, but that might be viewed as a reasonable risk considering the severity of Henry's condition. Also, by leaving the structures in one hemisphere intact, he would minimize the chance of destroying whatever the unknown functions of those structures were.

My grandfather chose a third option. He picked up his suction catheter, inserted it carefully into one of the trephine holes, and proceeded to suction out that hemisphere of Henry's medial temporal lobes. His amygdala, his uncus, his entorhinal cortex. His hippocampus. A good portion of all of those mysterious structures disappeared into the vacuum. Then he pulled the tool out of the first hole, cleaned it off, and inserted it into the second. Lacking a specific target in a specific hemisphere of Henry's medial temporal lobes, my grandfather had decided to destroy both.

This decision was the riskiest possible one for Henry. Whatever the functions of the medial temporal lobe structures were—and, again, nobody at the time had any idea what they did—my grandfather would be eliminating them. The risks to Henry were as inarguable as they were unimaginable.

The risks to my grandfather, on the other hand, were not.

At that moment, the riskiest possible option for his patient was the one with the most potential rewards for him. After years of straddling the line between medical practice and medical research in the back wards of asylums, of attempting to both cure insanity and gain an understanding of various brain structures, he was about to perform one of his medial temporal lobotomies on a man who was not mentally ill at all, whose only dysfunction was epilepsy. In the language of scientific research, Henry was a "normal," or at least much closer to being a normal than anyone who'd previously received one of my grandfather's limbic lobe operations. For four years, my grandfather had been conducting "a study of the limbic lobe in man," and so far he had only "small bits of passing data" to show for it. That afternoon, however, my grandfather's study was expanding to include a whole different class of research subject.

Imagine my grandfather peering into that second trephine hole, guiding his suction catheter deeper and deeper, his headlamp illuminating the intricate corrugations of the structures he was in the process of destroying. It's impossible to say exactly what thoughts drove him at that moment, what stew of motives. He had reason to believe his operation might help alleviate Henry's epilepsy. He also had reason to believe his operation might provide new insight into the functions of some of the most mysterious structures in the human brain. It's quite possible that he wasn't thinking much at all, at least not consciously. Years later, during a rare moment of introspection, he described himself as follows: "I prefer action to thought, which is why I am a surgeon. I like to see results."

He pressed the trigger on the suction catheter, and the remaining hemisphere of Henry's medial temporal lobes vanished into the vacuum.

As my grandfather made that final cut, Henry lay there, looking up at him. He could catch glimpses of his mask, of his surgical cap, of his headlamp. He could see his glasses, those thick-rimmed surgical loupes with their magnifiying lenses. He could hear my grandfather's breathing, feel his warm exhalations.

And maybe, just maybe, some sweat or blood or condensation

accumulated on a lense of those glasses, and maybe my grandfather asked a nurse to reach over and wipe it clean. Maybe that sight, which would have been one of the last ones ever processed by Henry's vanishing medial temporal lobes, somehow stuck with him, blurred and dreamlike and of indeterminate origin. Maybe that's why, for the rest of his life, Henry would tell people that he'd once dreamed of being a brain surgeon but had decided against it, because he wore glasses, and what if his glasses got dirty and a nurse attempting to clean them knocked them askew, causing him to make the wrong move, to cut too far, to go too deep.

If so, that meant that Henry was mistaken about what he'd seen, just as he was mistaken when he told people that he believed something had gone wrong with his own operation, that his own surgeon, Scoville, had made a mistake, had made the wrong move, and that was why he was the way he was. Because even if my grandfather did get something on his glasses, and even if they were then knocked askew by a nurse, that had no impact on the way the operation played out.

My grandfather didn't make any mistakes that day.

He took exactly what he wanted to from Henry.

He finished the operation.

He removed the tools.

He replaced the bone and stitched the flesh.

Six weeks later, he sent off a print version of his Harvey Cushing Society presentation to the *Journal of Neurosurgery* for publication. The paper contained one major addition to the remarks he'd made onstage back in April. His limbic lobe operations, he now wrote, had "resulted in no marked physiologic or behavioral changes, *with the one exception of a very grave, recent memory loss, so severe as to prevent the patient from remembering the locations of the rooms in which he lives, the names of his close associates, or even the way to toilet or urinal.*"

The italics were his, and that italicized clause transformed a forgettable, modest paper into one that will continue to be referenced for as long as we remain interested in how we hold on to the past. It became a cornerstone of the skyscraper that is modern memory science.

It was the birth announcement of Patient H.M.

It was also the obituary of Henry Molaison.

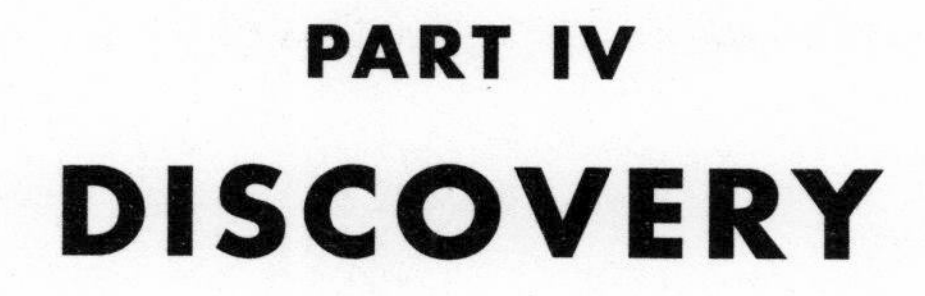

PART IV
DISCOVERY

TWENTY

WHERE ANGELS FEAR TO TREAD

"It's terrible how things accrue."

Those are the first words in the transcript of my first interview with Brenda Milner. When I read the transcript, I hear her voice, a crisp British accent, words tightly spaced, while a hazy picture of her office forms in my mind. That's what she was referring to: her office, and the clutter in it, how it had accrued over the years. The stacks of papers, the overstuffed bookshelves, the boxes full of files. It was one of those offices where it must be a struggle each morning to clear space for the new day's work. There were posters on the walls, and I think one of them had an animal on it, a *National Geographic*–style nature photo, though I can't remember what type of animal it was. There was one framed photograph on her desk, near her computer: a headshot of my grandfather in his light green surgical scrubs.

Our interview took place in 2010, and by that time things had been accruing in Milner's office for more than a half century. She was ninety-three years old, still a full professor at McGill, still teaching classes, still doing research, still living the messy, striving life of an active scientist.

Milner greeted me warmly. Somebody from McGill public relations accompanied me to her office, and before the interview began a photographer arrived and led us to another room, to take a quick shot of the two of us for a Montreal Neurological Institute newsletter. "Does he look like Scoville?" somebody asked Milner. She appraised me—I'm taller than my grandfather, with a bigger nose, closer-set eyes, and a lot less hair.

"No," she said, "not really."

I'd arrived in Montreal the night before and was disappointed to hear that I'd just missed a screening of the movie *Memento* that Milner had hosted for the McGill Film Society. *Memento*'s protagonist, an amnesic man attempting to solve a murder mystery, was inspired in part by Henry. I told Milner I liked the film, and she said she did, too, that in general it was one of the most realistic cinematic portrayals of amnesia she'd ever seen, even though, she pointed out, the script had bungled the definition of short-term memory. People like Henry and the man in *Memento* don't have bad short-term memory, she said. On the contrary, short-term memory is all they have.

I remember Milner sitting there by her cluttered desk, a tiny woman in a woolen skirt.

"Now," she said, "how can I help you?"

Until coming to the Neuro in 1950, Milner had not had a major interest in memory. She was deeply curious about the brain, of course, and all the mysterious ways its structure correlated with its functions. And she understood how lesion patients had contributed to our understanding of those functions: During her final examination in psychology at Cambridge University, as she sat in the historic Senate House, one of the questions asked for a summary of the current knowledge about the localization of function for sight and hearing and speech. That was in 1937, and though she doesn't remember her precise answer, she imagines she would have written about Phineas Gage, Monsieur Tan, and all of those other fascinatingly broken men and women. She would have had no reason to mention memory, though. In 1937, the general consensus was the same as it would be more than a decade later, when Milner began working with Penfield's patients: Memory was not associated with a particular structure in the brain but was instead distributed equally, equipotentially, across the whole cerebral landscape. According to that view, trying to find a seat of memory was the neurological equivalent of a snipe hunt: It didn't exist. Milner's own views on memory didn't change much until she sat down with Patient P.B. and Patient F.C., those two men with their two unilateral medial temporal lobe lesions, and told them some stories, and asked them to repeat them back to her. When they looked at her blankly, when it was clear they weren't retaining anything she told them, she realized that something significant was happening. She didn't believe in the seat of memory, that is, until she found it.

But had she?

If removing the left hemispheres of the hippocampi of these two men had caused profound amnesia, which it clearly had, then it should have done the same to the other patients who had received the same operation, which it clearly hadn't.

Milner and Penfield discussed that paradox at length. Eventually they arrived at some tentative conclusions.

Yes, Penfield had performed the same operation on the two amnesics that he'd performed on at least ninety other patients. And yes, the brains of those two patients, postoperatively, appeared to function very differently from the brains of the others who'd received the same lesions. And no, that didn't make any sense. Unless the brains of the two amnesics had been different to begin with.

They concluded that the only reasonable explanation for such different results from the same procedure was that the medial temporal lobes of the two amnesics must have already been damaged prior to the operations. The hemispheres opposite to the hemispheres that Penfield targeted must have been dysfunctional. Perhaps this damage had taken place during birth, if tongs had been used to extract them from the womb, or maybe they'd suffered undetected strokes later in life. This would have meant that Penfield's unilateral lesioning of their hippocampi would have been equivalent to bilateral lesioning. In effect, though Penfield had targeted only one hemisphere, he would have been destroying both.

If this turned out to be true, it meant that normal memory function depended on the hippocampus and other medial temporal lobe structures. Although Penfield and Milner both felt that they were on the verge of finally zeroing in on the seat of memory in the human brain, they also knew that they lacked proof. The brain imaging technologies of the day could not detect damage to the remaining hippocampi of the two amnesic patients. And even if Penfield had decided to reopen their skulls to look inside, the damage might not have been visible to the naked eye. To upend prevailing views of how memory worked would require more than just a hunch.

Were Penfield willing to test the theory by actually removing both hippocampi of a patient bilaterally and seeing what happened, he and Milner might have been able to obtain the evidence they needed. But Penfield would never do that. He was too wary of causing unnecessary harm. Despite having revolutionized the field of epilepsy surgery, he was fundamentally a conservative, cautious doctor. He viewed all novelty with skepticism. Psychosurgery, for example, appeared to him to be a technically interesting but therapeutically unsettling practice. In his long career Penfield had still never performed a lobotomy, and he sometimes obliquely criticized the doctors who did, making pronouncements about the "vainglory" of "young surgeons who have learned to use a scalpel so expertly that they can take anything out of anywhere without a fatality, to cut the pathways of the currents of intellect and leave a man who is still capable of walking."

Penfield's conservatism was both a boon to his patients and a barrier to progress. By the mid-1950s, several years had passed since Milner's testing of Penfield's first amnesic, and though they planned to write a paper based on these cases, they still hadn't published their findings. Their data felt incomplete, their evidence too easy to dismiss. They were, in the scientific sense, stalled.

Then, during a neurosurgical conference in New Mexico in May 1954, Penfield ran into just the sort of young, ambitious, mechanically gifted psychosurgeon he often accused of vainglory. My grandfather and Penfield discussed their respective experiences operating on the medial temporal lobes. Despite the fact that these operations had mostly been done for very different reasons—my grandfather performed them almost exclusively on asylum patients, Penfield almost exclusively on epileptics—what they found interesting was not their differing motives but their similar results. Penfield told my grandfather about the amnesia he and Milner had documented in Patients P.B. and F.C. and about their hunch that the limbic region must therefore be crucial to memory. Penfield was repulsed by psychosurgery, but he was fascinated when my grandfather told him that he, too, had encountered cases of postoperative amnesia. Not only that, but the younger surgeon's operations, unlike Penfield's, had all targeted both hemispheres. When Penfield returned to Montreal he told Milner what he'd learned.

Eventually, on March 22, 1955, he wrote my grandfather a letter.

"Dear Bill," he began, "Dr. Milner and I have been putting together our projected paper on loss of memory in relation to the hippocampal area. I have thought many times of our discussion out at Santa Fe, and it seems to me that the cases you referred to throw a very important light on the whole problem." He specified the most important cases as the three patients "in whom you made a removal bilaterally back to a distance of 8 or 9 centimeters in the temporal lobe," and asked whether it would be possible for Brenda Milner to travel to Connecticut to meet with them. "I remember that they were all psychotic," Penfield wrote. "I suppose you must feel as hesitant in regard to these cases as I do in regard to the two patients in whom I have produced a gross loss of memory. Actually, I should feel much worse inasmuch as the patients I operated upon were not psychotic and had a much better outlook on life than yours could possibly have had." Milner, Penfield continued, was willing to go to the patients "wherever they are."

"I have pulled all the pertinent charts and am delighted to have Dr. Milner come down and go over the cases," my grandfather wrote back. He corrected Penfield's assumption that all of his medial temporal lobe cases had involved psychotic patients, and then described Henry. "The only non-psychotic epileptic case will come into our office," he wrote. "He is one and three-quarter years post resection of the medial surface of the temporal lobes including the uncus, amygdala and hippocampal gyrus, and according to his mother, over the phone, 'his memory is absolutely no good; cannot even be sent to the store alone for purchases.'"

As for the asylum patients, he would ensure that the institutions granted Milner full access to them, though he warned that they might not be as easy for her to glean useful data from as Henry. "These cases," he wrote, "are all available for study but, of course, are complicated by other damage."

On April 25, 1955, Milner boarded the night train from Montreal to Hartford. She traveled light: a few changes of clothes, some toiletries, and a small collection of psychological tests. Of course the most important thing she carried didn't weigh anything at all. It was an idea, a theory, one that had been taking shape for years but had until now remained frustratingly hard to pin down. The train pulled out of the station and began pushing south, picking up speed, crossing the border and skirting the edge of Lake Champlain. The foothills of the Adirondack Mountains rose in the distance, and Milner tried to sleep in the hurtling darkness.

She met Henry for the first time the next morning, at Hartford Hospital. My grandfather introduced them, and Henry greeted Milner with a smile. He was twenty-nine years old, boyish, affable, polite. After the introductions, my grandfather left to perform a surgery, and Milner excused herself to prepare her testing materials at a table in a nearby office, leaving Henry in the hallway outside with Dr. Karl Pribram, the head of research at the Institute of Living. Pribram had come to the hospital because he, too, was curious about Henry. When Milner finished setting up, she found Pribram and Henry still engaged in conversation. She interrupted them, then led Henry into the examination room.

"What were you and Dr. Pribram talking about?" she asked.

Henry looked at her curiously. She was mistaken, he said. He hadn't been talking with anyone.

They sat at the table, and Milner pulled out a copy of something called the Wechsler Memory Scale. The WMS, as it was known, was the standard diagnostic tool used to test memory. She'd used the same test with Patients P.B. and F.C. It was published by a firm called the Psychological Corporation and was written and conceived by a Bellevue psychologist named David Wechsler, who was also the author of the most popular IQ test at the time. Finding Henry's "memory quotient" was a simple matter of presenting him with the tasks listed on the form, recording his answers, and then tabulating the results. The interpretation of a person's memory quotient and intelligence quotient were roughly analogous: A score of one hundred was considered average, while anything above or below one hundred would be considered superior or inferior to some degree. The test began with a series of very basic questions, which Milner posed to Henry one after the other, pausing between each to note his answer. Henry said he was twenty-seven, that the year was 1953, that the month was March, and that the president of the United States was Harry S. Truman.

Most of Henry's answers were not exactly incorrect, factually speaking. Instead they were simply wrong chronologically. Everything he said had been true, at some time or another, just not at the present. As she listened and took notes, Milner tried not to betray any surprise or shock that might influence Henry. (This was a standard diagnostic strategy: In a copy of the WMS testing manual that I acquired, another psychologist had written across the top of one page that while presenting the test it was a good idea to "pretend you are Jack Webb of *Dragnet*.") She proceeded through the questions without comment, even as the depths of Henry's amnesia became clear.

Milner then tested Henry on his ability to reproduce simple geometric drawings from memory and to remember unusual pairings of words, such as *cabbage/pen* and *obey/inch*. As Henry struggled, repeatedly coming up blank, Milner continued taking careful note of his errors, focusing on the protocols of the test, trying not to be distracted by the dawning sense that she was in the presence of an extraordinary patient.

Finally, she told Henry some stories, one about a young house-cleaner in Boston named Anna Thompson who'd been robbed of fifteen dollars, another about a ship that struck a mine near Liverpool during a perilous ocean crossing.

When she was finished, she waited a few moments, then asked him to repeat whatever he remembered back to her.

He looked at her curiously.

Henry scored a memory quotient of sixty-seven, the lowest Milner had ever seen.

Before the day was done, Milner also tested Henry's IQ. The tasks contained in the Wechsler-Bellevue Intelligence Scale were eclectic. There were specific questions meant to gauge one's general factual knowledge—Where is London? Who wrote *Faust*? What is the Apocrypha?—and broader, almost philosophical questions that required more elaborate and thoughtful answers: Why should we keep away from bad company? Why are laws necessary? Why are people who are born deaf usually unable to talk? One section asked for definitions of various words, ranging from the simple—*apple*—to the obscure—*moiety*—and another tested real-world arithmetic skills by asking, for example, how many pounds of sugar you could buy for a dollar, if seven pounds of sugar cost 25 cents. There were analogy tests, which asked in what ways oranges were similar to bananas, wagons similar to bicycles, and praise similar to punishment. Other tests involved pictures, not words, such as one that required the test taker to reassemble a jumble of paper fragments into their original form, a human head in profile.

Henry was good at all of them. More than good. He scored 118 points overall, far above average and well into the "superior intelligence" range. He was particularly good at the arithmetic, could work the answers out in his head quickly and fluidly, but he proved sharp and capable throughout. As Milner observed Henry tackling the IQ test, coasting through hard questions with ease, it was almost possible for her to forget the amnesia that lurked just below the surface of his bright eyes.

Henry was a tireless test taker. He focused on each new task with the same intensity, never complaining, never bored. At one point, during a pause in the testing, Milner excused herself and went to get a cup of coffee from the hospital cafeteria. It had been a long night on the train and a long day in the examination room. She returned a few minutes later armed with caffeine and ready to get back to work. Henry greeted her with a smile and a look of friendly uncertainty. He had no idea who she was, but he was pleased to meet her all over again.

The next day, Milner began visiting the psychotics in the state asylums. My grandfather provided her with a car and a driver, and phoned ahead to make sure she'd have access.

Her first visit was to Connecticut State Hospital, sixty miles west of Hartford, with its bucolic riverside setting and active dairy farm. She settled into an examination room and soon received her first patient, a woman with the initials M.B. Milner, as always, took careful notes. Patient M.B. was, she wrote, "a 55-year-old manic depressive woman, a former clerical worker." She'd been at the asylum since December 27, 1951, when she'd been described as "anxious, irritable, argumentative and restless, but well-orientated in all spheres." Almost exactly one year after her institutionalization, M.B. visited room 2200, where my grandfather performed a "radical bilateral medial temporal lobe resection." M.B.'s case files recorded that "postoperatively she was stuporous and confused for one week, but then recovered rapidly and without neurological deficit."

Even before Milner began the formal testing, it was clear to her that the asylum's conclusions about M.B.'s lack of deficits were wrong. The woman, Milner wrote, "had been brought to the examining room from another building but had already forgotten this; nor could she describe any other part of the hospital, although she had been living there continuously for nearly three and a half years." The tests Milner administered only verified these first impressions. "On the Wechsler Memory Scale her immediate recall of stories and drawings was inaccurate and fragmentary, and delayed recall was impossible for her even with prompting; when the material was presented again she failed to recognize it. Her conversation centered around her early life and she was unable to give any information about the years of her hospital stay." By the time M.B. was led back to her ward, Milner was convinced that she suffered from "a global loss of recent memory similar to that of H.M."

Over the next few days, Milner met with four more asylum patients.

One, a "paranoid schizophrenic with superimposed alcoholism," told Milner that she could "remember faces" but forgot "many daily happenings." Milner noted that during their conversation this patient "showed little knowledge of recent events." Another was a little less impaired, and "knew that her daughter had caught a 7 o'clock train to New York City that morning to buy a dress for a wedding the following Saturday. She could also describe the clothes worn by the secretary who had shown her into the office." A third patient fared poorly on memory tests, though she was aware that "she had been working in the hospital beauty parlor for the past week and that she had been washing towels that morning." The final patient was able to remember Milner's "name and place of origin 10 minutes after hearing them for the first time" but was "still subject to delusions and hallucinations" and "was found to be too out of contact for extensive formal testing."

Up until that week, Milner had never even been to an asylum. When she talked about these visits a half century later, the frustration of trying to gather useful data in those difficult environments still seemed fresh. The patients, she said, were mostly pathetic people from the back wards of the asylums, and she could, she said, "do very little with them. I could just establish that they had essentially the same kind of impairment. . . . But I couldn't do very much with them in terms of formal testing."

Still, it was clear that most of the asylum patients she met displayed some degree of memory loss, even if that degree was challenging to quantify. Most important, the severity of the loss appeared roughly proportionate to the amount of hippocampal tissue my grandfather had removed from their brains. That was the key.

At the end of the week, she boarded the train back to Montreal. The tentative ideas that she'd brought to Connecticut had grown into something stronger.

By that time, the bilateral medial temporal lobotomy that my grandfather developed, like his orbital undercutting lobotomy before it, had spread beyond Connecticut. Other neurosurgeons, in other states, in other asylums, had begun to experiment with it. Sometimes they would invite my grandfather to come demonstrate the operation.

Which is why, in May 1954, about nine months after operating on Henry, my grandfather traveled to Manteno State Hospital in Illinois, a little south of Chicago. By the time he arrived, the patient was already anesthetized and on the operating room table. My grandfather knew virtually nothing about him besides the obvious fact that he—unlike most of the asylum inmates my grandfather had operated on in Connecticut—was a man. He proceeded with the operation, cutting open the inmate's skull and removing most of his medial temporal lobes bilaterally in the presence of a young neurosurgeon named John F. Kendrick, who subsequently began performing the operation on his own.

After Milner's visit to Connecticut, it occurred to my grandfather that she might like to visit the man he'd operated on at the Manteno asylum. He wrote the asylum superintendent a letter, asking that he grant Milner access to the patient, then sent a copy of the letter to

Wilder Penfield, along with a note musing that the case "might prove of intense interest." Penfield provided a copy of both letters to Milner, along with a handwritten note:

"To Dr. Milner: Where angels fear to tread . . ."

Two months later, Milner arrived in Illinois. The patient in question had, she learned, recently been transferred from Manteno State Hospital to another nearby asylum called the Galesburg State Research Hospital. On January 12, 1956, Milner was provided with an examining room at Galesburg, and the man she would later describe in scientific papers as Patient D.C. was escorted to meet her.

"Did my grandfather ever feel guilty?"

Milner seemed surprised by my question.

"I don't think he felt guilty," she said. "I mean, we did not know what these structures did. And I don't really think he should have felt guilt about H.M., because H.M. was so desperate. He was having an absolutely miserable life." She paused. There was one patient in particular, it struck her now, who had weighed on his conscience, who had made him feel guilty.

"D.C.," she said. "The doctor in Chicago."

That was one of the first things Brenda Milner learned about D.C., sitting in that examination room at the Galesburg State Research Hospital reviewing his case history: He'd been a doctor. A practicing doctor, in Chicago. And then something had happened to him. A breakdown. The breakdown was precipitated, perhaps, by the loss of a lawsuit. Or maybe it was inevitable: He'd had a history of paranoid thoughts and violent outbursts, which started before he finished medical school. Regardless, the breakdown in question had been extreme: He tried to kill his wife, with an ax, unsuccessfully. That was in 1950, when he was forty-one years old. For the first four years of his institutionalization, he received many of the standard treatments from the armamentarium of contemporary asylums, including repeated rounds of insulin coma therapy and electroshock therapy. His condition did not appear to improve. And so eventually a call was placed to my grandfather.

The records indicated that Patient D.C.'s recovery from the operation was "uneventful," with "no neurological deficit," though at least one asylum employee noted that "since the operation he had been unable to find his way to bed and seemed no longer to recognize the hospital staff."

Milner administered D.C. the Wechsler intelligence test and found that he was extremely bright, even brighter than Henry, with an IQ of 122. Then she began testing his memory. Very quickly she realized that he "presented exactly the same pattern of memory loss as H.M." When she asked him where he was, he said he had no idea but explained that that was only natural, since he'd arrived there just the night before. He'd actually been there six weeks.

She ran him through the usual gauntlet of memory tests and tried a few new ones out. She asked him to draw pictures of a dog and an elephant. Then she put them aside. A few minutes later, she showed him the drawings. He was not a good artist. She asked him what animal the drawing of the dog was supposed to be. He squinted at it, then took a guess.

"A deer?" he said.

Then she asked him who had drawn the pictures.

He had no idea.

Those drawings, fifty years old now, are among the things that have accrued in Brenda Milner's overstuffed office, although she can't find them during my visit.

She does remember a phone call she made the day after she met D.C., though, before leaving Chicago.

"I called Dr. Scoville," she told me. "It was his birthday. His fiftieth birthday, I think. January the thirteenth, wasn't it?"

She called him, wished him a happy birthday, and told him about the tests she'd run on Patient D.C. and how they had revealed an amnesia just as profound as Henry's. Then she told him a little bit about D.C.'s history, specifically the fact that he had been a medical doctor.

"This was the thing that got to him," Milner told me. "You know, the professional thing. I'm sure if he had been something else than a doctor—even a high-level something else—it wouldn't have bothered him so much. A lawyer or something. But this was a doctor. And it shook him."

In May 1957, the *Journal of Neurology, Neurosurgery & Psychiatry* published an article by William Beecher Scoville and Brenda Milner titled "Loss of Recent Memory After Bilateral Hippocampal Lesions." The paper introduced Patient H.M. to the world and detailed the depths of his deficits.

The paper also described a variety of institutionalized men and women my grandfather had performed similar operations on—Patient M.B., Patient A.Z., Patient A.R., and Dr. Patient D.C., among others—but the clear focus was on Patient H.M., who combined a mind unmuddied by mental illness with bilateral medial temporal lobe lesions as extensive as any performed on the psychotics. The important takeaway was a broad one: "In summary, this patient appears to have a complete loss of memory for events subsequent to bilateral medial temporal lobe resection." By extension, H.M.'s case pointed "to the importance of the hippocampal region for normal memory function." My grandfather and Milner concluded with the stark, unequivocal, and deceptively revolutionary declaration that "bilateral medial temporal lobe resection in man results in a persistent impairment of recent memory" and that the medial temporal lobe structures must therefore be "critically concerned in the retention of current experience."

In other words, the location of the seat of memory, that ancient mystery, had been revealed. And it is that revelation, borne out by more than a half century of subsequent research, which made this paper the single most cited paper in memory science. It is in many ways the field's founding text.

Brenda Milner is ninety-three years old. Most of the people she's ever known are gone now, of course.

Wilder Penfield.

My grandfather.

Henry.

But sometimes the memories they created are so strong that the fact of their absence is hard to process.

"People ask if he was unhappy," Milner said to me at one point, talking about Henry. "I think he'd never known happiness. I mean, this is the thing about H.M. Any of these other patients, one would say, 'This thing that happened is catastrophic.' I mean if you've got a real life with a family, and you're interested in politics, and you're good at your job, and so on, and you suddenly can't remember what you had for breakfast, then this is a catastrophe. But if you had major convulsive seizures in spite of heavy medication for years and years and years, so that although you were potentially bright you couldn't think clearly, and it took so long to get through school, and you were rejected. . . . You can imagine that he did not really have a very good life with all the medication and all the seizures and the isolation. And since the surgery, since he is so amiable, and since he really likes doing tests, he likes doing something for science . . ."

She paused.

"I keep on saying 'likes.' I can't believe he's not here. I feel like I've lost a friend, and that is a funny thing, too, because you think of friendship as reciprocal, right? You're friends with someone and they're friends with you. And yet this is very one-sided. Because he doesn't know me."

TWENTY-ONE

MONKEYS AND MEN

I remember the bull's speed, and the sound of its hooves in the dirt, and the tendons tightening on the top of its neck as it lowered its head and prepared for impact. I don't remember the way it smelled, though I'm sure it did: They always do, a sour mix of sweat and manure. And I don't remember if it was grunting or making any other sounds. Sometimes they do, sometimes they don't.

It was the beginning of 2001, and for the previous couple of months I'd been hanging around the Academia Municipal Taurina, a bullfighting school in Guadalajara, Mexico. The students ranged in age from twelve to eighteen, and every day I'd come and watch them train, taking notes for a magazine story that I had no home for but was convinced could be great. These ambitious, talented, brave kids, trying to take their first steps in this risky business while navigating the usual pitfalls of teenagerhood: It was like *Death in the Afternoon* crossed with *Glee.*

Usually the kids trained without bulls. Instead they themselves would play the bulls, leaning down low and charging at each other, hard and fast, learning the intricacies of cape-and-foot work. Often they'd use a strange contraption, a custom-built wheelbarrow-like thing with real bull horns mounted on its front to make the fake charges a little more menacing.

Sometimes, though, they used real bulls.

Or, to be more precise, *becerros.* A *becerro* is a young bull, barely more than an adolescent at two to three years old and weighing around 300 pounds; they weigh a mere fraction of the full-grown *toros bravos* you usually find in bullrings, the NFL linebackers of the animal kingdom, which can easily top 1,500 pounds. Were one of these kids to get hit full-force by a *toro bravo,* that might be it for the kid. With a *becerro* there was more room for error, more room to make mistakes and learn. Which isn't to say that getting hit by a three-hundred-pound animal is pleasant. It isn't. As I was about to find out.

We'd driven in a caravan out to a ranch that afternoon, where one of the patrons of the school, a rich local farmer, had a small bullring on his property. There was a *becerro* waiting for us there in the ring. Jet-black, wiry, and wired. The bullfighting world is full of stories about the techniques used to prep bulls for training or matches. That they're amped up with amphetamines or have cords cinched tightly around their testicles. I don't know what had been done to this *becerro,* if anything, but it looked ready to fight.

I watched the kids take their turns with it. Mauro, Daniel, Rodrigo. One after another, they stepped into the ring. They weren't dressed in their formal bullfighting attire, those spangly, stiff, coruscating, pageant-ready suits. Instead they were dressed down, jeans or sweatpants, T-shirts, sneakers. They shared a capote. It was pink on one side, a sort of dull off-white on the other, heavier than you might expect, made of a coarse canvas. When it came his turn, each boy would step out from behind one of the chest-high wooden barriers arranged at four points along the inner circumference of the ring. There was enough room for a person to stand behind these barriers but not enough room for a bull, was the idea. They'd step out, often when the bull was facing away from them and on the other side of the ring. They'd take a few steps forward, then stop, straighten their backs. The top edge of the capote had a sturdy wooden rod fixed along its edge, and the student would hold it out, letting its full surface area billow below the rod, the pink side facing the bull. If the bull still didn't notice, the student would stomp his foot in the dirt, kicking up a tiny explosion of dust.

"Ay," he'd shout, his voice tight, and as commanding as he could muster. "Toro!"

Eventually, the bull would turn.

They were good, these kids. They'd stand their ground, shoulders pulled back, chins up, their posture designed to project confidence, not fear, no matter what they felt inside. They'd shake the capote, once, twice, three times—"Toro! Toro! TORO!"—keeping it perpendicular to the horns of the bull, the bull that was by now eyeing them intently.

And then it would charge.

The color of the capote didn't matter. It's a myth that red enrages. Bulls are color-blind. What they see is movement, what they hear is noise, what they feel is threatened, what they want is contact, and, with it, conquest. The capote took up more space, on a two-dimensional plane at least, than the kid did. It moved more, too. The capote was the flag—whipping up and down, back and forth—and the kid was the flagpole.

The bull went for the cape. Why wouldn't it?

And the kid, like Lucy snapping up the football just as Charlie Brown attempts to kick it, lifted the capote in a flourish, twirling in time with the bull just as it attempted to make contact. Usually the bull followed the capote as it moved, rotating around the axis of the kid, bucking its head impotently against the fabric. One pass. Two. Maybe three. And then the kid stepped away, took shelter behind another one of the barriers, and the next kid readied himself, waiting for the bull to settle a bit before he picked up the borrowed capote and took his own steps into the ring.

They trained like this for forty-five minutes while I stood behind one of the barriers, my notepad propped against the top of it, my ballpoint scratching out words, my skin reddening under the blue Mexican sky, trying to take it all in. It was almost lunchtime when the last kid—I think it was Mauro—had his time in the ring. And then I don't remember who it was that suggested I give it a try.

I took the capote. I stepped into the ring. I walked forward a few steps. I tried my best to pull my shoulders back, to stand straight, to stick out my chin. The bull was in the opposite corner, looking away from me. I held the capote out.

"Ay! Toro! Toro! TORO!"

It turned. It seemed to hesitate for a few moments, looking at me with its blank, dull eyes. I could see its flanks heaving. It was tired. It must have been frustrated, too. It hadn't connected, not even once.

I shook the capote again, and it charged.

I remember the approach, the lowered head, the clomping hooves, the way my hand whitened around the rod that supported the cape. I remember shaking the cape with increasing urgency. I remember wondering why the bull didn't seem to be paying any attention to the cape, why all it seemed to see was me. A good matador, when the bull is about to make its pass, stands even straighter than straight, thrusting his hips forward while he pulls back his shoulders, an act of postural symbolism so obvious it needs no explanation. My hips receded as I stuck the capote out as far as I could, away from my body, shaking it harder and harder, trying to make the bull change its unchangeable mind.

Its forehead connected with the side of my hip, lifting me off the dirt and into the air. I remember twisting, trying and failing to get my feet under me. I hit the ground hard. It was on me. I remember holding one of its horns, gripping it tighter than I'd gripped the capote, holding it away from me so that it wouldn't come close to my head, my neck. I heard the kids come rushing, yelling at the bull, one of them smacking its flank, trying to distract it. After maybe five seconds it worked: The bull turned its attention away from me, took a tentative lunge toward Mauro. I scrambled to my feet and ran behind one of the barriers, laughing with excitement. I had so much adrenaline in my veins that I didn't notice the swelling hoofprint embossed on my stomach till later.

So why had the bull connected with me and not the others?

An obvious answer, and a correct one, is that I'm a terrible matador. But there's another answer, one equally true, and to find it we return to Henry's story.

The lines dividing human and animal research during the 1950s were blurry almost everywhere, but nowhere more so than at my grandmother's asylum. In 1949, the same year the superintendent, Charles Burlingame, built my grandfather the first operating room devoted exclusively to psychosurgery, he also, in the same building, opened a monkey research laboratory. Burlingame hired a young, ambitious neurosurgeon turned neuropsychologist named Karl Pribram to head the laboratory and become the institute's director of research. Pribram arrived at the asylum straight from John Fulton's laboratory at Yale, and his mandate was described in the institute's annual report that year: "Thus, follow-up studies can be carried out on several problems arising from psychosurgical (lobotomy) work on humans, making possible a more rational approach to this controversial mode of therapy in psychiatry."

Even before Pribram left Fulton's primate laboratory to start his own at the Institute of Living, his interests, like those of many of his contemporaries, had begun to migrate from the frontal lobes deeper into the brain, toward the temporal lobes. He and an even younger researcher, a visiting PhD student from McGill named Mortimer Mishkin, set out to pick up where Paul Bucy and Heinrich Klüver had left off the decade before, investigating the behavioral and neurological effects of bilateral temporal lobe ablations. In their first efforts, they requisitioned ten animals: one adult male chacma baboon, one female guinea baboon, and eight young rhesus macaques. They tested each animal extensively on a variety of tasks. Pribram then opened their skulls, one by one, and removed various portions of their temporal lobes before retesting them. They were generally kept alive for approximately four to eight months postoperatively, and efforts were made to control and keep tabs on most elements of their lives, including what they ate and their body temperatures; Pribram, however, noted that although "we believe that all spilled food was recovered and weighed we cannot be certain of this. In the same manner, although every precaution was taken to accustom the animals to the rectal temperature-taking procedure, it cannot be said with certainty that the relaxation of the animal was the same pre- and postoperatively." After sufficient data was collected, they were killed and their brains analyzed to provide a precise measurement of the extent of their lesions.

What they found in many ways confirmed what Klüver and Bucy had found. For example, the animals with temporal lobe lesions seemed willing to eat almost anything. Pribram and Mishkin would soak pieces of potato and cotton ball in foul-tasting quinine and offer them to the monkeys. The lesioned monkeys would chew whatever was proferred without hesitation. They would also eat meat, which these vegetarian primates would have otherwise avoided. This newfound adventurousness extended beyond their palates, too: They seemingly had appetites for all sorts of things they didn't care for before. The researchers would offer the monkeys items considered noxious, such as razor-sharp pieces of metal and burning pieces of paper, and the monkeys would repeatedly pick them up and examine them without hesitation, despite the damage such items caused them. "A stimulus object is considered noxious if it visibly injured the animal's integument by cutting or burning," they wrote. "The number of times in a session that an animal would approach, accept, and examine such an object is recorded." On several occasions, after a monkey took the burning paper, "the animal's whiskers would catch fire," and in general, "in spite of the obvious discomfort these noxious agents seemed to cause, the animal would return over and over again to expose himself to injury."

Pribram and Mishkin also noted that the animals showed the classic "tameness" and "lack of fear" that Klüver and Bucy had described in their animals. They tested this in part by yelling at the animals in a threatening manner, and on several occasions crawling into the cages of the once ornery, now placid primates. "Without prior planning, the authors independently felt it safe to enter the animal's cage and 'petted him for a considerable time.' When the observer placed his hand in the animal's mouth it was chewed very gently." They also placed nonlesioned monkeys in the cages with the lesioned ones to see what would happen. "When attacked by a larger animal he would not attempt to escape but would sit quietly ducking the debris thrown at him, wincing or grimacing briefly when hit or bitten."

Along with these rehashings of Klüver and Bucy's original protocols, Pribram and Mishkin also administered some new tests. Chief among these was the so-called delayed-response task, which was supposed to test their memories. The task was taught to all of the animals prior to their surgeries to make sure they had a basic understanding of the method. It worked like this: The animal would watch as a researcher hid a single peanut under one of two identical upside-down cups. The researcher would then lower a screen in front of the animal and the cups. The screen would remain in place for about fifteen seconds, then the researcher would raise it, and the animal would be allowed to reach for one of the cups. If he chose correctly, he'd get to eat the peanut. Prior to operation, the animals were each run through the task one hundred times and chose the peanut-concealing cup 85 percent of the time. Then, after large portions of their medial temporal lobes were removed, they were administered the task again.

They did fine. As Pribram and Mishkin wrote in their paper, "performance in the delayed-reaction test was unimpaired." Which is to say that as far as they could tell, when primates lost their hippocampi and other nearby structures it did not damage their ability to remember things. Whatever was happening to the monkeys as a result of the lesions, it did not seem to be affecting their memory systems.

But then, H.M.

By the time my grandfather operated on Henry, Pribram had already established his laboratory at the Institute of Living, and although he and Mishkin were both still affiliated with Yale, they lived and did most of their work in Hartford. Henry's operation was performed at

Hartford Hospital, not the Institute of Living, but Pribram heard about it and was immediately intrigued. In fact, in the original letter my grandfather wrote to Wilder Penfield inviting Brenda Milner to come study Henry and his other medial temporal lobe cases, my grandfather added that Pribram "also would like to study these cases which, of course, is quite all right with me."

Pribram never ended up conducting any studies involving H.M. or the other patients. His specialty was the study of nonhuman primates, and maybe he simply found the transition from macaque to Homo sapiens too difficult. But although Pribram and Mishkin didn't study H.M. directly, they and countless other brain researchers had to grapple with what H.M. meant. If human and nonhuman primate brains were similar, functionally speaking—which after all was the animating principle behind monkey research—then why were monkeys with medial temporal lobe lesions apparently left with intact memory systems? How did they remember which cups those peanuts were hidden under?

Soon researchers everywhere were poking at this apparent discrepancy, trying to replicate H.M.'s lesions in primates and seeing if doing so would induce amnesia. My grandfather was among them: Along with a neuropsychologist named Robert Correll, he established a small colony of macaques in a lab at Hartford Hospital. For a while, during his off-hours, my grandfather fell into the habit of visiting the lab and doing to a macaque exactly what he had done to Henry, removing its medial temporal lobes bilaterally. He and Correll attempted to test the memories of the primates pre- and postoperatively, presenting them with various tasks, rewarding them with bits of food or banana-flavored pellets. His results, however, were largely the same as Pribram and Mishkin's had been: The lesions seemed to leave the macaques' memories unaffected, at least their memories as measured by delayed-response studies. My grandfather attributed this failure in part to "the absence of a generally accepted operational definition of memory." Eventually he sacrificed all the monkeys, made slides from their brains, and moved on.

As for Karl Pribram, he left the Institute of Living in 1957 and transferred to Stanford University, where he received a joint professorship in the departments of psychology and psychiatry. Pribram's protégé, Mishkin, left the Institute of Living for the National Institute of Mental Health, in Bethesda, Maryland, where he established his own primate lab, one of the largest in the country. At the NIMH, Mishkin continued to focus on trying to document H.M.-like amnesia in monkeys. He would lesion the monkeys' medial temporal lobes, then run them through a series of tests, then sacrifice them to precisely measure the dimensions of their lesions.

Lesioning, testing, sacrificing. Again and again.

It took two decades before he found what he was looking for.

To get to Mishkin's office at the NIMH, I had to pass through multiple levels of security, first showing my passport to a guard behind a glass-shielded desk, then passing through a metal detector manned by another guard, then discussing the reasons for my visit with another guard before I was issued a temporary visitor's pass on a lanyard to hang around my neck. Mishkin's office was almost a quarter mile from the entrance to the sprawling NIMH campus, and as I walked to it I thought of a book I'd loved as a kid, *Mrs. Frisby and the Rats of NIMH,* a novel about a group of laboratory animals that are granted extraordinary intelligence, which they use to escape from their cages so they can live in their own world rather than "on the edge of somebody else's, like fleas on a dog's back."

There was another security guard in the lobby of Mishkin's building, and I showed him my visitor's pass and my driver's license and told him I had an appointment. He got on the phone.

"Dr. Mishkin?"

A minute or so later, Mishkin emerged. He looked much the same as the latest picture I'd seen of him, shaking hands with President Obama in 2010 during the ceremonial presentation of the National Medal of Science. During the event, Obama had praised Mishkin for his "contributions to understanding the neural basis of perception and memory in primates." Mishkin, who had a firm handshake and an interesting, interested face, with bright eyes and a wide smile, led me down a few hallways and through a heavy door into his office space.

"Did all of this security happen after 9/11?" I asked.

"No," he said. "It happened after I opened my lab."

Animal rights activists, he explained. All that lesioning, testing, and sacrificing made him and the rest of his laboratory staff, not to mention the lab itself, a target for those who thought humans should keep their hands off their nearest animal kin.

Mishkin wanted me to know that he hadn't always worked with animals. The first study he ever published, in 1950 while he was still at McGill, was a study of lobotomy patients. "But it was nonsense," he said. "We were using inkblots to study schizophrenic patients who had lesions to their brains. My god. Inkblots! It was pretty incredible that we were doing things like that. And we knew that it was stupid. But that's what was being done at the time. I actually refused to continue, it was so silly." The basic problem, as he saw it, was that nobody knew the functions of the structures they were destroying. Neurosurgeons may have thought they knew, but neurosurgeons were not scientists, even if they sometimes pretended to be. Mishkin knew enough to at least know what he didn't know. He knew it was a terrible thing to mutilate a human brain on the basis of nothing more than a glorified hunch. Animals were another matter. Whatever you thought about the ethics or morals of animal experimentation, the rationale for it boiled down to a simple statement, one you either agreed or disagreed with: Better them than us.

Mishkin's office was large and cluttered. There was a phrenological skull on one shelf, next to an assortment of model brains. I noticed that along with the neuro stuff there were also a number of items related to astronomy, including a beautiful framed photograph of an earthrise as seen from the surface of the moon. Mishkin told me that astronomy had always been an avid but amateur interest of his. I took a seat across from him. There was a small framed H. G. Wells quote on the wall to my right: NO PASSION IN THE WORLD, NO LOVE OR HATE, IS EQUAL TO THE PASSION TO ALTER SOMEONE ELSE'S DRAFT. He saw I was staring at it, then pointed out the joke to me: One of the commas, the one after HATE, had been penciled in.

"I added that," he said.

That's science, in a way. Each new generation takes the drafts created by the men and women who came before it and revises them, sometimes making small tweaks, adding commas or dropping clauses, sometimes making more drastic changes, discarding chapters altogether or writing entirely new ones. They're all editors, and all of them, the good ones at least, have to be passionate about altering someone else's draft if they're ever going to make any progress.

In Mishkin's career, one of his most significant edits revised our understanding of monkey amnesia and began to bring it in line with our understanding of the human amnesia suffered by Patient H.M. The discovery hugely strengthened the scientific community's confidence in its understanding of memory mechanisms. He made the discovery here, at the NIMH, with another series of macaques. After replicating H.M.'s operation by lesioning their medial temporal lobes bilaterally, removing structures including the majority of their hippocampi and amygdalae, he ran the macaques through a new variation of the delayed-response tests that, for decades, had been coming up empty. This variation was called a delayed-nonmatching-to-sample task, and it differed from previous tasks in an important way. Now, instead of Mishkin requiring the monkeys to remember which object concealed food, he concealed the food under an entirely new object each time. So to earn the peanut, the monkey would have to remember which object he had chosen before, then choose the other one. This test protocol was designed to overcome the basic problem with the previous tests, which was the suspicion that a monkey could learn to associate a particular object with a positive reward in an instinctual way without employing what we think of as memory. By contrast, in this new task, as Mishkin wrote, "because the food is always associated with a novel object, the ability to link a specific object to a reward has no bearing on performance. The reward serves merely as an incentive; the test measures recognition memory specifically."

And it worked. Meaning that after the lesioning, whatever memory systems enabled normal monkeys to complete the task previously no longer worked. "Their scores fell nearly to chance," Mishkin wrote. "It seemed, then, that we had created a true memory loss."

Things had come full circle. Primate research had inspired my grandfather to create Patient H.M., and Patient H.M. had inspired this new round of primate research, which after years of false starts finally began to validate and reinforce the discoveries made about Patient H.M. In the never-complete manuscript that is our knowledge of self, Henry Molaison was acquiring an ever more central role.

Mishkin demonstrated the historic test to me at his desk, using my cellphone as a stand-in for a peanut.

I waited behind the barrier and watched as the guys corralled the bull. Using whoops and slaps, they guided it through a door someone had opened in the ring, into a narrow pen. It was done for the day. Nobody else wanted to take another turn in the ring with it. This was partly because they'd all gone a few rounds already, but it was also because of what had just happened to me, and what that meant.

Imagine a bullring as an experiment.

The bull is the test subject, and its task is to choose between the bullfighter and the bullfighter's capote. If it chooses correctly, it will receive the reward of satiated rage, horns against flesh. If it chooses incorrectly, it will receive nothing but an impotent thrust against a billow of air. (Although, of course, during a real bullfight, not just a practice session, if it chooses incorrectly it dies. But it doesn't know that, so that fact is irrelevant to how the experiment plays out.) Today, during the first part of the experiment, the bull chose wrong, over and over again. It charged the capote, the bigger target, which swayed and shimmied enticingly.

Once, twice, three times. A dozen, even. Each time it came up empty.

Then me.

The bull turned and saw that dual spectacle, the flagpole and the flag, and was forced once more to make a choice. It charged. Straight at me, straight for me, straight into me. The flag went flying, and so did I. The bull finally made the right choice.

Now again, I'm a lousy bullfighter. I don't stand as still as I should and I don't make the capote undulate as appetizingly as I could. But even if it weren't me in that ring, even if it were Mauro or one of the other students, someone who actually knew how to *torear,* eventually the bull would have wised up. This is a known fact in the bullfighting community: On any given afternoon, you can only train with a given bull for a certain number of hours before it is, in a sense, spoiled. Even these broad-shouldered brutes adapt to the task at hand, learn to ignore the capote, and target the man.

This is basically a variant of Mishkin's delayed-nonmatching-to-sample task. The bull isn't being tasked with learning to associate a certain choice with a positive outcome. Instead it's being tasked to do something that requires a more complex form of memory: It has to remember that a certain choice it made earlier was unsuccessful, and that it had therefore better make the opposite choice. As enticing as the capote is, the bull has to go against all instinct, girded and guided only by memory, and choose the bullfighter instead.

It did. It chose me. It found success. And once it did so and was rewarded with the satisfaction of trampling me in the dirt, the experiment had effectively come to an end. Because now, were it to choose again, its choice would be simpler. Learning to hit the bullfighter is only difficult until it has done so. Then it's easy, just a matter of strengthening the association between the choice it made and the reward it received. That sort of association doesn't require any meaningful form of complex memory any more than does the drool from a Russian dog hearing a dinner bell.

So the bull remembered. That much seems clear. It had access to the events of its past and made a decision based on it.

But here's another question, one harder to answer:

Does the bull remember me in the same way that I remember the bull?

Even now, more than a decade later, I can see it thundering toward me. I can remember the heat of the sun, the sound of the hooves, the feel of the impact. I can remember a scene, incomplete but not insubstantial.

This ability to create scenes from our pasts may or may not be uniquely human. And we may never know for sure. Elisabeth Murray, one of Mishkin's younger colleagues at NIMH, wrote an article a few years ago, the title of which sums up the problem: "What, If Anything, Can Monkeys Tell Us About Human Amnesia When They Can't Say Anything at All?" You could substitute mice or rats or sea slugs or any of the other animals we've tried to substitute for ourselves in memory studies. Eventually, any translational research project—that is, one that attempts to translate our understanding of nonhuman memory systems to human ones—crashes into a wall of unreliable interpretation.

Even if that bull could call me to mind, what then? What could it do with that scene? Could it play with it, muse over it, find the connections between it and other scenes? That's what we do constantly. I call that memory up, that charging *becerro,* and other memories sprout from it like spokes from a hub. I think of a series of pictures from an old family album. Black-and-white. My grandfather again, leaning against a wall, a deeply tanned man beside him. In the crook of my grandfather's arm: a capote. He tried bullfighting, too, during a trip to Spain to attend a medical conference. There are more pictures following it: He's in the bullring, the cape unfurled, the bull charging. I don't know whether he was any better than me, or whether his bull ultimately connected, but he was certainly better dressed. He was wearing a good suit, the jacket off, the shirt still tucked in. He didn't look as afraid as I remember myself being. He was nothing if not daring. Another spoke of the wheel: I remember reading the transcript from a panel discussion on medical ethics. The discussion took place in the seventies, and my grandfather was a panelist. So was Karl Pribram, incidentally, and a man named José Delgado, a onetime colleague of both of theirs at Yale. Delgado is famous for an experiment he did with bulls; that's why this leaps to mind. In his experiment, Delgado implanted a device into a bull's brain. It was a remote-controlled device, and he claimed that when activated it would remove all of a bull's aggression instantly. He demonstrated it in a bullring sometime in the 1960s and invited the press. You can find a clip on YouTube. The bull charges, and Delgado waits until the last possible minute, at which point he presses a button on the remote control and the bull comes to a shuddering halt, looking suddenly confused and uncertain. Delgado liked demonstrations like this and liked saying the most outlandish things. During the symposium, he boasted that the neural implants he was developing promised to revolutionize evolution itself: "The question," Delgado said, "rather than, 'What is man?' should be 'What kind of man are we going to construct?'"

My grandfather responded with uncharacteristic humility.

"With all due respect to Dr. Delgado," he said, "I work almost wholly on humans, and we are more aware of the disastrous effects that sometimes occur in neurosurgery."

Mishkin told me he had to leave soon to attend a talk by a colleague. The talk was called "FMRI Correlation Maps from Spontaneous Recordings of Single Neurons," and he tried to explain to me what that meant, but I got a little lost. Earlier, when we were talking about amnesia, he'd complained that his own memory was beginning to fade and that he suffered from anomia, an inability to recall the names of everyday objects. I hadn't noticed. Instead speaking with Mishkin reminded me of something Brenda Milner had told me when I asked what accounted for the enduring sharpness of her own remarkable brain. "Curiosity," she said. Every day she woke up with a genuine passion to discover something new, and that's what kept her mind so quick.

I asked Mishkin if he felt that same driving, undying curiosity. He smiled, nodded.

"It is a really interesting adventure, what we are into," he said. "Trying to discover how the brain works! How long is it now? Over sixty years. There's nothing more interesting."

He laughed and pointed at his earthrise poster.

"Well, except the cosmos," he said. "The two great mysteries: the cosmos and the brain!"

TWENTY-TWO

INTERPRETING THE STARS

In a quiet conference room in a busy laboratory complex in San Diego, I placed the tip of a ballpoint pen on a piece of paper. A young German research psychologist named Ruth Klaming was at my side, a stopwatch in her hand, telling me what to do. A small metal stand hid my hand from my direct view, though I could see it in a mirror that was propped on the table in front of me. The piece of paper had a single large five-pointed star printed on it. The star had a double outline, and there was approximately a centimeter between the inner line and the outer one. My task, as Ruth explained it, was to trace a line all the way around the star, starting at the top and moving counterclockwise, between the inner and outer borders, trying not to touch either. Ruth pressed start on her stopwatch and I began. I looked at my hand in the mirror and moved it, one short tick, and watched as the pen in the mirror moved in entirely the wrong direction, toward the outer border. Then I stopped. I prepared to move the pen again but when I started and saw the line moving once again in the wrong direction, this time straying outside of the star, my hand froze. I sat there gripping the pen, staring at the mirror, trying to work out the puzzle. Obviously the problem was the mirror. I needed to move my hand in the opposite direction to what it was telling me was the right direction. In other words, if the image in the mirror indicated I should move the pen down toward me, in reality I needed to move the pen away from me. I steeled myself—it was almost physically uncomfortable to go against the visual cues provided by the mirror—and pulled the pen shakily down and to the right for an inch or two until it reached the star's first juncture, its second vertex. Then I froze again, trying to figure out which way to move my hand next. I got it wrong again, breaching the border, before recovering and moving slowly down toward the third point. This process repeated itself for the rest of the journey around the star until I reached the spot where I'd begun. Ruth pressed a button on the stopwatch.

She told me my time, not even bothering to pretend that my score was good. Then she removed the paper and replaced it with a fresh one, and I started over. Four more pieces of paper, four more stars. My times improved slightly with each one, until by the fifth star I felt almost comfortable, almost confident, almost able to overcome the odd disjoint between what I saw and what I felt.

Ruth gathered my papers and filed them in a binder, then placed a fresh sheet in the device. She left the room for a minute and I moved to another seat at the table. When Ruth returned she was accompanied by a man I'll call Jason, and Jason took my former seat. He was wearing baggy jeans and a red T-shirt with the lightning bolt symbol of the Flash printed on it. Around his waist he had a fanny pack, which was where he kept his smartphone along with a modded Game Boy filled with classic Nintendo games. Jason was in his late twenties and looked younger than that, with a wispy mustache and a hairstyle verging on mulletish.

Ruth explained the test to Jason and gave him a pen, and he placed the point at the top of the star. Then she clicked the stopwatch and he began. His first try was faster than mine—and his second even faster. He improved steadily until by the fifth try he'd made it around the star blazingly fast. Ruth congratulated him, gathered up all the stars, and filed them away in her binder.

We ate lunch in the same conference room, sandwiches and soda. Ruth was there, as was Jason's mom.

I introduced myself to Jason. He asked me where I was from, and when I told him I was living in the Yukon Territory, he nodded.

"Do the people up there constantly try to deceive you?" he asked.

I looked puzzled, and he paused a beat before delivering the punch line with a wry smile.

"Or do *you con* them?"

I laughed and then changed the subject, asking Jason about his modded Game Boy. He showed it to me and explained how it worked. It contained a sort of extra-large bootleg cartridge that gave him instant access to some of his favorite games: *The Legend of Zelda, Super Mario Bros. 3,* several of the *Final Fantasy* RPGs. He put the Game Boy down and we ate in silence for a minute or two. Then his mom spoke up.

"That's Luke," she said to Jason, inclining her head in my direction. "Do you know where he's been living?"

Jason gave me a searching look, as though fishing for some sort of prompt. I waited. Finally he smiled, and I knew he was about to crack another joke.

"Planet Earth," he said.

When I told him the Yukon Territory, he nodded, and a second later I saw the hint of another smile.

"Do the people up there constantly try to deceive you?" he asked.

When Jason was in his teens, he contracted a severe brain infection, which marched through his medial temporal lobes, destroying, one after another, most of his hippocampus and amygdala and uncus and entorhinal cortex. Once the infection had run its course, the lesion it left behind was uncannily similar to the one my grandfather made in Henry. And, like Henry, Jason was profoundly amnesic. His memory tended to last for exactly as long as he kept an object or idea or name or face in his attention. Once his attention shifted, whatever was there before disappeared. Jason coped with his amnesia like many amnesics did: Unable to draw on the past, they pay acute attention to the present, searching for cues and clues to help them make sense of their surroundings. This surface quickness can make it hard to see the underlying dysfunction: If I hadn't known who Jason was before I met him, I probably wouldn't have suspected anything was the matter with him until he repeated the Yukon joke.

After lunch, Ruth ran me through the mirror tracing test again. I did better right from the start. My hand just seemed to know which way to go and how to compensate for the visual discord. I was still a little clumsy, the line a little shaky, but the earlier paralysis was gone. By the end of the second round of five stars, I'd made it around much quicker than before, with fewer errors.

Then I watched Jason take the test again. He sat where I'd been sitting.

"Have you seen this test before?" Ruth asked him.

"No," he said.

Ruth jotted down a note, then explained the test protocol to him again, just as she had an hour earlier. Jason picked up a pen, placed the point at the top of the star, peering intently into the mirror. Ruth tapped the stopwatch, and he began.

As before, Jason completed the task more quickly and accurately than I had, and by the fifth star he was able to lay down one of the highest scores Ruth had ever seen.

Jason had absolutely no memory of the first time he took the test, and by the time he left the conference room he would have no memory of the last time he took it. How was he able to improve his performance on a task without remembering having performed it?

To answer this question, it's necessary to jump back a half century to another testing room, and another bundle of scribbled-on stars.

After the publication of the 1957 paper she wrote with my grandfather, Brenda Milner continued making trips from Montreal to Hartford as often as she could, riding the night train, spending a few days at a time with Henry, plumbing the depths of his amnesia.

Initially those depths seemed boundless. For Milner, it was always a little shocking to see that blankness in Henry's eyes whenever they met, that utter lack of recognition. She was getting to know him so well, but to him she was always a stranger.

She ran a huge variety of tests on him, tests of intelligence, vocabulary, facial recognition, and more. Once, during an informal quiz, she asked Henry to remember the numbers five, eight, and four, then left the testing room for twenty minutes. When she returned, she asked him if he remembered the numbers.

"Five, eight, four," he said to Milner's surprise.

"Oh, that's very good! How did you do that?"

"Well, five, eight, and four add up to seventeen," Henry answered. "Divide by two, you have nine and eight. Remember eight. Then five—you're left with five and four—five, eight, four. It's simple."

Milner then asked him if she remembered her name.

"No," he said apologetically.

"I'm Dr. Milner," she said, "and I come from Montreal."

They then chatted about Canada for a minute or two.

"Now," Milner said finally, "do you still remember the numbers?"

"Number?" Henry said, "Was there a number?"

In other words, he'd played with them, keeping the numbers at the top of his mind, not allowing them to slip into the abyss. Milner's shock—was his amnesia not as grave as she'd thought?—turned to understanding as she realized that he hadn't actually been able to remember the numbers she'd given him, since the numbers had never really left his present moment. Two minutes later, after they'd moved on to other subjects, she asked him if he still remembered the numbers, and he gave her his usual blank look.

And then, one afternoon, Milner sat Henry down at a table and placed in front of him a sheet of paper with a double-bordered star on it. It was a task that had recently been invented, and as far as Milner knew it had never been administered to an amnesic. She was curious to see how Henry would fare on it but expected him to do poorly, allowing her to measure yet one more dimension of his memory deficits.

Just like Jason, just like me, and just like Milner herself, Henry struggled on the first star but by the fifth had improved. This did not surprise Milner. Henry had done the first five stars right in a row, which meant that, much like the numbers five, eight, and four had on that previous occasion, the task had never had a chance to slip from the top of his mind.

The next day, Milner placed another blank star in front of him.

"Do you remember seeing this before?" she asked him, and he shook his head.

"Nope," he said.

She explained the task, and he did it. Well. Very well. Just as well as the fifth star the day before. Henry seemed surprised at his own facility with the task.

"Gosh, that wasn't as hard as I thought it would be," he said.

Milner, on the other hand, was more than surprised. Watching Henry's hand trace an assured path around that star, a star he had no conscious memory of, was the most memorable moment of her scientific career. It hit her with the force of a revelation.

Henry couldn't remember taking the test. But he'd improved on it nevertheless. Milner knew this didn't mean that Henry actually retained some dim memory of his previous circuits around the star. She was confident by now that the events of Henry's life were indeed gone almost as soon as they happened.

What it meant, she realized, was that although Henry's brain lacked the ability to record the specific experiences in his life, it apparently was nevertheless able to retain some aspects of those experiences at an unconscious level. In other words, there were apparently at least two different and separate memory systems in the human brain. The system responsible for laying down specific occurrences, or episodes, was hopelessly compromised in Henry. The other system, the one that allowed one to acquire and improve upon learned skills, appeared to be intact. Henry might on a day-to-day basis never be able to remember having done something, but he was apparently able to remember how he did it.

Milner's initial studies of Henry had already formed the foundation for modern memory science, zeroing in on the seat of memory in the human brain. Now her further studies of Henry unexpectedly led to a second revolution: the notion that the brain contained at least two distinct and independent memory systems, one that was intact in Henry and one that was not.

In years to come, people would apply various names to these systems—procedural versus declarative, implicit versus explicit—but at that moment, watching Henry's hand trace that unexpectedly assured line around that mimeographed star, Milner had no name for it. She just watched in silent wonder.

Jason and I took a walk near the laboratory. He lit up a Kool and we wandered down a quiet street. Henry was a smoker, too. Things are different now, though, and it isn't every day you accompany someone on a smoke break. I found myself wondering whether there might be some sort of connection between amnesia and smoking. Whether amnesics might be more likely to pick up the habit. I used to smoke, and the thing about smoking is that it's something you do, no matter how many times, in the moment. It's a meditative, quiet, undemanding act. The problem is, eventually the ability to look behind or beyond the present makes smoking less appealing. I quit, finally, because I remembered myself with clearer lungs and could imagine them clogging more and more as years went by. People like Jason and Henry have limited access to their past selves, and most scientists think they can't really project themselves into the future, either. So why wouldn't they make every passing, soon-to-be-lost moment count, make it as pleasurable, as quickening and sharp, as possible?

Why wouldn't they smoke?

Their vices and their brain lesions weren't the only things they had in common. Jason was around the same age Henry was when Henry's story was about to be revealed to the world, and now Jason, too, was on the verge of becoming a Patient with a capital P. Some scientists were preparing a paper about Jason, and its working subtitle was "A Modern-Day Patient H.M."

Still, to be an amnesic in the 2010s was very different from being one in the 1950s. Take smartphones, for example. Every smartphone user recognizes how the devices become memory supplements, shoring up the leaky containers in our heads that would otherwise let facts and names slip away. For a modern amnesic, however, a smartphone can become not so much a supplement as a substitute, a prosthetic hippocampus. Jason keeps his phone always on hand and is always using it to take pictures. He downloads the majority of these pictures to his laptop, storing them away and clearing space, but he also always keeps a few dozen on his phone.

While he puffed on his cigarette, he pulled out his phone and opened the photo app. He scrolled through the pictures, looking at some of the events of the past few days, the airplane view of San Diego, a sunny beach, the outside of the laboratory, Ruth the psychologist. And a few pics from home, of pickup trucks and friends. He looked at the pictures and described them to me, then got to the last one and scrolled to the right and the first picture popped up again and he described it to me again, just as he did before.

Jason is more independent than Henry was. He lives on his own. This has caused complications: Some old friends of his took advantage and repeatedly stole small amounts of money from him, transgressions he would always forget so never have to forgive. Each time they'd show up at his door he'd let them back in. When we met, his mother was in the process of installing a high-tech surveillance system so that she could see a video feed of anyone at his front door and have final say on who gets to enter. Still, he has a great degree of independence. GPS has even made it possible for him to drive on his own. Once he has a destination programmed, he'll follow it to the end, even if by the time he arrives he's forgotten why he set off on his journey in the first place. The moment-by-moment mechanics of driving, all those little on-the-fly decisions, are no problem for him. And he even reads novels, something Henry never did. He enjoys Stephen

King. As he reads, he writes up compressed plot summaries every couple of pages in the margins. The fact that he can do this is amazing. But these are compensatory devices and habits, coping mechanisms, not solutions. The central problem they're designed to tackle remains: Jason cannot remember what happens to him.

We talked about videogames, about his favorite brand of cigarettes, about the cars he wants and a fight he got into while he was in high school. It was in the hallway, and he got suspended. We talk about an old girlfriend of his and about his favorite bands. I liked Jason. He was a good guy, open and enthusiastic, pleasingly geeky. We shared some of the same interests. It felt like we were connecting, bonding somehow.

But we weren't. Or he wasn't. Because by the end of the day, Jason wouldn't remember me. I would leave his mind without a trace.

There's something liberating about talking with an amnesic. I suddenly realized that I could tell him anything at all, my deepest secrets, my most embarrassing hopes, and he would listen and respond, giving me input, perhaps even advice, but would never be able to tell anyone else. Jason was like a priest that way, or a therapist, or a diary, only even more secure.

There was something heartbreaking about it, too. This young, open man, taking in the world around him, with nothing ahead of or behind him, walking forever on a long, long tightrope between a clouded past and an unreachable future.

He finished his cigarette and lit another and offered me one for the second time, and for the second time I told him I didn't smoke.

"We have to get back to the laboratory," I said.

He let me lead him there, not really understanding where he was going, not asking any questions, smoking the moment's first and last cigarette.

TWENTY-THREE

THE SON-OF-A-BITCH CENTER

My grandfather always scheduled his lobotomies on Saturday mornings, is how Dr. Dennis Spencer remembers it. A little past nine A.M. They were kind of "surreptitious," that's the word Spencer uses. My grandfather didn't place these operations on the regular schedule at the hospital, and they always happened on those weekend mornings when the neurosurgical suite was quiet, staffed with just my grandfather's longtime OR crew as well as any neurosurgical residents, like Spencer, who happened to be on hand. Usually the patient would be wheeled to the operating room straight from the Institute of Living next door. That's where most of them came from, back in the mid-1970s, though Spencer recalls that my grandfather also had some sort of deal with a psychiatrist in New Jersey and got plenty of referrals from him. Spencer watched my grandfather perform a lot of lobotomies during his residency, but the one he remembers best, the one he wants to tell me about, was one my grandfather performed on a woman in her early forties who had Tourette's or OCD, something like that. Whatever it was, it made her say, "son of a bitch, son of a bitch, son of a bitch" over and over and over again.

This was 1973 or 1974. Spencer had been a resident for two or three years. He was a student at the Yale School of Medicine and spent most of his time there, in New Haven, but he also regularly made the one-hour drive to Hartford to work with my grandfather. The two hospitals—Hartford Hospital and Yale–New Haven Hospital—had an unusual arrangement back then, sharing residents drawn from a combined pool of neurosurgery students at Yale and the University of Connecticut. By the time Spencer began his residency, that arrangement was beginning to strain at the seams, and within a few years it would collapse altogether. The problem was that the cultures of the two hospitals were very different, a direct by-product of the huge differences between the two chiefs of neurosurgery. At Yale, the chief was a man named Bill Collins. He was a solid, careful, smart surgeon, though even his admirers, Spencer among them, would admit he was maybe conservative to a fault. He didn't like taking risks. For him, neurosurgery was about finding the best, most efficient, safest solution to whatever the problem at hand was, then sticking to it. My grandfather . . . my grandfather wasn't like that.

Spencer started hearing stories about my grandfather as soon as he arrived at Yale. Wild stories. Stories that were hard to believe until you met him, like the one about my grandfather and Enzo Ferrari. Here's how it goes: Back around 1958, my grandfather was attending a medical conference in Europe when he decided, as he often did, that he needed a new sports car, stat. He drove to the Ferrari factory in Maranello, Italy. At the time, the company's founder, Enzo Ferrari, would only sell his vehicles to men he thought were good enough drivers to handle them. My grandfather and Ferrari set out on a test-drive together, and after a white-knuckle half hour on the hairpin highways near the factory, Ferrari turned to his prospective customer, his face drained of color but his voice firm.

"Dr. Scoville," Ferrari said, "if I sell you this car you'll be dead within the year."

My grandfather returned home with a Mercedes Gullwing instead.

Hard to believe, right? Hard to believe until the first time you saw my grandfather come rushing into the neurosurgery ward yelling at the nurses and the secretaries and anyone who'd listen that he'd just led some Connecticut state troopers on a high-speed chase and he thought he'd lost them, but that if they came looking for him to tell them that he was in the OR performing an emergency craniotomy or something.

Wild Bill, that's what some of the residents took to calling him.

Was it a surprise that Wild Bill and his counterpart at the Yale hospital, staid Bill Collins, didn't get along? Collins thought my grandfather was a bad influence on the Yale residents, and my grandfather thought Collins was too careful for his own good, that nobody made progress by doing the same thing over and over. Neurosurgery, for my grandfather, was an adventure, an ever-changing one, and should never be made into some by-the-book slog.

Collins wasn't the only guy my grandfather didn't get along with. As soon as Spencer began his residency, people warned him about a long-standing, bitter turf war being waged between my grandfather and James Foster, Hartford Hospital's chief of general surgery. Foster had his hands on the hospital's purse strings, and as far as my grandfather was concerned he was always trying to limit his goals for the neurosurgery department, tamping down the expansion of equipment, of staff, of space. Limiting *him,* in other words. He hated to be limited. He fought hard against Foster, demanding and often getting his way: The neurosurgery department had grown a huge amount since my grandfather founded it three decades before. Still, he always wanted it to be bigger, better.

For my grandfather, the conflict between him and Foster wasn't just professional, wasn't even just personal: It was literally a matter of life or death. Exactly what that meant was another story, one that began at a party, a big blowout at my grandfather's house. He hosted these parties every year for all the neurosurgeons and neurosurgical residents from both Hartford Hospital and Yale–New Haven Hospital. (Not that Collins would ever come, of course. They'd get wild, these parties. Collins wouldn't have approved.) Anyway, this particular party took place just a few days after my grandfather had bought himself a new motorcycle, a souped-up BMW. His wife was out of town. He wanted to show it off. He got on, and one of the residents got on behind him, and my grandfather started riding the two of them, fast, all over his big front lawn. Then he decided to try doing a wheelie. He wasn't used to popping wheelies with the weight of somebody else riding behind him, so the front wheel went up, then it just kept going up. Up and over. Bike flipped backward, high speed, dumping the two of them onto the lawn. No helmets, of course. The resident was fine, my grandfather was knocked unconscious, and the bike landed on top of him. Picture a lawn full of inebriated neurosurgeons standing around in sudden shock looking at Wild Bill sprawled there with a head injury.

He came to after a minute or so, and then the ambulance arrived and carted him off to Hartford Hospital, with a whole armada of surgeons following closely behind. He was wheeled straight to a private room, and he said he was feeling fine, but somebody noticed a suspicious swelling in his stomach area. A ruptured spleen is what they suspected. Word spread, and a few minutes later James Foster himself came into the room, looked him over.

"Bill," Foster said, "you've got a busted spleen. I'm gonna have to operate."

My grandfather, lying there in that hospital bed, looked up at his nemesis.

"Hell no," he said.

Foster's eyebrows rose, and my grandfather continued.

"I know what you're planning," he said. "You're planning to let me die on your table. You want me out of the way. That's not gonna happen. My spleen's fine. My spleen stays where it is."

They went back and forth, arguing heatedly, my grandfather refusing to budge, refusing to consent. Eventually Foster left, shaking his head, and told all the neurosurgeons waiting outside that they *had* to convince my grandfather to consent, that he'd die otherwise. And so they marched in, one after another, trying to persuade my grandfather to let his enemy save his life. Nothing worked until finally Ben Whitcomb, my grandfather's second-in-command and longtime best friend, pleaded with him. You've gotta do this, Bill, he said. You've gotta let him operate.

My grandfather looked at Whitcomb with hurt and confusion in his eyes.

"I don't know what he paid you, Ben, to make you betray me, but you're my best friend in the world and if I can't trust you I can't trust anybody."

He consented and was wheeled off to surgery. Foster opened him up and removed his spleen, which was indeed ruptured, and it all went smoothly; a couple of hours later my grandfather was in the recovery room, waking up, still groggy, when one of his residents, Norman Gahm, came in holding a glass jar. The jar was full of formaldehyde, and there was something dark and fibrous floating in it, with a long ragged split down the middle.

"See?" Gahm said. "It was ruptured!"

And yeah, that story might sound hard to believe, too, until you saw the glass jar, with its cargo still floating within it, on a shelf in the office of Gahm, who held on to the damn thing for decades: Wild Bill's ruptured spleen.

Telling the spleen story reminds Spencer of *another* great story, about the only other time he knows of that my grandfather was himself operated on. It was to treat a slipped disk in his back. Now, my grandfather developed a lot of the basic tools that spinal surgeons use when they're treating slipped disks. The Scoville retractor. The Scoville clip. He also pioneered the so-called keyhole approach, through a tiny incision, which became the standard. As it happened, keeper-of-the-spleen Norman Gahm was scheduled to perform the disk operation on him, and on the day of the operation my grandfather got to the hospital early and started ordering people around. What he did was he got them to set up this elaborate system of mirrors in the OR so that later, while he was lying on the table, his back cut open and his spine exposed, with only a local anesthetic, he could look over his own surgeon's shoulder. He supervised his own operation. Spencer never heard of anyone else doing that, before or since.

The thing was, he could get away with it. With everything. Partly it was the era—those were different times—and partly it was that Spencer didn't think there was another neurosurgeon in the state of Connecticut as skilled in the operating room as Bill Scoville. Those hands of his: quick, adroit, steady, creative. Spencer loved to watch him work.

So on that morning in question, the one he wanted to tell me about, after the lady was rolled into the OR and my grandfather was getting all prepped and ready to go, Spencer was eager to watch him do what he was going to do. You could call him wild, you could say he took chances others didn't, but before you criticized him too harshly you needed to take a look at him scrubbing up. Spencer never saw a surgeon take as much care as my grandfather did scrubbing up. He didn't just wash his hands, he dipped them in a tub of hydrochloric acid! Spencer had never seen anything like it. Must have been hell on his skin. Also, his surgical mask. Same mask that everyone else wore, paper-thin, with the elastic around the back of the head. But if you looked closely you'd see that underneath the mask my grandfather had two rubber tubes, one in each corner of his mouth, and they snaked out from his mouth and curved around behind the mask like snorkels. So that when he breathed, even through the mask, those exhalations didn't go anywhere near the exposed insides of his patients. That level of care was rare.

Sometimes things still went wrong, of course. One of the first things Spencer learned when he began his Hartford Hospital residency was that all of my grandfather's residents were expected to regularly head over to the Hartford veterans' hospital nursing center and check in on a guy there named Gunner. Gunner had been a patient of my grandfather's with a ruptured disk. The operation itself was usually easy enough, but in Gunner's case my grandfather decided to perform it in an unusual way, with Gunner sitting in a chair instead of lying on a stretcher. Which would have been fine except that an air embolism got into his bloodstream and rose up and blocked off Gunner's basilar artery, causing a massive stroke in his brain stem and obliterating his motor functions. From then on all he could do was move his eyes, right for yes and left for no. He still could think fine, but he was totally locked in. My grandfather would always send residents to check up on him, to make sure he was being tended to. And my grandfather would go himself, too, sometimes. He'd bring flowers.

But back to the story he wanted to tell me. The lobotomy story. So there the woman was, lying on her back, muttering, "Sonofabitchsonofabitchsonofabitch." And there my grandfather was, scrubbed and ready to go. Spencer watched as my grandfather prepared, watched as he began, watched him slice two little slits right above her eyebrows, watched him pry them open with retractors, exposing the skull beneath. Those eyebrow slits were a bit of a novelty, incidentally. A little tweak: Rather than slicing that half-moon arc across the top of their heads and rolling their foreheads down, like he used to, he'd started making just those two little slits, slits that would be concealed by their eyebrows. That was another example of his constant tinkering, of his constant attempts to improve even his oldest procedures. His orbital undercutting lobotomy *was* a very old procedure by then, in the mid-1970s. A quarter century old. Not only was it old, it was rare. People weren't performing many lobotomies by then. They'd fallen out of favor, become stigmatized. *One Flew Over the Cuckoo's Nest, A Clockwork Orange, Planet of the Apes:* Popular culture, and popular opinion, had turned against it. Even within the medical community, it had become a sort of black sheep. That was another reason for the friction between my grandfather and the Yale chief of neurosurgery: Collins hated lobotomies, and he hated the fact that my grandfather was making his Yale residents, like Spencer, accomplices to them.

As for Spencer, he didn't know what to think. He was just there to absorb, to learn. He stood on that stool behind my grandfather, watched him bring his trephine drill down onto the woman's forehead, watched him cut out those two plugs of bone and put them to the side, watched him pick up the suction catheter and the flat brain spatula and lean in and begin the work. He watched my grandfather lesion the right hemisphere of the woman's brain first, and he listened as she continued her incessant cursing, lying there under the bright lights: "sonofabitchsonofabitchsonofabitch." Watched as she kept it up throughout the first half of the operation and that first hemisphere. Watched as my grandfather carefully withdrew his tools, then inserted them into the second hole in her head and started in on the left hemisphere. My grandfather carefully measured how far in he'd inserted the suction catheter, then began to slowly pivot the tool, starting to make the second lesion. She was still cursing as he began to cut—"sonofabitchsonofabitchsonofabitch"—and then suddenly—"sonofabitchsonofabitchsonofabi . . ."—she stopped. My grandfather paused, waited to see if she'd start in again. She didn't. Then he turned and looked back at Spencer.

It's been four decades since that day in the OR, and though Spencer went on to have a long and illustrious career, eventually becoming the head of the Yale department of neurosurgery and the president of the American Association of Neurological Surgeons, he'll always remember exactly what my grandfather said to him then, back in that long-gone era when giants like Wild Bill still roamed the earth, leaving all those great stories in their wake.

"Spencer," he said, "we've just found the son-of-a-bitch center!"

TWENTY-FOUR

THE MIT RESEARCH PROJECT KNOWN AS THE AMNESIC PATIENT H.M.

My mother remembers spending many nights sitting by her bedroom window holding an empty tin can to her ear. Her bedroom was on the second floor of their house, 334 North Steele Road, in a tree-shaded neighborhood of West Hartford, Connecticut. She had removed the top of the can and nailed a hole through the bottom of it, then inserted a thick string, which she tied into a fat knot so that even when she tugged on it, it wouldn't come loose. The string extended from the bottom of the can and stretched across North Steele Road to a house directly opposite hers and into another second-floor bedroom window, where one of her best friends, Suzanne, would be speaking into a tin can of her own. The two houses were identical to each other, designed by the same architect, mirror images on opposite sides of the street. Suzanne's words vibrated along the taut string, then resonated in the can and thrummed in my mother's ear.

They became neighbors in first grade and stayed neighbors through high school, and neither Suzanne nor my mother remembers exactly how old they were when they began making those tin can telephones, or when they outgrew them. They don't remember the specifics of their conversations. They might have talked about friends, movies, music. My mom's dog, Wiggles; Suzanne's dog, Skippy. The latest radio episodes of *The Shadow* and *The Lone Ranger.* Maybe, eventually, boys.

Two young girls in suburban postwar America, proto-texting deep into the night.

In the winter, on weekends, my grandfather would often take Suzanne and my mother and my uncles Peter and Barrett along on ski trips, driving to and from the mountains at his usual breakneck speeds. Suzanne's house and my mother's were identical, but their fathers were very different: Suzanne's was an unassuming engine-parts salesman; my grandfather was a dashing neurosurgeon with a rotating fleet of sports cars. From an early age, Suzanne decided she wanted to be a doctor when she grew up. She began assembling her own first-aid kits and taking them off to summer camp with her so she could tend to any skinned knees or bee stings she encountered among her camp mates.

Whatever admiration Suzanne felt for my grandfather, there is evidence it was not mutual. Once, during high school, my uncle Barrett got into a fistfight in their backyard with another boy. A crowd of children gathered, and my grandfather watched, too. Suzanne was there. She was rooting for the other boy. When Barrett lost, my grandfather grabbed Suzanne by the arm and dragged her off the property, telling her she was no longer welcome in his home. It was the only time my mother ever saw her father engaging in an act of physical aggression against another person.

In 1955, two years after Henry's operation, my mom and Suzanne both graduated from the private Oxford School in West Hartford, and the following fall both of them enrolled at Smith College. Suzanne studied premed, but her first chemistry course dissuaded her from her childhood goal of becoming a medical doctor, so she shifted to psychology. My mom studied English. Postcollege, the paths taken by the two of them diverged. After a stint as an Avon Lady, my mom discovered a passion for early childhood education and opened a series of progressively minded daycares, models of socioeconomic and racial integration. Suzanne, meanwhile, stayed in the world of higher education and became a university professor and a PhD. But they remained friends. They both eventually settled in the Boston area, where I was born. Growing up, I saw Suzanne now and then. My mom would have her over for dinner, or we'd all go out to eat. Suzanne had gotten married and divorced by then and changed her last name to Corkin. She had three kids, two boys and a girl, all roughly my age.

I knew Suzanne was my mom's oldest friend, but I didn't know much about her beyond that.

I know a lot more now.

I know, for example, that when my grandfather died and my mother was helping settle his estate, she chose to give Suzanne an unusual memento: the anonymous human skull that he'd kept in his home office.

I also know that Suzanne inherited much more than that skull from my grandfather.

Decades before my grandfather died, she'd already taken possession of his most famous patient.

After graduating from Smith, Suzanne Corkin moved to Montreal to pursue a doctorate in psychology at McGill. It was 1960, and shortly after she arrived she read a recent paper in the *Journal of Neurology, Neurosurgery & Psychiatry* and realized that the William Beecher Scoville who co-authored it was the same William Beecher Scoville she'd grown up across the street from, the dashing neurosurgeon who was the father of one of her oldest friends. The other author, of course, was Brenda Milner. A year later, Corkin had the opportunity to join Milner's laboratory at the Montreal Neurological Institute, and she took it. She began working on her PhD thesis there, an exploration of the effects of different types of brain lesions on somesthetic function, otherwise known as sense of touch. Most of Corkin's research subjects were patients of Wilder Penfield's, men and women he'd operated on to treat their epilepsy. Corkin would run them through various tasks, such as trying to identify common objects—a comb, a bottle cap, a book of matches—simply by handling them, or trying to see at which moment, when she moved two sharp points slowly together on their skin, the two points became indistinguishable from each other. Like her mentor Milner, Corkin became a meticulous investigator, keeping careful notes, always trying to determine whether some sensory deficits might have been caused by aspects of the patients' personal histories rather than their brain lesions, such as one subject whose dulled sense of touch Corkin assumed was "probably attributable to calluses on both thumbs, a result of the patient's janitorial work."

Although memory wasn't Corkin's focus, one of the tests in her arsenal did relate to memory. It was called a tactually guided maze, and it was supposed to measure a research subject's ability to learn to navigate a labyrinth by touch alone. The maze was cut into a sheet of aluminum over a wooden box and hidden behind a curtain. Subjects would reach their hands through the curtain and use a stylus to try to get from the beginning to the end of the maze, and Corkin would ring a bell every time the stylus hit a dead end. Using a stopwatch, she would time how long it took subjects to reach the finish, then she would run them through the maze again and again, seeing if they eventually got faster and made fewer errors. Her main intention was to measure the subjects' ability to learn the route, but of course this was also a sort of memory task: If the subjects got better at solving the maze, it was because they were remembering the correct path. Most of Penfield's patients had no trouble with the task. Whatever small unilateral lesions he'd made in their brains did not at all compromise their ability to learn by touch or any other sense.

Then, in May 1962, Henry visited the Montreal Neurological Institute. Milner had organized the visit, and it was the first time she had an opportunity to observe Henry outside of my grandfather's office in Hartford. Henry rode the night train up with his mother, Lizzie. They stayed for a week, spending their nights at a rooming house near the Neuro. Lizzie, who'd spent most of her life in sleepy Connecticut and rural Louisiana, seemed intimidated by the bustle and tumult of Montreal. Milner encouraged her to get out and explore the city, but Lizzie chose not to and instead spent most of her days sitting on a bench in a hallway at the Neuro, waiting for the scientists to finish doing whatever they were doing to her son. Milner worried that Lizzie was going to get bored, but she had no such worries about Henry. For one thing, it seemed unlikely that when every minute was entirely new to you, detached and separate from the endless chain of minutes that preceded it, boredom was even possible. For another thing, they kept him busy.

During that week, Henry submitted to a huge number of tests. Many were repeats of ones he had taken before, though he didn't recall having done so. Milner and her graduate students retested his IQ, his immediate recall of numbers, and his ability to trace a star in a mirror, among other things. Milner wanted to see if the passage of time had caused any changes to Henry's condition. It had not, in any significant way. Milner also allowed her graduate students their own time with Henry, to present him any new tasks that they thought might be illuminating. Henry was not quite famous yet, but his unique importance as a research subject was becoming clear. Even at the Neuro, where working with lesion patients was routine, Henry stood out, his lesions bilateral, his amnesia deeper, his scientific utility unquestionable.

Corkin met Henry for the first time that week. She shook hands with him and his mother in a hallway at the Neuro, then led Henry back to a testing room. They made small talk, chatting about Hartford, their mutual hometown. Corkin's mom had once attended St. Peter's school, just like Henry, and Corkin was born at Hartford Hospital in 1937, just sixteen years before Patient H.M. was born, in a manner of speaking, in my grandfather's operating room. Henry was eleven years older than Corkin, and the two came from vastly different backgrounds—Henry was firmly working-class, while Corkin would describe her own upbringing as privileged—but there were common threads of experience. After talking for a little while, Corkin sat Henry down at a table and pulled up a chair across from him. On the table was her tactually guided maze, although a screen was placed in front of it so Henry could not see it. Corkin explained the procedure.

She gave Henry the stylus, then took hold of his hand and brought it to the start of the maze. She gently guided it through to the finish, to give him a basic orientation, then brought his hand back to the start, released it, and picked up her stopwatch and her bell. Henry began.

The first time he navigated the maze he made almost exactly eighty errors, each eliciting one ring of Corkin's bell. The second and third time he made slightly fewer errors, and on his tenth attempt he made additional improvements, committing only seventy-two errors. The eleventh time, however, his score regressed: He made the wrong move more than eighty times. Corkin tested him on the maze over a period of two days, observing him over dozens and dozens of attempts, and by the end he had shown no net improvement in terms of the number of errors he was making. Henry's amnesia evidently completely prevented him from memorizing the correct path through the maze. This was not surprising. When reviewing the data later, however, Corkin noticed something: Although Henry's overall rate of errors never improved, the time it took him to complete the maze did. On his first attempt, it took him about ninety seconds. On his final attempt, it took him less than forty. So although he clearly hadn't learned the maze itself, he had learned something. Specifically he became more adept at the physical procedure of navigating the maze, of guiding a stylus around a hidden track. Like most people, he moved the stylus in a slow, cumbersome way at first, but by the end he was moving it quickly and confidently. This was in some ways similar to Henry's results on Milner's mirror tracing test: It was another proof that his procedural memory, his memory of *how* to do things, was intact. The exciting difference, however, was that Corkin's tactually guided maze actually demonstrated both Henry's intact procedural memory and his demolished episodic memory *simultaneously.* For the first time, in one elegant experiment, Henry's strengths and Henry's weaknesses were both laid bare.

For a young graduate student like Corkin, it was a thrilling moment. It was also a pivotal one: Although she didn't know it at the time, and Henry never would, those first sessions were the start of the most intensive relationship between a research subject and a researcher in the history of science. Corkin delivered her PhD thesis—"Somesthetic Function After Focal Cerebral Damage in Man"—in July 1964, and although Henry made only a brief cameo in that thesis, he was just beginning a starring role in Corkin's career.

After graduating from McGill, Corkin moved from Montreal back to New England, accepting a position in the psychology department of MIT. The department was brand-new: A German immigrant named Hans-Lukas Teuber had founded it that same year, and Corkin was one of the first people he recruited. Teuber came to MIT by way of Harvard and New York University and had built his reputation on the careful study of lesion patients, though in his case most of the lesions he was studying were the products of war, not surgery. Most of his research subjects up to that time had been veterans, men who'd sustained penetrating head wounds of one sort or another, and Teuber had proved brilliant at teasing out the various ways their damaged brains affected them and extrapolating what that said about how normal brains worked. As he put it, he considered brain injuries "experiments of nature" and was a passionate advocate for the value of "studying the disturbances of brain function as a clue to normal modes of central nervous system functioning." He aimed to import and instill this passion for the lesion method to his new department at MIT.

Corkin was a natural fit. Like her new boss, she, too, had amassed a great deal of useful experience working with lesion patients. Unlike Teuber's patients, most of the research subjects Corkin had worked with were not "accidents of nature" but instead the willful products of surgery, and one of them, Patient H.M., was already clearly among the most important lesion patients in history. There was a word that scientists had begun using to describe him. They called him *pure.* The purity in question didn't have anything to do with morals or hygiene. It was entirely anatomical. My grandfather's resection had produced a living, breathing test subject whose lesioned brain provided an opportunity to probe the neurological underpinnings of memory in unprecedented ways. The unlikelihood that a patient like Henry could ever have come to be without an act of surgery was important. As Corkin herself explained years later, it would be hard to conceive of, for example, the soldiers Teuber was accustomed to studying, men who'd been shot in the head, winding up with brains similar to Henry's: "To get a pure one would be rare. Because think about what it would take to blow out both hippocampi. You'd be dead. I think it would be most compatible with not being alive."

Teuber recognized Henry's importance, which meant he also recognized the importance of Corkin's access to Henry, not to mention her personal relationship with Brenda Milner and William Beecher Scoville. By hiring Corkin, Teuber was acquiring not only a first-rate scientist practiced in his beloved lesion method but also by extension the world's premier lesion patient. Indeed, shortly after Corkin signed on at MIT, Teuber drove to my grandfather's office in Hartford for the first of many visits. Henry was waiting for him there, and Henry and Teuber drove back to Cambridge together. They parked outside a new redbrick building called the Clinical Research Center, and Teuber led Henry inside to a small bedroom where he would live for the next two weeks. During the four decades that followed that first visit, Henry made fifty-four individual trips to the Clinical Research Center, sometimes staying for as long as a month. A mountain of clinical data began piling up, the beginning of what would become the largest amount of data ever gathered on a single research subject. Still, for Henry, his new home never became any less bewildering than it was for him that first week, when three times he rang for a nurse in the middle of the night and posed an apologetic question.

"Where am I?" Henry asked. "And how did I come to be here?"

Hans-Lukas Teuber drowned during a vacation in the Virgin Islands in 1977.

By that point, Brenda Milner at McGill had largely ceded control of Henry-related research to MIT. Doing so made sense on a number of levels. For one thing, Hartford, where Henry lived, was much closer to Cambridge than to Montreal. For another, Milner had in many ways moved on. Her early work with Henry led directly to two groundbreaking discoveries: one, that memory function was localizable to a particular part of the brain, and two, that the brain contained at least two different and independent memory systems. That, for Milner, was enough. She was still deeply curious, still actively engaged in research, and still working with brain-lesioned patients at the Neuro, but her interests were increasingly migrating forward from the medial temporal lobes to the frontal lobes. She'd discovered Patient H.M. and made him famous, which made her famous in turn. Now she could let him go. Truth be told, it wasn't all that hard, on a personal level. Henry's memory deficits, his incessant repetition of the same stories, his inability to remember Milner or anyone else, made it difficult to establish anything but a superficial bond with him. "We found ourselves beginning to regard him the way you would regard a pet," Milner once told an interviewer. "He lost his humanness. You can't build a friendship or any sort of human affection for the person."

After Teuber's death, Corkin became the lead investigator for Patient H.M., a position she would hold until his death. In a book she wrote about Henry, Corkin explained the situation like this: "By the late 1970s I had become the primary point of contact for anyone who wanted to access him for research. Hans-Lukas Teuber died in 1977, and Brenda Milner moved on to other research topics while still maintaining a strong interest in Henry. I had inherited him as a patient." She took her new responsibilities toward Henry seriously, and by all accounts displayed a genuine interest in his well-being. She would send him postcards on holidays, flowers on his birthday. She made sure he never lacked crossword puzzle books. She took good care of him, in her own way.

And Henry, in his own way, took good care of Corkin. Inheriting the world's most important human research subject was good for

Corkin's career, scientifically and professionally. After Teuber's death, MIT invited her to head up her own laboratory, and although thousands of different research subjects eventually participated in studies at the new Behavioral Neuroscience Laboratory, Henry was clearly the lab's most prized asset. Many of the graduate students and postdocs who applied for positions with Corkin did so in the hopes of getting a chance to work with the famous Patient H.M., and over the years more than 22 percent of the papers that came out of the lab were about Henry. Those papers were also the ones that gained the most attention and interest, both among the wider scientific community and the general public. This attention and interest all raised the profile of Corkin and her laboratory, which received millions of dollars in private and federal funding. "I came to realize Henry's limitless worth as a research participant," Corkin would later write, adding that "Henry was certainly a boon to my lab's reputation."

She became Henry's gatekeeper, fielding all requests from outside researchers who wanted to work with him. She said that she "felt strongly that Henry should not be made available to every person who wanted to meet him" and that if she had allowed "all interested researchers to test and interview him, the resulting free-for-all would have been a constant drain on his time and energy." Even within the lab, Corkin imposed strict rules related to the interactions that people were allowed to have with Henry. She forbade photography and video. To this day, no video of Henry has been made public. Jenni Ogden, a neuropsychologist from New Zealand who spent years working in Corkin's lab, recalls that she once surreptitiously took a few snapshots of Henry while she was running some tests with him in 1986. Years later, long after she'd left the lab to start her own at the University of Auckland, she was having a friendly catch-up chat with Corkin when she mentioned having taken those pictures.

Corkin went silent on the other end of the line.

"I want you to send those pictures to me," she said finally. "And I want you to destroy the negatives."

Similarly, when Endel Tulving, one of the world's leading memory researchers, met Henry he asked Corkin whether he could tape-record their conversation, and she refused. "It's just silliness," Tulving said later.

Howard Eichenbaum, a renowned Boston University neuroscientist and a member of the American Academy of Arts and Sciences, told me that there were two ways to interpret Corkin's zealous guardianship of Henry. Eichenbaum usually works with rats, but during the 1980s he and Corkin collaborated on a few papers involving Henry. "Certainly I think she would say that she was hugely protective of Henry," he said, "but there are two things that are going on." He explained that when researchers like Corkin gain privileged access to an important human research subject like Henry, they know that "this is their resource, this is where they are going to get their data from." He contrasted animal-focused researchers such as himself, who "can always order some more rats," to human-focused researchers such as Corkin, who can never hope to find another Henry. "Everyone who has an amnesic patient guards the access to them," Eichenbaum said, "because *they* want to do the experiments on them. They can't do all those experiments at once, so it's going to play out over years, and they don't really want to share." Eichenbaum said it was inevitable that Corkin and her colleagues would develop possessive feelings toward Henry. "They kind of *own,* in some sense—although obviously you don't own people—that resource. They have the contact for that resource, so they want to hold on to that resource to do the experiments herself."

According to Eichenbaum, Corkin's fierceness as a gatekeeper was understandable. After all, he said, "her career is based on having that proprietary access."

Memory scientists often speak of the important difference between knowing that a certain fact is true and knowing how you came to learn it. For example, here's a simple question: What's the capital of France? The answer probably leapt to your mind in an instant. Now, here's another question: When exactly did you learn that Paris is the capital of France? If you're like most people, you have no idea. That particular fact twinkles in your mind amid an enormous constellation of other facts, most of them forever disconnected from the moment they first sprang to life. That store of mostly disconnected facts is known as your semantic memory.

Your semantic memory is contrasted with your episodic memory, which is your memory of fleshed-out narratives rather than merely facts. When you engage your episodic memory, you engage in a form of mental time travel, bringing yourself back to a particular place and time, reimagining a scene you've already lived. When you engage your semantic memory you are doing the mental equivalent of flipping through an encyclopedia or photo album, plucking out bits of information whose origins might be unclear. Incidentally, semantic memory can also be autobiographical. You might know for certain that you are capable of walking on your hands without remembering the day you first learned how to do so. In neuroscientific shorthand, this is known as the difference between "knowing *that*" and "knowing *how.*"

One of the most remarkable things that came to light during Suzanne Corkin's research with Henry had to do with this distinction between episodic and semantic memory. Prior to Corkin, the basic understanding of Henry's amnesia could be summarized as follows: The lesions in his brain prevented him from acquiring any new episodic or semantic memories while the episodic and semantic memories he'd acquired prior to the operation were left more or less intact. As it turned out, this was incorrect. Through a series of experiments, Corkin and her colleagues demonstrated that virtually all of Henry's episodic memories, even the ones that had been created prior to the operation, either no longer existed or were completely inaccessible to him. He didn't just have no postoperative episodic memories, he had no episodic memories, period. Instead of being able to do what the rest of us can do—use our minds to reexperience and reexamine many of the stories of our lives—Henry was left with the ability only to rifle through his mind's files of disconnected facts, never knowing the context or origins of any of them, never stringing them together into real, living narratives. As Corkin put it, Henry's entire past, even predating the operation, had become "semanticized."

Although MIT is an institution where scientists have access to all the best and most expensive neuroscientific tools, from MRIs to PETs to EEGs, the way Corkin and her colleagues proved that Henry lacked an ability to access episodic memories was decidedly low-tech. They interviewed him. For hours and hours, they sat and asked him about his past. Henry could easily recall his date of birth—two twenty-six, twenty-six—but now they asked if he could describe a specific birthday party, any birthday party ever. It turned out he could not. He could pull up facts about himself, about his family, about the world—but he couldn't string these facts together into episodes, narratives, stories. This basic deficit, and his seeming efforts to compensate for and talk around it, was heartbreaking.

"What is your favorite memory of your mother?" Corkin once asked Henry during an interview in 1992.

"Well, I—that she's my mother."

"But can you remember any particular event that was special—like a holiday, Christmas, birthday, Easter?"

"There I have an argument about myself about Christmastime."

"What about Christmas?"

"Well, 'cause my daddy was from the South, and they didn't celebrate down there like they do up here—in the North. Like they don't have trees or anything like that. And, uh, but he came north even though he was born in Louisiana. And I know the name of the town he was born in."

Another researcher once asked Henry if he'd ever fallen in love. He told her that he had.

"Okay, tell me about it."

"Well, just how you felt and everything and the ways it could be. And they would fall for you. And you still don't know."

"Can you tell me about when you first felt that you were falling in love with somebody? One specific event?"

"No."

"No, you can't think of that? Can you think of one specific event lasting for several hours from your early childhood? Can you come up with anything like that?"

"No, I can't."

Henry's almost complete lack of episodic memories is difficult to relate to. We are a storytelling species, and we spend a great deal of our time stringing the facts of our pasts into narratives in the present. A mind unable to do so can seem like a fundamentally alien mind. Forgetfulness is one thing—we've all got porous memories—but the deficits Henry endured are something else altogether. It may in fact be impossible to know what it was like to experience life through Henry's eyes, from within Henry's mind, but it's important to remember that most of the episodes from our own lives become semanticized over time, bleached of their context. My own episodic memory, for example, has countless pits and gulfs and vacuums.

I have no idea when I first heard of Patient H.M. Maybe it was from my mother. Maybe when I was a kid she told me that her father had once performed an experimental operation and that his patient had gone on to become an important research subject. Or maybe I learned about him some other way. I know that by the time I went to college I had a vague understanding of H.M.'s story. I dated a psychology major for a while, and I remember trying to impress her with my family connection to one of the central figures in her textbooks. Still, the specific moment that the seed planted itself in my mind is gone, and all that's left is the disconnected certainty that at some point I must have learned the basic story of Patient H.M. I know that I did, but I don't know how.

My memory of what happened when I first began to pursue Henry's story for myself is much richer.

In March 2004 I went to Chicago for the annual meeting of the national City and Regional Magazine Association. I'd been working as a staff writer at *Atlanta* magazine for a year or so, churning out features about local hip-hop stars, neo-Nazi detectives, and Jimmy Carter. On the final day of the meeting there was a big banquet and awards presentation, and I won some. It turned out that one of the judges was an editor at *Esquire*. He called me up a week later and asked if I had any story ideas.

This was, for me, a big opportunity, a chance to break out from local to national magazines. I spent a couple of days thinking up pitches. The first one I came up with was about a guy named William Furman, who was both an important historical figure and virtually anonymous. Furman was a onetime death row inmate whose case was overturned, leading to the temporary abolishment of the death penalty in America. He was a murderer who had paradoxically saved the lives of hundreds of other people slated to die.

Terry, the editor at *Esquire*, rejected the pitch. He did so nicely, though, and offered some advice: "Think about stories that you are uniquely qualified to deliver—either because of your passion, your access, your perspective, maybe even your own experience. Think about stories that are potentially rich, with multiple layers that will resonate with readers."

A few days later, a new story idea occurred to me, one that seemed to tick all of Terry's boxes. A story with a personal thread, but one that might resonate with other people. A profile of a man I'd always been deeply curious about. In a sense, he was a man very much like William Furman: a historic but anonymous figure.

And, crucially, he was a man I imagined I could gain unprecedented access to.

After all, my mother's oldest friend was his gatekeeper.

I sent Terry a pitch about Patient H.M., describing him as "perhaps the only man alive that literally lives in the moment. The moment has lasted fifty years." The pitch was straightforward, presenting the basics of Henry's story, and my grandfather's, as I understood them at the time.

Terry quickly sent me back a seven-word response: "I love this idea. Let's do it."

So I sent Suzanne Corkin a letter. A real one, on paper, express mail. I told her that I'd like to profile Patient H.M. for *Esquire* and that the magazine had expressed interest in my doing so. I also enclosed a copy of the pitch I'd sent to Terry. The next day I used the tracking number to make sure the letter had been delivered. It had. Then I waited.

I didn't hear back. Not that day, or the next day, or the day after that.

I emailed Suzanne, and then I phoned her, at work and at home. I got her cell number from my mom and left a voicemail there, too.

Finally, a week after I sent the letter, my own cell rang.

I recognized Suzanne's voice, though her tone, cold and sharp, was new to me.

"I really wish you'd approached me before sending that pitch off," she said. "Would it be possible to get *Esquire* to return the pitch to you, or destroy it?"

I'd last seen Corkin about a year before, when she and her son and my mom and I went out to dinner at a Mexican restaurant in Cambridge. I don't remember much about the night, except that Corkin's son was planning to go out tango dancing a little later and that Corkin had sent back both a drink and an entrée because neither had been made to her liking. If we'd had a conversation during that dinner, it would have been a superficial one. She was my mom's friend, not mine.

But still. She may not have been a friend, but she'd always been friendly.

On the phone, though, that didn't seem to be the case. She told me, in the same cold tone she began with, that it would be impossible for me to meet Henry. First of all, there was the matter of his name: Henry. I had referred to him in my pitch by his first name and that bothered her. I had to understand, she said, that Patient H.M.'s real name, first and last, was a secret. She acknowledged that his first name had been made public multiple times already, including in her own classroom lectures. For example, one book about Henry called *Memory's Ghost,* released just a few years before, regularly referred to him as "Henry M." But the fact that I'd used his first name still bothered her. Also, she said my pitch had "lots of errors." It bothered her that I'd focused on the hippocampus, for example, when "the latest research shows that it's not just the hippocampus involved."

"That's why I won't be able to share any of his records with you," she said, "and for the same reason, you can't meet with him. Not that you would get much from such a meeting. I think he may have had a stroke recently."

She gave a curt apology for taking so long to respond to my letter, said she was busy and had to go. Then she hung up.

Over the next several weeks, I exchanged a series of emails and phone calls with Corkin, trying to convince her to change her mind and let me meet Henry. I also sent her another letter, making it clear that I was going to start reporting this story with or without her cooperation. After receiving that letter, Corkin suddenly seemed to relent. She called and told me that it might in fact be possible for me to meet with Henry after all. She'd discussed the matter with Henry's conservator, or court-appointed guardian, and the conservator had given provisional permission. Corkin said she'd need to be present for the interview and to discuss it with the MIT lawyer first, and perhaps have me sign something before the meeting. She even gave me a tentative window during which the interview might take place: November or December, during Henry's next visit to the laboratories at MIT.

This was followed by more silence. I repeatedly followed up, trying to lock down a date, and she repeatedly told me that she hadn't yet had a chance to speak with the lawyer but would soon.

I got the feeling that she might be stalling, hoping I'd go away once enough time and, presumably, deadlines had passed.

Then, more than two months after I'd first pitched the story, Corkin sent me an email containing a legal contract.

Confidentiality Agreement

> The undersigned, Luke Dittrich (hereinafter called "RECEIVING PARTY"), in consideration for the use of certain information, knowledge, related to an M.I.T. research project entitled "The Amnesic Patient H.M.," conducted by Suzanne Corkin, Ph.D. (hereinafter called "INFORMATION") made available to it by the Massachusetts Institute of Technology (hereinafter called "M.I.T."), hereby agrees as follows:
>
> 1. RECEIVING PARTY agrees to keep in confidence and not to use the INFORMATION for its commercial benefit (except for technical evaluation internal to RECEIVING PARTY). RECEIVING PARTY further understands that the INFORMATION includes personally identifiable health information, through a patient interview, which is subject to the Health Insurance Portability and Accountability Act (HIPAA) and agrees that it shall keep in confidence and not disclose any part of the INFORMATION to a third party or parties in perpetuity.

The contract's next item conceded that my contractual obligations to MIT would not extend to information that MIT had not provided me or that I possessed prior to signing the contract. The item after that warned that if I had any employees, they, too, would be bound by the contract to "protect the confidential and proprietary nature of INFORMATION." Then the contract ended with a final point:

> 4. M.I.T. understands that RECEIVING PARTY will be free to publish the results of the technical evaluation after providing M.I.T. with a thirty- (30) day period in which to review each publication to identify any inadvertent disclosure of M.I.T.'s INFORMATION. If any INFORMATION is found in such publication,

RECEIVING PARTY agrees to remove such INFORMATION prior to publication.

The contract struck me as bizarre and somewhat unsettling. There was the description of Henry, a human being, as "an MIT research project entitled 'The Amnesic Patient H.M.'" There was the nebulous definition of the "INFORMATION" about said project/person, which I would apparently be allowed to receive but never use, and the demand that MIT have final editorial say over any "technical evaluation" that I wanted to publish. There was the strange use of the word *proprietary* in regard to Patient H.M. I wasn't a lawyer, though, so I forwarded the document to *Esquire*'s lawyer, who bounced it right back, telling me not to sign, since it was clearly "incompatible with journalism."

I thought back to something Corkin had said to me during that curt conversation we had after she received my initial letter. She was explaining why my use of Henry's first name had upset her.

"We never refer to him as anything other than H.M.," she said. "That's important. Because otherwise, people could find him."

It was time to see if she was right.

On November 8, 2004, I took an early flight from Atlanta to Hartford, determined to find Patient H.M. My plan was to first figure out his last name and then, armed with that information, to find Henry himself. (What I'd do after that, I had no idea.) In my notebook, I made a list of all the stray bits of biographical detail I'd been able to gather about him so far. The scientists who wrote about him were careful not to reveal too much, and in general they were successful. There wasn't a single paper, out of the hundreds that had been written about him since 1957, that revealed much in the way of identifying information. But if you read a lot of them, you could piece together a fragmentary portrait.

I knew he was born on February 26, 1926.

I knew his father's name was Gustave, and that Gustave was born in Thibodaux, in Lafourche Parish, Louisiana, in 1892.

I knew his grandfather had been a deputy sheriff in that same town.

I knew Henry had attended St. Peter's Church, on Main Street in downtown Hartford, and was baptized and received his first communion there.

I knew that he had once worked at the Underwood Typewriter factory.

That was about all I knew.

I arrived at Bradley International Airport before noon, picked up my rental car, and then headed out to start reporting.

I began at the Hartford Bureau of Vital Records, told the woman behind the counter that I was trying to find Henry's last name, and gave her the information I had. She told me their databases weren't set up like that. She suggested I contact the Connecticut Department of Public Health's records department. I told her I'd already called it the day before and had come up empty.

She gave it some thought.

"Maybe you should try the Connecticut State Library?" she said.

The Connecticut State Library sat on Main Street across from the state capitol. It had a wide façade flanked by big marble columns topped with classical sculptures of robed women, and the entrance doors were copper-plated, with handsome red leather panels. I asked a security guard where the library's history and geneaology department was, and he pointed me down a set of curved marble stairs leading to the basement.

The department was an expansive series of rooms at the end of a long, cramped, white-tiled hallway. There were old card catalogs, microfilms of newspapers, bound volumes of censuses, annual reports, property transactions, family histories. A compendium of local lives and the traces they left. I took a walk around. There were a handful of other people there that day, sitting at reading tables, hunched over documents. I had no idea where to start. The department's chief archivist, Mel Smith, sat behind a bleak, fluorescent-lit desk near the entrance. I noticed a fistful of orange number-two pencils in a wire-mesh cup beside a Dell computer that looked old even in 2004. I introduced myself, told him what I was looking for, and laid out the facts I knew for sure. He leaned back in his chair and tapped his fingers on his armrest. Then he swiveled to a computer workstation.

"What did you say his father's first name was?" he asked.

"Gustave," I said.

Mel pulled up a program that gave him access to a digitized version of the 1930 Federal Census.

"And his last name began with an M?" he said.

"Almost definitely," I said. (I did worry that the "M" in "Patient H.M." might be a red herring, meant to throw people like me off.)

He typed a few words into a search box.

"And where did you say Gustave was born?"

"Louisiana."

He typed in one more word and pressed enter with a little flourish. He tilted the screen so I could see it, and we both leaned in close.

We were looking at a scan of a photograph of a page from the census. The page was a yellowed old-fashioned spreadsheet, dense with cramped blue-ink handwriting. The handwriting, however, had been fed through an optical character recognition program, which made it searchable, just like digital text. About two-thirds of the way down the page, a cell on the far-left corner was highlighted.

There was only one Louisiana-born Gustave residing in Hartford in 1930. He'd lived on Main Street, with his wife, Elizabeth.

They had a single child, a son.

Their son was four years old. He'd been born on February 26, 1926.

His name was Henry.

Henry Molaison.

There are rare moments in reporting when something trips a wire in your head and you feel a genuine physical rush, almost a dizziness, like taking a first drag on a cigarette when you haven't had one in years. I left the library coasting on that feeling. Walking down the steps outside, I stopped and turned and took another good look at the façade. Above the columns, on a wide frieze flanked by the statues of the robed Muses, there were three words etched into the marble in foot-high letters: KNOWLEDGE. HISTORY. JUSTICE. I made some notes, already imagining how I would write this scene.

I tried to find Henry Molaison.

I thought with his name it would be easy.

I started, of course, with Google.

"Patient H.M." gave me thousands of results.

"Henry Molaison" gave me nothing. No address, no history, nothing at all.

I knew he had a conservator, a court-appointed guardian, and that the records relating to his conservatorship should be held at the probate court of whatever municipality he resided in. I visited several probate courts around Hartford, came up empty.

One paper mentioned that he lived in a nursing home somewhere in Connecticut, so I made a list of homes in the Hartford area, and I started calling them.

Haven Health Center: *No, we don't have anyone by that name. Sorry. Bye.*

Riverside Health and Rehabilitation Center: *How do you spell that last name? . . . No.*

Ellis Manor: *Molson? How do you spell it?*

Trinity Hill Care Center: *You know, I can't give out any information on our patients. There's a federal law, it's called the HIPAA law.*

Second call to same place, different person answers: *What's the name again? . . . No, we don't.*

Avery Nursing Home: *Happy holidays! It's Carol speaking. . . . Hang on. . . . Mmmm, okay, he's not on the list.*

And on and on. Every nursing home in Hartford, East Hartford, West Hartford, Newington, Bloomfield . . .

Nothing.

I didn't find Henry. I went back to Atlanta, disappointed. I continued to press Corkin, to see if there was any way she would arrange a meeting without her preconditions. She continued to reply, but slowly, telling me she needed to discuss things with the MIT lawyers. Finally I received an email from Corkin informing me that she was "adamant" about being able to correct whatever I wrote, and that if "I cannot read the manuscript in advance of publication, it's a no go."

The only way she was going to give me access to Henry, in other words, was if she could control the story I told about him.

PART V

SECRET WARS

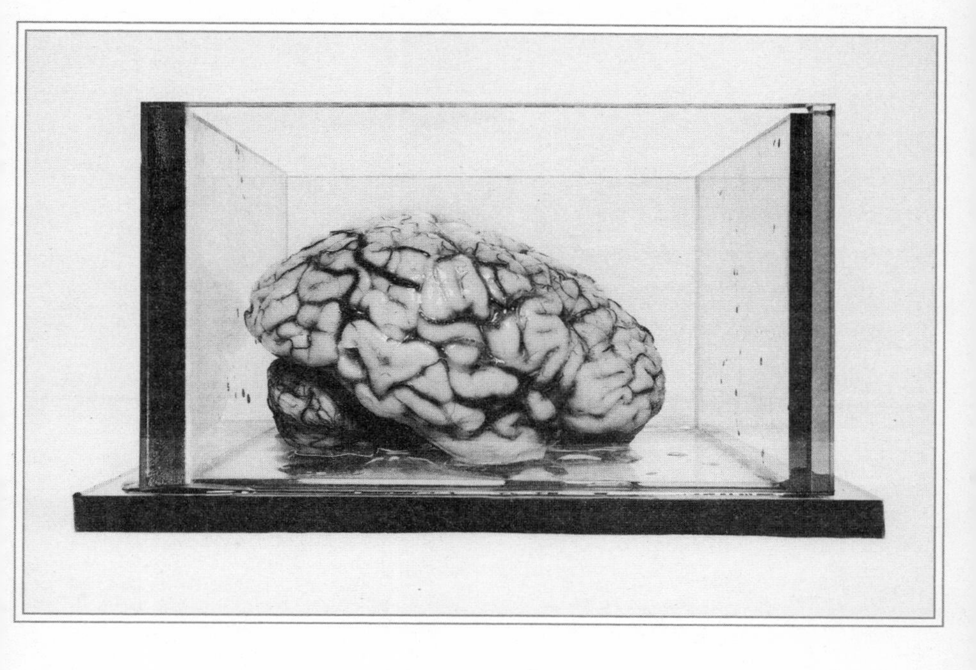

TWENTY-FIVE

DEWEY DEFEATS TRUMAN

Dr. William Marslen-Wilson and Patient H.M., MIT Clinical Research Center, May 1970

Marslen-Wilson: Do you remember anything about Pearl Harbor?
H.M.: Well, it happened on a Sunday.
Marslen-Wilson: Ah. What date?
H.M.: Well, uh, December seventh.
Marslen-Wilson: What year?
H.M.: Well, I think of 1941 right off.
Marslen-Wilson: That's right. Um, do you remember how you yourself heard about Pearl Harbor? Where you were, or anything like that?
H.M.: No, I don't. I . . .

He paused, gathering his thoughts. He had been fifteen years old when the Japanese bombed Pearl Harbor.

H.M.: I say I don't . . . in a way, I think of . . . that my father told . . . was telling my mother it. . . . He had heard about it before . . . and there . . . I have an argument with myself right off. . . .

The longer he thought about it, the more details of the day came trickling back.

H.M.: Daddy had gone down the street, to get the car in the garage. And saw the men that were at the gas station there, where they had the garage and everything. And they talked about it and told him. Because they'd been up all night in a way, and they heard about everything, and they were able to tell him. And then he told us about it. And we actually turned the radio on.
Marslen-Wilson: And where were you? Where was this? At home?
H.M.: Yes, this was at home.

Wilson asked Henry about other milestones of the war.

Marslen-Wilson: Do you remember, er, VE Day?
H.M.: Well, Victory-in-Europe Day?
Marslen-Wilson: Do you remember the celebration? Where were you when that happened? Do you remember anything about it?
H.M.: I don't remember exactly where I was, and then again I have an arg . . . uh, a feeling that I was in, uh, we were down, I say Hurd Park . . . and we were on a picnic. And the people heard about it there, on their car, because their car had a radio, our car did not, and they told everybody. And everybody hollered and jumped around.
Marslen-Wilson: What was this for?
H.M.: For VE Day.
Marslen-Wilson: I see. What about VJ Day? Do you remember how you heard about that?
H.M.: Well . . .

He paused.

H.M.: No . . .

He paused again, staying silent for about ten seconds. He searched his mind for a memory. He came up empty.

H.M.: The actual VJ Day I don't remember.
Marslen-Wilson: Do you know who is president now?
H.M.: No, I don't. To be truthful to you, I don't. Because I was wondering right along then. I said, well now, who is president now? Trying to put two and two together. And, well, some of the answers that I was giving, that were correct, and some of the way you were wording things, so I was associating them together in a way, and then maybe come up with it, the name and everything.

He paused.

H.M.: And I think of Dewey. Uh, was a governor of New York State.
Marslen-Wilson: And he may be president now?
H.M.: And he may be president now. Thinking of the black suit that he wears and the white shirt. That's what I think of.
Marslen-Wilson: What's he look like?
H.M.: Well. Five-foot-eight, about. Six or eight. Five-foot-six or eight. And, uh, his round head more . . . hair naturally, covering his head . . . and . . .
Marslen-Wilson: This is Dewey?
H.M.: Hmm?
Marslen-Wilson: This is Dewey?
H.M.: This is Dewey. And. And. Had a very large, uh, a large head . . . a larger head than you'd expect for a person of his size.
Marslen-Wilson: What about, er, Nixon?
H.M.: Well, I think of him being a president and then a vice president.
Marslen-Wilson: Yes, what happened to him? Is he still president?
H.M.: And . . . I think he is. Still president.
Marslen-Wilson: Does he have a wife?
H.M.: Pat Nixon.
Marslen-Wilson: Any children?
H.M.: And, er, children. I think of four boys right off.

Marslen-Wilson: Four boys?
H.M.: I think of boys. Four boys. And I think he has girls. I mean, three boys, anyhow. Not four boys, three.

Wilson paused. At the time of the interview, in 1970, Richard Nixon *was* president of the United States. He had also, years before, been vice president. His wife was named Pat. Henry's comments about Nixon's children were completely wrong, though. Nixon had two daughters, no sons. There was, however, a different prominent political family that *did* have three brothers and one sister. Four brothers, if you counted the one who died in action during World War II.

Marslen-Wilson: Do you know of any, anybody called Kennedy who has been president?
H.M.: Ye . . . Well, he's been a president, and I believe, well, uh . . . Trying to think of his first . . . Robert, isn't it? Robert Kennedy?
Marslen-Wilson: Yes, what happened to him? Is he still president?
H.M.: No, I don't believe he is. . . . I think of him being shot.
Marslen-Wilson: He was shot? Where? When?
H.M.: The date I cannot tell you. And, er, I think of Ohio right off.
Marslen-Wilson: I think it was further south and west than that.
H.M.: When you say that then I thought of Alabama.
Marslen-Wilson: No, I think it was further southwest.
H.M.: Southwest. Isn't . . . it's around Ohio?
Marslen-Wilson: Much further south and west.
H.M.: Much further so . . . Well, uh, I think of Reno right off?
Marslen-Wilson: Reno?
H.M.: Yes, Nevada.
Marslen-Wilson: You think he was shot there?
H.M.: And I don't think he was shot there. . . . I think it was a town just outside that, though.
Marslen-Wilson: Yes, who was he shot by?
H.M.: Uh, well, the guy that pulled the trigger, naturally.
Marslen-Wilson: That's right. [Laughter.]

H.M.: I don't remember his name.

MARSLEN-WILSON: What, what were the circumstances of the shooting?

H.M.: Well, he was . . . I think of him on a re . . . on a review . . . and . . . er . . . the assassin was I think of two stories up, in a window, shooting down. . . .

MARSLEN-WILSON: Yes. What do you mean, "He was on a review"? He was walking somewhere?

H.M.: He wasn't walking, he was in a car. He was riding in a car.

MARSLEN-WILSON: Was there anyone in the car with him?

H.M.: And I think of his wife right off.

MARSLEN-WILSON: Yes. What was she called?

H.M.: I think of Pat. And. There was a general in the car with him.

MARSLEN-WILSON: A general?

H.M.: Yes, a general.

MARSLEN-WILSON: Was he shot, too?

H.M.: And there I have, er . . . I know a bullet passed through him. And, uh, I . . . I say him . . . Being of course both males there, in the backseat. That were sitting . . . Er. And went into . . . Pat Nixon.

MARSLEN-WILSON: She was shot?

H.M.: By the bullet that went through. One of them, or ricochet. No. I just thought: "ricochet," right off. That came right like that. [He snapped his fingers.] I think it was a bullet that went through one of them and then ricocheted off, or something.

MARSLEN-WILSON: Who did it hit?

H.M.: Well, went through one of them and hit Pat.

MARSLEN-WILSON: Pat who?

H.M.: Nixon.

MARSLEN-WILSON: Uh-huh. Do you . . . the assassin, you don't remember? If I told you his name began with O . . . The man who shot President Kennedy?

H.M.: I think of O'Hara right off.

MARSLEN-WILSON: O . . . S . . .

H.M.: Oswald.

Memory scientists divide amnesia into two basic types: anterograde and retrograde. Anterograde means "moving forward" and retrograde means "moving backward." It was obvious since almost the moment he emerged from the operating room that Henry suffered from profound anterograde amnesia: He could hardly remember any new events as he moved forward through time. But he also, though less obviously, suffered from severe retrograde amnesia. Many of the years immediately preceding the operation were blank to him, and interviews helped paint a picture of just how far back Henry's retrograde amnesia stretched. His personal recollection of major world events appeared to be more or less intact up to 1944 and the German surrender. His personal recollection of the Japanese surrender the following year, however, appeared to have dropped away. He could remember where he was at the beginning of the war but not the end.

This fit nicely with the dominant theory of how experiences are turned into permanent long-term memories. According to that theory, memory traces—the first products of lived experiences—exist in a fragile, impermanent state for years, a limbo of sorts. These traces reside in various parts of the brain, but if they're going to stay accessible down the road, the hippocampus must work on them, strengthening them, sending neural impulses to them and receiving impulses from them for years, until one day they become strong enough to live on their own, at least somewhat independent of the medial temporal lobes.

Henry's answers often supported that framework: He remembered what researchers expected him to remember, forgot what they expected him to forget.

But researchers struggled to make sense of the clear exceptions to Henry's amnesia. How did he know anything at all about the Kennedys and the Kennedy assassination? Or Richard Nixon and his wife, Pat? They offered tentative explanations. Maybe some vestigial portion of his hippocampus allowed particularly vivid events or people to stick. Or maybe other nearby brain structures had picked up the slack, taking over some of the functions of Henry's medial temporal lobes.

Nobody knew for sure.

Sometimes Henry just remembered things he had no business remembering.

Other times, he remembered things that never happened at all.

Marslen-Wilson: Do you know who any of the people there are?

Wilson held up a photo.

H.M.: I think of the Rolling Stones right off.
Marslen-Wilson: Rolling Stones? Who are the Rolling Stones?
H.M.: Rolling Stones were the singers.
Marslen-Wilson: Yes. What sort of music?
H.M.: Jive music.

Wilson pointed at the photograph again.

Marslen-Wilson: They're not the Rolling Stones.
H.M.: They're not the Rolling Stones. . . . No . . . I didn't think they were because I thought there was five . . . Rolling Stones.
Marslen-Wilson: That's right.
H.M.: And I can only see four there.
Marslen-Wilson: Their name begins with B.
H.M.: Uh. B. I know they're brothers. They're naturally four brothers. Because there's four of them there . . . and . . . But I can't think of their names.
Marslen-Wilson: Beatles. Did you ever hear of them?
H.M.: B-E-A-T-L-E-S, the Beatles.
Marslen-Wilson: That's right. Do you listen to them?
H.M.: No. I haven't.
Marslen-Wilson: But you know how to spell their name?

Henry went quiet for a few seconds, his gaze shifting to one side. He was having a petit mal seizure. Maybe he scratched at his leg, maybe his jaw moved up and down, back and forth. Then he stopped seizing and came to.

H.M.: Hmm?
Marslen-Wilson: You know how to spell their name?
H.M.: Whose names? See, I'm sorry. I went into a spell. Just as you were asking the question and everything. I sort of, popped right out.
Marslen-Wilson: What do you . . . What happens . . . What do you feel when that happens?
H.M.: Don't feel anything in a way and . . . it's just that . . . well . . . I guess you could say . . . like waking up in the morning, when an ordinary person wakes up in the morning they come . . . and . . . to a realization of things . . . that I guess you could call it . . . more realization of . . . and . . . well . . . the big question mark that you have yourself, of course. You wonder what the heck it is now, because you know that you've been awake before, because you remember that in a way, but . . . not in the way you . . . but . . . not remembering in a way, but you know that you've been awake, then that little blank spell and you wonder what has been going on there. Because there might have been something, an awful lot. And you don't remember, and like you, even though you heard . . . you couldn't tell if . . . was . . . what the person said or anything like that . . . you wonder . . . yourself.
Marslen-Wilson: Well all I asked you was what was . . . uh . . .
H.M.: The names.
Marslen-Wilson: I can't remember myself now! Oh yes, the Beatles.
H.M.: Beatles . . . I was, uh, thinking of names, trying to think of the . . . uh . . . I was thinking of first names of the Beatles and . . . can't think of them.
Marslen-Wilson: You thought of another singing group, didn't you?
H.M.: I can't think of it, no.

Wilson held up another photograph, this one of Neil Armstrong walking on the moon.

Marslen-Wilson: Now, what's this a picture of?
H.M.: Well, it's a man walking on the moon.
Marslen-Wilson: You reckon?
H.M.: Um, that'd be the first photograph of a man walking on the moon.
Marslen-Wilson: Do you know the name of anybody on the moon . . . went to the moon . . . what do they call them?

Wilson waited for a while, then gave Henry the answer.

Marslen-Wilson: Astronauts.
H.M.: Ah, the astronauts. And. Right as you say astronauts, I think of, uh, how there are some going around . . . and some guys that, uh, four of them, trying, call themselves the Astronauts, and they are trying for . . . They call themselves the Astronauts. Singing.
Marslen-Wilson: The Astronauts?
H.M.: Yes.
Marslen-Wilson: Singing?
H.M.: Well, they call themselves the Astronauts. They are not . . .
Marslen-Wilson: Ah, they're a singing group?
H.M.: They're a singing group.
Marslen-Wilson: And what sort of music do they play?
H.M.: Well, jive music.

Wilson held up a photograph of Jack Ruby shooting Lee Harvey Oswald. It was the famous photograph taken just a second after the gunshot, Oswald facing the camera, Ruby lunging toward him, the pistol aimed at his gut.

Marslen-Wilson: Do you know who that is?
H.M.: Well, I think of Frank Sinatra. Right off in the middle. That's being shot.
Marslen-Wilson: Do you think it must be in a movie, or what?
H.M.: Ah, uh. And I think of Boston.

Marslen-Wilson: Frank Sinatra being shot in Boston?
H.M.: I think that's it. And. Uh. I have an argument with myself, though. Of Boston and . . . or was it out through . . . was it . . . not out through . . . But was it in Dallas? And then . . . Then Detroit comes into the picture.
Marslen-Wilson: Where who was shot?
H.M.: Um. Sinatra, I think . . . The man in the middle.
Marslen-Wilson: Who's he being shot by?
H.M.: And I think of . . . thought . . . I think I said that right . . . I . . . Probably wrong. The Oswald. That's what I thought. I thought of Oswald.
Marslen-Wilson: Where? What? Do you think that's Oswald?

Wilson pointed at Oswald.

H.M.: No. That one.

Henry pointed at Ruby.

Marslen-Wilson: So Oswald is shooting Frank Sinatra?

Henry nodded.

H.M.: Frank Sinatra. In Detroit.

The questioning went on for days, years, decades.

Graduate students would cycle in and out of MIT's Clinical Research Center, cherishing the time they got to spend with the increasingly famous Patient H.M. Jenni Ogden, for example. A visiting neuropsychologist from Auckland University who spent 1985 to 1986 in Corkin's lab, Ogden recorded several of her conversations with Henry, brought the tape back with her when she returned home, held on to it for decades, and still listens to it now and then, like a favorite album. Some of the recordings document an unpublished experiment she conducted with Henry, when she tried to determine if he was capable of parroting different emotional states. (She asked him to repeat the phrase "We're going to the movies" in just about every way imaginable: sad, angry, anxious, happy, amused. He did fine.) The parts of the tape Ogden likes best, though, are the less structured parts, like the bit when she and Henry just sat and chatted about Elvis.

OGDEN: Do you know what Elvis Presley was? Or is?
H.M.: Well, he was a recording star. And he used to sing a lot.
OGDEN: He did indeed! What sort of things did he used to sing?
H.M.: Well, jive.
OGDEN: Jive, yeah. Do you like to jive? Or did you like to jive?
H.M.: No.
OGDEN: Why not?
H.M.: I didn't . . . Well, I liked to listen. That was all.
OGDEN: You did. He's a good singer. Do you think he's still alive, Elvis Presley?
H.M.: No, I don't think so.
OGDEN: Have you any idea what might have happened to him?
H.M.: Well, I believe that he got the first bullet. I think. That was for Kennedy, I think it was.
OGDEN: And you remember Kennedy?
H.M.: Yeah. Robert.
OGDEN: Robert Kennedy. What was he?
H.M.: Well, he was the president. I think about three times.
OGDEN: The president about three times.
H.M.: Yeah. He was appointed president, too. And he got a bullet.
OGDEN: So what was that all about?
H.M.: Well, they were trying to assassinate him.
OGDEN: And did they? Did they kill him or not?
H.M.: No, they didn't.
OGDEN: So is he still alive?

H.M.: Yes, he's still alive. But he got out of politics. . . .
OGDEN: What's that thing over there?
H.M.: Well. I think, a videograph.
OGDEN: A videograph? I'll give you a . . . What's another word for it? Anything else you can think of?
H.M.: Well, you can see . . . You can put questions in it, and you have a *yes* or *no* on it, too.
OGDEN: That's right. You certainly do. That's called a co . . . ?

She paused, waiting for him to complete the word.

H.M.: Cardiograph?
OGDEN: No. Com . . . ? Com . . . ?
H.M.: Compressor?
OGDEN: No. You sometimes get it. Com . . . ?
H.M.: I can't think of the word.
OGDEN: Computer.

Henry's failure to come up with the word was not a surprise. Computers like the one sitting on a desk in the examination room weren't invented till long after 1953, and Henry's vocabulary had more or less ceased to expand since his operation, though occasionally he could come up with a word if prompted with the first syllable or two. Sometimes researchers asked Henry to provide definitions for modern words or terms, and his answers were almost always wrong. He defined boat people as "people who cater bon voyage parties," granola as "a portable keyboard wind instrument," apartheid as "the separation of young cows that have not yet given birth to calves," and brainwash as "the fluid that surrounds and bathes the brain."

OGDEN: Can you tell me who the president of the United States is at the moment?
H.M.: No, I can't.
OGDEN: Who's the last president you remember?

H.M.: I don't . . . Ike.
OGDEN: Ike. Now, if I tell you that the president now used to be a film star, does that help? Not a very good film star, but he used to be one a long time ago. I think he used to be a film star in Westerns. And now he's the president of the United States. Rea . . . ?
H.M.: Reagan.
OGDEN: Reagan! Very good. Do you remember he used to be a film star?
H.M.: Well, yes.

They talked for a little while about other movie stars that Henry remembered. Gary Cooper. Myrna Loy. Jimmy Stewart.

OGDEN: What about Frank Sinatra?
H.M.: Well, he did a lot of singing and he was in films and on the stage and radio and records.
OGDEN: Do you think he's still alive, Frank Sinatra?
H.M.: There I don't know.
OGDEN: He is. He is still alive, I think.

Ogden changed the subject then, deciding to see what Henry could tell her about someone who was not a public figure but who nevertheless loomed large in Henry's life.

OGDEN: What . . . Who, or what, is Sue Corkin?
H.M.: Well. She was a . . . well, a senator.
OGDEN: A senator?
H.M.: Yeah.

In 1973, Suzanne Corkin tried to figure out if Henry had difficulty detecting ambiguity. She presented him with a sentence: "The marine captain liked his new position." The sentence could be interpreted in at least two different ways, and she asked Henry whether he could describe those two meanings to her.

H.M.: The first thing I thought of was a marine captain, he liked the new position on a boat that he was in charge of, the size and kind it was, and that he was just made a marine captain, and that's why he liked the position, too. Because he was above them. And of all, most of all . . .
Corkin: So you're saying that he liked his job, in other words?
H.M.: He liked his job.
Corkin: Okay. Now, there is another meaning in that sentence. Can you tell me what it is?
H.M.: I just gave you two.
Corkin: Those are both really the same. Because they were both related to his job. There is another meaning.
H.M.: Well, 'cause he was on a new boat, you might say a new boat, he was made captain of a new liner or whatever it is and it's different than what he had before. He might have had a, a, a . . .
Corkin: You mean his job was different?
H.M.: Yes, he might, he has people . . .
Corkin: That's the same meaning that you told me.

They go back and forth for a while, until it's clear that Henry isn't going to come up with the alternate meaning Corkin is fishing for. So she tells it to him. She explains that apart from the interpretation Henry is making—the Marine captain liked his new job—you could also interpret the sentence as meaning that the captain liked his new position, in a literal, physical sense. That he had just sat down, for example, and liked the feeling of being seated.

H.M.: Oh.
Corkin: Okay? Do you see how those are really rather different meanings?
H.M.: They're different.
Corkin: One has to do with his job and the other is if he is sitting, standing, or whatever.
H.M.: The position he's in.

Corkin: The position of his body. Okay, you see? Do you understand how the very same words can mean two rather different things, two different interpretations depending on how you read it? Okay?

What Corkin said was true, of course: Words have fluid meanings, and can always be interpreted in different ways. This was certainly true in the case of Henry's own words, which were pored over by generations of scientists, scrutinized like the words of a prophet, wielded in support of different and sometimes contradictory belief systems.

Donald MacKay, a cognitive psychologist and psycholinguist, studied Henry's words as intently, and as fiercely, as anyone. He began working with Henry as an MIT graduate student in the late 1960s and early 1970s and continued to do so after he left MIT to take a position with the UCLA department of psychology. He wrote more than a dozen journal articles that explored Henry's way of speaking and claimed to find a variety of persistent deficits in Henry's ability to formulate coherent and contextually appropriate sentences. MacKay's criticisms could be opaque to nonlinguists—one typical paper spent pages chronicling the subtle differences between Henry's "major violations of miscellaneous conjunction constraints" and his "major violations of copular conjunction complement restraints"—but as a whole the articles provided compelling evidence that Henry did have an unusually hard time expressing himself logically and grammatically.

For example, in the 1990s, Lori James, one of MacKay's graduate students, showed Henry a picture that depicted one man rock climbing while two other men looked on. One of the men on the ground was pointing at the rock climber. James asked Henry to formulate a sentence describing the scene, and to use the words *fall* and *leg* in that sentence. Normal controls tended to produce a sentence along the lines of "The man is telling him not to fall and break his leg."

H.M.: Seeing how somebody's climbing that mountain, they are discussing it themselves 'cause stuff he should take.

James: Mm-hmm. So just try to make up a sentence using these two words.
H.M.: David wanted him to fall and to see what lady's using to pull himself up besides his hands.
James: So can you make one sentence up? Using both words.
H.M.: Well, I see that Dave did past and he's going up fast.
James: So, you just need to make up a sentence using these two words. So make up a sentence using the two words.
H.M.: Um, well, he's got a pack and so does each one of those.
James: Yeah, I see that. But again you just need to use these two words to make a sentence up.
H.M.: Just to see how he's legs, see. How he's using his legs to climb.
James: I know. But you're ignoring my question, aren't you?
H.M.: Well, both of them.
James: I know, but I just want you to say a sentence using these words.
H.M.: Well, how they have to fall, uh, climb, easing up.
James: So, what are the two words?
H.M.: Fall and leg.
James: So, can you make up a sentence about this picture?
H.M.: Jay had to use climb, too.

Not all researchers agreed with MacKay that Henry's "errors in novel spoken discourse were so severe as to render his output incoherent and incomprehensible," but most people who worked with Henry noticed that he had at the very least some language problems. There was disagreement over the origins of these problems, however. MacKay believed that they were likely the result of Henry's brain lesions, while Corkin speculated that Henry's "mild language disorder . . . might have preceded the operation, and could be related to substandard education and low socioeconomic background."

Then, in 2005, a group of researchers from Duke University published an article that presented a dramatically contrary view: They concluded that Henry didn't have any language problems at all. After interviewing Henry for several hours over three days, they failed to see "the language deficits noted previously. Instead, H.M.'s level of oral usage was remarkably competent." They noted that their findings "contradicted other studies of H.M.'s language" and singled out MacKay's work, declaring their own to be a "more ecologically valid analysis of H.M.'s language skills," in part because they had conducted their interview in Henry's "familiar home environment" rather than "in unfamiliar laboratory studies."

They included several excerpts from their interviews with Henry to support their claim, including this exchange, which came after they asked Henry whether there was anything at all he'd like to tell them.

H.M.: Well, I know of one thing: What's found out about me will help others be.
Researcher: That's right. You're a hero! Did you know that? You're a national hero. Did you know that you are famous?
H.M.: No.
Researcher: Yeah, you're famous. You are! Are you glad? Is that nice to know?
H.M.: Well, it's nice to know, in a way.
Researcher: Not everybody gets to be famous, sir, but you are!

MacKay responded to the Duke paper with what passes in academia as a full-frontal assault, taking issue with what he saw as the "procedural flaws" throughout, from "statistical errors" to an "inadequate control group" to the fact that the evidence they'd presented consisted of a few short excerpts from an unpublished transcript covering five to six hours of conversation. "A selective focus on examples favoring the no-major-errors hypothesis is problematic," MacKay wrote, "because science can only progress as an empirical enterprise by seeking *counterexamples* and analyzing them in detail." The most lacerating portion of MacKay's paper, however, was when he used the work of the Duke researchers against them, arguing that even their own cherry-picked excerpts contained evidence of major verbal problems on Henry's part. He pointed out that in the exchange above, for example, Henry's statement that "what's learned about me will help others be" doesn't make grammatical sense. MacKay also pointed out that, grammatical or not, Henry's seemingly altruistic statement was one that he had been repeating ad nauseam for decades, in almost any context. MacKay cited examples of Henry giving basically the same response to questions ranging from "Are you happy?" to "How are you feeling?" to "Where do you think you are?" to "What aspect of remembering are you wondering about?"

"Repeating the same response to so many different questions," MacKay wrote, "seemed abnormal."

Everyone who worked with Henry grew familiar with those sorts of stock phrases and anecdotes, the building blocks of his conversations. Again and again he would tell the scientists about how what they learned from him would help others, about camping in Vermont, about crossword puzzles, about his desire to be a neurosurgeon, and he'd pepper these chestnuts with all his familiar verbal tics: *in a way . . . an argument with myself . . . I guess . . . right off . . .* As Henry grew older, and more famous, he became in some ways like an aging rock star, periodically hauled onstage to deliver his greatest hits to his fans, the neuroscientists who studied him. Sometimes even his fans grew sick of hearing those same old songs. Alice Cronin-Golomb, one of Corkin's graduate students in the late 1980s, was often tasked with driving Henry back and forth between Cambridge and Hartford, and in an essay she wrote about the experience she recalled how she was maybe "the only person to actively try to keep his memory from working."

"During those drives," she wrote, "there was a highway sign for Chicopee Falls, which would always cue him to say 'Chicopee Falls? I had an aunt in Chicopee Falls!' And I'd be listening to the same story time after time. One time I just couldn't bear the thought of hearing the story again, so when I saw the sign before he did I yelled, 'Henry! Look over there!' and pointed in the direction opposite the sign so he wouldn't get that cue."

As time passed, even as Henry's fame grew greater, even as the articles about him accumulated, even as the spats over his strengths and weaknesses flared, there was one other way in which he resembled an aging rock star: His best days were long past. Scientifically speaking, the basic discoveries that Brenda Milner made during her first afternoons with Henry were unquestionably more important than any discoveries made by the legions of other scientists who worked with Henry during the six decades that followed.

But still they came, making their pilgrimages to see Henry, asking the same questions again and again and arguing about what his answers meant.

Sometimes Henry's interviews were loose, informal, rambling, and other times they were tightly controlled. He'd be presented with a task and expected to do it. There were very few tests from the researchers' arsenals that were not applied to Henry at one time or another. Every aspect of him was scrutinized, and the contours of his deficits were mapped out with increasing precision, day by day, year by year.

Even his sense of humor was put under the microscope. In 1997, a researcher spent an afternoon sliding eight-by-eleven photocopies of old *Far Side* and *New Yorker* cartoons across a table to Henry, asking him to explain what made them funny. One cartoon showed a businesswoman in a boardroom, standing in front of a job-performance chart, speaking to a group of colleagues.

"The beatings will continue until morale improves," she was saying.

"It's about this woman talking to the ward there," Henry said. "And the secretary is sitting down, writing. And then the woman is supposed to be listening to her, listening to what this woman is saying. And, uh, that picture they've got in the background there, it's just a picture. But they're a business, in a way, in the area. Or in, maybe, a distant view, because the mountain area in the back. Well, she's making a comment there, that the beatings will continue until morality improves. And then, she said 'morale improves.' Morality in a way. Instead of L it's a T. It should be a T."

Henry paused, studying the picture for a few more moments before continuing.

"And their window frame is slanted," he said.

The researchers concluded that Henry did not have normal capacity to comprehend or construct jokes, and that his sense of humor appeared to be severely damaged. "Henry doesn't have what it takes to be humorous," one of them later told me. The researcher acknowledged that there was anecdotal evidence that Henry occasionally cracked jokes, but pointed out that those anecdotes were usually predicated on certain assumptions. For example, Suzanne Corkin often told people about a remark Henry had made one day when she'd complimented him on his passion for crossword puzzles.

"You're the puzzle king," she'd told him.

"Yes," Henry had responded, "I'm puzzling."

While Corkin interpreted Henry's response as a joke, it could also be interpreted in other ways. Maybe, for example, he'd misheard her and was simply parroting back what he thought she'd said. One word, as Corkin herself had pointed out, can mean two rather different things.

Joke or no joke, what Henry said was true: Even decades into his career as a research subject, he remained a deeply puzzling case.

Reading Henry's interview transcripts could be like staring at clouds. The Kennedy assassination might come up repeatedly in a single conversation, but it would present itself in slightly different permutations: Sometimes Franklin Delano Roosevelt was shot in Dallas, sometimes it was Pat Nixon sitting in that fated convertible, in Ohio. Sometimes Kennedy rode with a general beside him, sometimes he rode with Elvis. Time and people and places faded in and out, innefably intertwined, clear for a moment before dissipating again.

The most compelling moments were always the rare ones when

Henry would try to explain what it was like to *be* him. He'd struggle to articulate it, to describe what the world looked like from within his fractured mind. He never quite succeeded, since his amnesia wouldn't let him hold on to the ideas long enough to get them out. He'd seem on the verge of a breakthrough, of a definitive statement, and then his train of thought would derail, and he'd start all over again.

Henry was studied more than any other human research subject in history, but there were things about him that would remain a mystery, not just to the scientists but to Henry himself.

Dr. William Marslen-Wilson and Patient H.M., MIT Clinical Research Center, May 1970

Marslen-Wilson: What do you do back in Hartford?

H.M.: Well . . . I learned how to rewind electric motors.

Marslen-Wilson: Recently, after your operation, what do you do?

H.M.: I don't know.

Marslen-Wilson: Do you stay at home all day?

H.M.: I know I don't go . . . I don't believe I go out. To work or labor or anything. Then I must stay at home. Or. There I have an argument with myself, too. Do I? Why? Why do I stay home? And . . . what the . . . Like I said, that argument with myself, in a way. And I wonder. But I know that, well, whatever's being done is done right.

Marslen-Wilson: But why do you think you stay home?

H.M.: Well, whatever's learnt is learnt. And that's more important.

Marslen-Wilson: Yes, but why do you, when you're at Hartford . . . why do you stay at home and not go out to work or anything like that?

H.M.: Well, because I would forget, when getting through work, to come home. And the way home.

Marslen-Wilson: So you'd do too much work?

H.M.: Or. I would forget just the job that I was going into, and maybe not arrive at the job.

Marslen-Wilson: Why do you think this is?

H.M.: Well, well, I think of an, ah, operation. And then I have an argument with myself right there. Did the knife slip a little? Or was it a thing that's naturally caused by it, naturally, when you have this kind of an operation?
Marslen-Wilson: That caused what?
H.M.: This, uh, well, loss. Or you could say, loss of memory, in a way. But not, uh, the reality.
Marslen-Wilson: Not?
H.M.: Well, it's, uh, you can be realistic in a way, but . . . You really think things out more. And get all of them.
Marslen-Wilson: All of them?
H.M.: Get all the ends, and then put them all together, and then always think about and decide, then. Instead of figuring along one way, only, in a way. You figure them all around, and then go through it again.
Marslen-Wilson: This is what you do?
H.M.: That's what I'm thinking. That's what I'm thinking of.
Marslen-Wilson: You think this is something you do different from other people?
H.M.: Because, well, most people, when they just think, they think things through. Once. And they are able to pick out what they have thunk. I say *thunk.* That's not a word. But, uh, what they have thought, and they know. They're able to pick it out, from memory and everything. Where you run through it, and then go through it again. You run through it, and you find out what's good, but you go through them all again. . . .
Marslen-Wilson: So you mean by not having a very good memory, you can't improve the way you do things?
H.M.: Well, by not remembering the things, you can't improve them. And you can't remember either way—the good way and the bad way—and you can't put them together and figure them out that way.
Marslen-Wilson: Oh, I see. So you can't tell whether something's good or bad, I mean, a job you've done is good or bad, because you can't remember which is the good way and which is the bad way?

H.M.: Yes . . .
MARSLEN-WILSON: Do you think about this a lot?
H.M.: In a way, yes.
MARSLEN-WILSON: In what sorts of situations does this come up?
H.M.: Well, I don't think of any particular situation . . . and then you have the argument with yourself. Then you wonder to yourself, well: One way is better now . . . then you argue, of course, that argument. And . . . you wonder. Really wonder, to yourself, which is which. Because maybe the way that you said, that he said, is the way you thought. And vice versa. You know, just vice versa, in a way? And, just the way it turns around. And the way, well, you wonder to yourself. . . .
MARSLEN-WILSON: Okay, well . . .
H.M.: That's what I think of right off, too: Is the best thing to do . . . Is the best . . . Is the right thing right?

TWENTY-SIX

A SWEET, TRACTABLE MAN

One day in late December 1974, an acquaintance of the Molaisons named Lillian Herrick stopped by the home that Henry shared with his mother at 63 Crescent Drive, in Hartford. She found Henry's mother, Elizabeth, lying on the floor, "completely out of it." It wasn't clear what had happened, and Henry, sitting nearby, appeared oblivious to his mother's distress. There was a terrible stench, since neither Henry nor Elizabeth had showered or washed for days. Herrick called an ambulance, and the doctors who examined Elizabeth determined she showed signs of dementia. Herrick was a retired psychiatric nurse from the Institute of Living who now earned additional income taking in elderly people at her own house. Arrangements were made, and the following month Henry and his mother moved into Herrick's home, a large three-story house on New Britain Avenue in West Hartford.

As soon as they moved in, it became clear to Herrick that Henry and his mother had a difficult, unpleasant relationship. They fought a lot. Henry, Herrick told one of the MIT researchers, was "not nice to his mother. She nagged him, he retaliated." This retaliation sometimes became physical. Henry would kick his mother in the shins or hit her on the forehead.

This was not the first time Henry had demonstrated a violent streak. Four years before, for example, on a Sunday afternoon in the spring of 1970, after yelling at his mother repeatedly to leave him alone and stay out of his way, Henry slammed his fist against a door so hard that he broke his hand. Not long after that, Henry had an even more extreme outburst. At the time, Henry had been spending his weekdays at a place called the Hartford Regional Center for the Mentally Retarded, a state-funded organization that provided menial jobs to people who might otherwise have trouble finding employment. Sometimes Henry would mount key chains on cardboard display cases, and other times, such as on the day in question, Henry's task was to fill plastic bags with a specific quantity of uninflated balloons. It was a simple thing, but his amnesia sometimes made even simple things—such as keeping count of how many balloons he'd packed—frustratingly difficult. Maybe that's why he suddenly leapt up from his workbench and started yelling at nobody in particular, saying, according to one account, "that he had no memory, was no good to anyone, was just in the way. He threatened to kill himself and said he was going to hell and would take his mother with him." The people at the center tried to calm him down, but Henry kicked at them and shoved them away. He ran to a wall and started smashing his head against it, until a doctor arrived and injected him with a sedative.

After seeing how difficult Henry's relationship with his mother was, Herrick decided that the best way to deal with the conflict was to separate them. She moved Elizabeth to an upstairs bedroom, while Henry stayed downstairs. The separation seemed to work. After about six months, "he quieted down," Herrick said. She also tended to Henry's hygiene, trying to instill better habits. She made sure he remembered to shower, to brush his teeth, to comb his hair, and to be ready to go at nine-thirty A.M. on workdays. "He never protests," Herrick said, "but he wouldn't do these things if you didn't get after him." She left little notes for him around the house. There was one on the television reminding him he had to turn it off by nine-thirty P.M. every night. Herrick also made sure that whenever the folks from MIT wanted to continue their experiments with Henry, they could. She'd pack his suitcase, drive him to Cambridge, and drop him off at the Clinical Research Center.

. . .

Henry's occasional violent episodes in Hartford contrasted with the way he behaved in Cambridge. According to Suzanne Corkin, Henry was always, to her, a "sweet, tractable man." Here's a sampling of a few of the other ways Corkin has described him in writing:

"He was a pleasant, engaging, docile man with a keen sense of humor, who knew that he had a poor memory and accepted his fate."

"When my colleagues and I interacted with Henry he was always friendly but passive."

"At the CRC he was always docile and friendly."

"For someone with such a severe memory problem, Henry was surprisingly easygoing. He was cheerful and never seemed uncomfortable or nervous."

"During our conversations with him, he seemed happy and content; he smiled often and rarely complained."

Henry was, in other words, the perfect research subject. Docile, passive, uncomplaining. Researchers noted that if you asked him to sit somewhere, he would remain there indefinitely, that he would only speak when spoken to, that he would almost never complain of hunger or thirst or pain. Corkin tended to attribute this to some sort of innate tractability. "From what we know of Henry," she wrote, "he had always been an agreeable, passive person."

There was, however, another explanation for Henry's behavior, one which Corkin downplayed but other researchers did not. Simply put, primates that undergo bilateral medial temporal lobotomies similar to Henry's, such as the macaques that Klüver and Bucy operated on, always became tamer, more tractable, more docile. Most of the scientists who worked with Henry assumed there was a relationship between Henry's cooperativeness as a research participant and his operation. When I asked the neuroscientist Howard Eichenbaum whether he believed that Henry's brain lesions contributed to his passivity, he was certain of it: "What was removed along with his hippocampus and his cortical areas was the structure of his amygdala," he said. "And it's known that those kinds of lesions in animals make them very passive. And it's generally thought that his sort of passiveness, and maybe some of his other features, like his lack of emotionality and pain and hunger and so on, were all due to the amygdala."

Henry's passivity, then, was not surprising. What was surprising was the fact that sometimes he was not passive, that sometimes, despite his surgically created tameness, he would lash out. What explanation could there be for those outbursts, for those sudden storms?

To understand the answer, it's important first of all to point out that Henry did in fact often feel anxious and worried and unhappy while at MIT. During some of his long stays at the Clinical Research Center, researchers would administer tests that revealed high levels of internal strife. To take one example, on August 10, 1982, they presented Henry with something called the Beck Depression Inventory. It was a questionnaire, multiple choice. Here are a few of the statements Henry circled when asked to describe how he felt at that moment:

"I feel that the future is hopeless and that things cannot improve."

"I feel that I am a complete failure as a person."

"I am dissatisfied or bored with everything."

"I feel guilty all the time."

"I feel I may be punished."

"I am disappointed in myself."

Other questionnaires, given to Henry at other times, revealed a similar state of mind, such as one in which he was asked to document his internal feelings by ticking "yes" or "no" next to a series of descriptors. Henry ticked "yes" next to "enraged," "terrified," "frightening thoughts," and "cannot relax."

None of those personality questionnaire results have ever been published, although in one paper, from 1996, Suzanne Corkin did mention that Henry had been administered the questionnaires, and she summarized her interpretation of Henry's answers by saying that they provided "no evidence of anxiety" or depression. In general, Corkin tended to put the sunniest possible spin on Henry's condition, depicting him as a sort of avatar of enlightened contentment. "We can be so wrapped up in memories that we fail to live in the here and now," she once wrote in a discussion of Henry, adding that "Buddhism and other philosophies teach us that much of our suffering comes from our own thinking, particularly when we dwell in the past and in the future," and noting that "dedicated meditators spend years practicing being attentive to the present—something Henry could not help but do." She went on to speculate about "how liberating it might be to always experience life as it is right now, in the simplicity of a world bounded by thirty seconds."

Those questionnaires, by contrast, indicate that rather than feeling liberated, Henry, when he engaged in introspection, was sometimes anguished by his absent past, by his muddled present, by his unimaginable future. It's true he didn't often complain about that anguish, but those passive, uncomplaining tendencies are known by-products of the sorts of trenches my grandfather cut in his brain. As for why his outbursts always occurred while he was in Hartford rather than at the Clinical Research Center, bear in mind that whenever Henry was at MIT, he was in a place that was and would forever remain alien to him. He would have been kept occupied by a procession of strangers, plying him with an endless battery of stimulating tasks. The whole environment—a university bustling with scientists and brimming with sophisticated testing equipment and technology—was unlike anything he'd ever experienced in his preoperative life. It was a place stripped of the people and environments that connected Henry to his past, a place apart. While at MIT, Henry appears to have experienced feelings of worthlessness, and confusion, and hopelessness, but those feelings may have been less acute than in Connecticut, where Henry would have been constantly reminded of the eternal limbo to which my grandfather's operation had sentenced him.

Imagine what it was like for Henry to see his mother as she aged, to witness the relentless march of time across her face as her hair grayed and her wrinkles deepened. Each and every time he saw her, he would have to grapple with her instant transformation from the young woman he remembered to the older woman she had become, while grasping at the blank abyss of lost years that separated the two.

Or imagine what it was like for Henry during his workdays at the Hartford Regional Center for the Mentally Retarded, a smart man surrounded by strangers who weren't. He was doing work he was overqualified for—packing balloons, mounting key chains—but that his amnesia nevertheless made difficult. Every slippery moment he spent at the center would remind him of the terrible and mysterious fact that his life had come to a standstill.

It is impossible for anyone to ever know what it was really like to inhabit Henry's mind and to live in Henry's world. There is no evidence, however, to support the conclusion that it was anything like nirvana.

In 1978, Henry's mother's increasing dementia forced her into a nursing home. Henry remained behind at Lillian Herrick's home until Herrick herself became ill with cancer. In December 1980, Herrick decided she could no longer care for Henry, and she moved him into Bickford Health Care Center, a nursing home owned by one of Herrick's brothers. Henry was fifty-four years old, decades younger than most of the other patients there, and he still believed, of course, that he was decades younger than his actual age. Once again, he was in an environment that couldn't help reminding him of the fundamental vacuum at his core and the missing chapters of his life. And once again, he would have outbursts. He threw things, he yelled, he threatened to jump out of the window.

In 1982, he had a particularly violent episode, taking a poorly aimed swing at an employee of the nursing home, hitting a wall with his balled-up fist. Staff called the police, and two officers arrived on the scene. They decided not to make an arrest. Instead a nurse gave Henry antianxiety meds, and he eventually fell asleep. The following day, someone asked him what he recalled of the night before.

"I don't remember—that's my problem," he said. "Sometimes it's better not to remember."

TWENTY-SEVEN

IT IS NECESSARY TO GO TO NIAGARA TO SEE NIAGARA FALLS

The neuroanatomist Jacopo Annese usually drove a red Porsche 944, but on a summer day in 2006 he was a passenger in an unmemorable sedan, watching the redbrick husks of old paper mills glide by his window. He was on Main Street in Windsor Locks, Connecticut, a few blocks from Bickford Health Care Center, where he was going to meet Henry for the first time. Bickford Health Care Center was a single-level complex that used to be a motel. A little canal ran alongside it, as did some railroad tracks. It was west of a highway. Annese was surprised, to be honest, that Suzanne Corkin, who was driving, didn't make him wear a bag over his head. Even before he met her, Annese knew by reputation how protective of her prized test subject Corkin was, vetting researchers exhaustively, demanding the signing of nondisclosure contracts, disallowing tape recorders, that sort of thing. She'd built a good portion of her career on her access to Henry and wouldn't let just anybody in.

But when Annese requested that she set up this meeting, when he told her that he'd like to see Henry at least once while Henry was still alive, she consented.

The car pulled into the parking lot of the nursing home and nosed into an empty space. Annese and Corkin got out and walked inside together into the lobby. Annese got his first glimpse of Henry there: an old man, overweight, sitting in a wheelchair. Henry looked up at them with a dull expression on his face and no hint of recognition in his eyes. He responded to their greetings, but just barely. He was still on massive daily doses of anticonvulsants, as well as antipsychotics, anxiolytics, antidepressants, blood thinners, and various other medications including Xanax, Seroquel, Oleptro, Mellaril. He'd fallen and broken his ankle at least twice in the past two decades and had his hip replaced in 1986. His bones were brittle. He'd had a stroke two years before. Profoundly amnesic since 1953, Henry now also suffered from dementia, which brought with it a general blunting of his intellectual faculties, a blunting only exacerbated by all his medications. He had become what Corkin described as a "pharmacy in a wheelchair." Incidentally, Henry's increasing decrepitude had itself suggested some new experiments to Corkin. During another meeting, she'd quizzed Henry on how old he thought he was. He guessed that he was perhaps in his thirties. Then she handed him a mirror.

"What do you think about how you look?" she asked while he stared back at his own wrinkled, uncomprehending face.

"I'm not a boy," he said eventually.

Corkin and Annese wheeled him to the cafeteria, where Henry drank a smoothie from a straw. Annese was a voluble, extroverted man and tried to engage Henry in conversation, but Henry hardly responded. He sat and drank mostly in silence. Eventually somebody took the smoothie away, and Annese and Corkin got up and wheeled Henry back to his room. They said goodbye to him, speaking loudly and firmly to make sure he understood. His hearing was going, too. He'd suffered from severe tinnitus since 1986, probably a side effect of Dilantin, and it sometimes got so bad—a relentless, inescapable, excruciating buzzing sound at all hours of the day and night—that he would plead with the staff at the nursing home to bring him a gun so he could blow his brains out.

Henry's brain, of course, was what this visit was all about. It was why Corkin had allowed Annese to come here, why she had parted the veil. She needed a man of Annese's particular skills. Annese, for his part, was grateful to Corkin for letting him in, glad to get a chance to meet Henry and spend a little time around him. Ever since graduate school, he'd found anonymous cadavers the hardest to harvest. It made it much easier if you knew the person as a person before you dealt with the person as a corpse.

Scientists had been trying to get a look inside Henry's skull almost ever since the moment my grandfather replaced the bone plugs and sealed it back up. In the first three decades following the operation, Henry received numerous CT scans and X-rays, and although these different technologies each had individual strengths and weaknesses, none were entirely satisfactory. They provided the scientists what was, at best, a blurred view of Henry's brain, like looking through a lense smeared with Vaseline.

Magnetic resonance imaging, a technology developed in the late 1970s and widely used at MIT throughout the 1980s, promised a clearer picture, but Henry didn't receive his first MRI scan until 1992. The delay was due to a fear that putting Henry in an MRI machine might cause him serious injury or even kill him, owing to the fact that MRI machines work through the use of extremely powerful electromagnets, and my grandfather had left behind several metal clips inside Henry. These clips were used to pinch shut veins and membranes in Henry's brain and to help determine the depths of his lesions in postoperative X-rays. When a person walks near an MRI machine holding a set of keys, those keys might be wrenched away and rocket at high velocity toward the interior of the machine. If the clips inside Henry's head were magnetic, something similar might happen to them, with obviously catastrophic consequences. Even if the clips were not magnetic, metallic objects of all sorts heat up when placed within the intense fields of an MRI machine, and if Henry's clips became too hot, that, too, could cause problems.

On the other hand, there was a good chance that the clips were safe. A neurosurgeon who had worked with my grandfather told Corkin that he believed the clips my grandfather used at the time he performed Henry's operation came from a manufacturer named

Codman & Shurtleff, the same company that built my grandfather's custom trephines. A call to Codman & Shurtleff revealed that the clips they sold back in the early 1950s were typically made of either silver or tantalum, which are both nonmagnetic, and a review of journal articles indicated that nonmagnetic clips were unlikely to heat up to dangerous temperatures during MRI scans. There may have been a risk, then—it's hard to imagine there was not—but the team decided that if there was, it was one worth taking. So in May 1992, at Massachusetts General Hospital's Martinos Imaging Center, Henry lay down on a stretcher and was wheeled into the hollow, super-magnetized core of a 1.5-tesla MRI machine, where the steady thrum of rotating magnets filled his ears, loud enough to drown out the persistent buzzing of his tinnitus. He expressed no discomfort: The clips inside his head, apparently, were staying put and staying cool. Corkin would later say she hadn't thought there'd been a risk at all, even a slight one.

Behind a heavy door in an adjacent room, Corkin and her colleagues stared at a computer screen and watched with excitement as ghostly cross sections of Henry's brain began to appear, providing the best view anyone had had of it since my grandfather made his cuts four decades before. During those four decades, scientists performed hundreds of experiments with Henry, amassing what was already the most extensive amount of clinical data of any human research subject ever. The through line of all this research was an attempt to connect what they learned about Henry's amnesia and other deficits to what they knew about Henry's brain lesions. Finally they had a chance to see and measure those lesions rather than just depend on my grandfather's best guess about their size and scope or squint at the blurred images produced by more primitive neuroimaging technologies. After decades of measuring the effects that my grandfather's operation had on Henry's mind, they had their best opportunity yet to look at the lesions themselves, the causes of those effects. The biggest initial surprise of the scans, incidentally, was that the lesions appeared to be substantially *less* extensive than my grandfather had estimated in his postoperative report. Whereas he had described destroying all of

Henry's hippocampal structures bilaterally over a distance of approximately eight centimeters, the MRI images indicated that there were at least two centimeters of preserved hippocampal tissue on both sides, a little more in the left hemisphere than the right. The scans also revealed that Henry's cerebellum, the part of the brain sitting just above the stem and responsible for general motor skills and coordination, was shrunken and in poor shape, presumably a result of the high levels of antiepileptic drugs Henry had been taking during most of his life. In addition, my grandfather's suction catheter destroyed most of Henry's olfactory cortex, which might have explained his pronounced and extensively documented difficulty in distinguishing between different scents.

As excited as Corkin and her colleagues were to see these scans, however, they were also somewhat frustrated by them. MRI was the state of the art in neuroimaging technology at the time, but it had limitations. The resolution was still far from ideal: A human brain typically contained about 1 billion neurons, while an MRI scan image had a resolution of only about 65,000 pixels. Areas that encompassed entire constellations of neurons and axons and dendrites were reduced to a few pixels.

You could spend a half century testing somebody—examining, poking, prodding, feeding—and come up with all sorts of theories to explain your findings. You could even put that person in an MRI machine and study the blurred images that appeared on your screen. But the brain, nestled in its fortress of bone, doesn't give up its secrets easily.

Eventually, Corkin knew, to truly understand the dimensions and effects of the cuts that my grandfather made in Henry's brain, another cut would have to be made.

On November 13, 1992, six months after that first MRI scan, Henry received two visitors at Bickford. One of them, Edward McGuire, was an attorney. The other was a man named Thomas F. Mooney. The purpose of their meeting was to help McGuire determine whether Mooney was a good candidate to become Henry's conservator, a role that would give Mooney legal control over Henry's medical care, among other things. A conservator is required when people have a debilitating condition such as dementia, schizophrenia, mental retardation, or, as in Henry's case, profound amnesia, which makes it impossible for them to make informed decisions regarding their own well-being. In most such cases, a parent acts as a de facto conservator. Henry's father had died decades before, and his mother had died in 1980. For the twelve years prior to this meeting, Henry was without a conservator or legal guardian of any sort. Suzanne Corkin, as well as some of the lawyers at MIT, were concerned about this.

Central to their concerns was the principle of informed consent. This was a principle that had evolved slowly over the course of the twentieth century, in fits and starts, often spurred forward by horrors such as those documented by the Nuremberg trials of the 1940s or the revelations about the Tuskegee Syphilis Experiment in the 1960s. By 1992, there was a clear consensus, not to mention a legal mandate, that required scientists who worked with human subjects to ensure that those subjects fully understood and agreed to any experiments they took part in. When research subjects had a condition that made providing their own consent difficult, that consent had to be obtained from a parent, conservator, or other legal guardian.

During the first two decades of experiments with Henry, consent was usually provided by Henry's mother, Elizabeth, or his father, Gustave, with Henry co-signing. By 1974, Gustave was dead and Elizabeth had dementia, so Lillian Herrick, the woman who'd been boarding Elizabeth and Henry at her home, started signing Henry's consent forms whenever she dropped him off at MIT. This unusual arrangement—Herrick was not Henry's conservator but, in a sense, his landlady—continued until Henry moved out of Herrick's home and into the Bickford center in 1980. From 1980 until 1992, the only person who signed Henry's consent forms was Henry himself.

Was Henry, on his own, capable of providing informed consent?

On the one hand, he was a bright individual, with an above-average IQ and intact reasoning abilities. On the other, the missing parts of Henry's brain affected him in ways that many experts would say made true consent impossible. For one thing, Henry could only hold on to the present moment for extremely short periods of time. This meant that a researcher attempting to gain consent from Henry faced the challenge of introducing herself, introducing the proposed experiment, explaining its methods, elucidating any possible risks, explaining the concept of informed consent, and then getting him to sign the consent form before any of that information flew out of his head. In addition, as soon as Henry signed the form he would forget everything about it, including the fact that he had consented at all, which meant that he would, some would argue, need to continually reconsent, over and over, as the experiment unfolded.

When it came to informed consent, however, an even more fundamental problem than Henry's amnesia was his passivity. Despite his occasional outbursts, Henry was, as Corkin and other researchers described him, almost always a docile and tractable man, and this docility and tractability surely was a result of his brain lesions, specifically his missing amygdala. My grandfather's operation had made Henry neurologically predisposed to consent to anything.

Many of the researchers who worked with Henry at MIT didn't appear to know much about what process was followed to obtain informed consent in his case. Instead they simply trusted that Corkin, the principal investigator, had made sure to follow proper protocol. "I'm sure Sue did," the psychologist Nancy Hebben told me. "But I didn't have to specifically do something. There must have been a consent form of some sort." Hebben was in charge of running the tests in the early 1980s on Henry's pain thresholds, tests that left little mysterious burns on his chest and forearms that Henry would later ask his nurses to explain. When I asked if she believed that Henry was capable of giving informed consent for those experiments, or any others for that matter, Hebben shook her head. "My guess is it would not have been Henry signing. How could you consider that informed consent? Because he wouldn't remember. I mean, he could understand. He wasn't a stupid man. So you could explain it and he could say, Oh yeah, I agree to do that. But then he wouldn't remember that he had agreed to do it. . . . I mean, how could he possibly give informed consent?" Hebben assumed Corkin had arranged for somebody else to provide consent on Henry's behalf, but she was wrong. For at least a decade, including the period when Hebben did most of her work with him, Henry was the only person signing his consent forms.

By 1992, as Corkin described it later, Henry's "difficulty in retaining new information raised the nagging question of how we obtained informed consent." In July of that year, Peter Reich, the chief of the psychiatry department at MIT, was asked by Corkin to interview Henry, and afterward Reich wrote a reassuring letter in which he said that Henry "appeared to understand the nature of the procedures he has been undergoing as a CRC subject and expressed his willingness to undergo these procedures." This was not enough, however, to allay everyone's concerns. The standards governing human research, not to mention governing the funding of human research, had changed considerably in the decades since "the MIT Research Project Known as the Amnesic Patient H.M." began. By the time Reich interviewed Henry, Corkin had already initiated the search for a conservator, someone who could provide the rock-solid informed consent that Henry himself could not.

Thomas Mooney was the second person to apply. A year earlier, in 1991, Corkin had arranged for a physician named John J. Kennedy to submit his conservatorship application to the Windsor Locks Probate Court, which oversaw such matters. Kennedy was on staff at the Bickford Health Care Center and was Henry's personal physician there. The application required that Kennedy describe the "mental, emotional and/or physical condition which prevents respondent from performing necessary and proper functions for his or her well-being," and Kennedy wrote that Henry "has had a lobotomy and is presently mentally unable to perform daily functions properly." The application also asked Kennedy to list Henry's closest relatives as well as any other "interested parties." In Connecticut's probate practice book, an interested party is defined as any "person having a legal or financial interest" in the proposed conservatorship. Apart from himself, Kennedy listed only one such party: Suzanne Corkin. He also wrote that Henry had "no close relatives." After submitting his application, Kennedy received a confidential memo from the probate court, and shortly after that he withdrew his application. It's unclear why, though his conservatorship might have been legally problematic at best, since a clear conflict of interest exists when a person's physician is given complete authority over the patient's healthcare decisions.

A year later, at Corkin's request, Thomas Mooney—the son of Henry's old landlady Lillian Herrick—submitted his application. It was drafted on the same boilerplate form as Kennedy's had been, though some of the information was different. Henry's debility was described in greater detail as follows: "Henry G. Molaison has a neurological condition known as global amnesia, a severe memory impairment that prevents him from being mentally able to care for himself." Also, while Kennedy's application claimed Henry had no close relatives, Mooney's application claimed that he himself was Henry's cousin. One thing remained the same in both applications, however: Suzanne Corkin was listed as the only "interested party" apart from the proposed conservator himself.

During the meeting between Henry, Mooney, and the attorney McGuire, Henry charmed McGuire. "I was impressed with Mr. Molaison's congeniality and gentleness, his sense of humor and obvious native intelligence," McGuire wrote afterward. Corkin had briefed McGuire extensively on Henry's case prior to the meeting, speaking with him on the telephone and sending him a letter and a copy of a recent journal article, so McGuire was aware that Henry was, as McGuire wrote, "a unique person. He is apparently the most thoroughly studied and described neurological patient in the world." McGuire remarked that during the meeting, Henry "spoke clearly of recollections from his youth, including his schooling, places where his family had lived, the injury when he was 7 years old which may have, at least in part, precipitated his epilepsy, and his work experience. His particular amnesia was also very evident, as he repeated his recollections several times, obviously not recalling that he had just said the same thing. However, each time he repeated a recollection, the details were the same. Mr. Molaison conversed easily and comfortably. When I asked him if he would like Tom Mooney to have the responsibility of making sure that he is well taken care of, his reply was that he would be glad to have the help, as long as it would not be too burdensome for Tom."

In paperwork filed with the court, Mooney was referred to alternately as Henry's cousin and his nephew. Henry, when asked by McGuire, said he thought Mooney might be his second or third cousin. In fact, if Mooney and Henry shared blood ties, they were thin ones, undetectable even by following Henry's family tree for several generations. Nevertheless, from McGuire's account of the meeting, it's clear he had been led to believe that Mooney was, as he put it, "Mr. Molaison's only known living relative." This was an important point, one that would make Mooney a suitable candidate for conservatorship. A person's next of kin, after all, might be expected to take a sincere interest in that person's well-being. The problem was, it wasn't true, although it's unclear whether Mooney realised this or not. At the time of their meeting, Henry had at least three first cousins—Frank Molaison, Marjorie Ramsdorf, and Myra Crowley—living in Connecticut, all more closely related to Henry than Mooney. None of those cousins had been contacted.

At one point during the meeting, Henry told one of his oft-repeated stories about a childhood trip he'd made to Buffalo. Mooney asked him if he'd seen Niagara Falls during his Buffalo trip.

"No," Henry said, "Niagara Falls cannot be seen from Buffalo. It is necessary to go to Niagara to see Niagara Falls."

Mooney and McGuire laughed, and four days later, on November 17, 1992, the Windsor Locks Probate Court officially appointed Tom Mooney as Henry's conservator.

. . .

Mooney stayed Henry's conservator for the rest of Henry's life. It was not, as Henry had fretted it might be, an overly burdensome position for Mooney: All of Henry's day-to-day needs were handled by the nursing home, and Mooney had few responsibilities and did not appear to go one step beyond what was required. Indeed, when the probate court conducted a mandated follow-up review of Mooney's conservatorship four years later, in 1996, an attorney named Mary T. Bergamini submitted the following report:

"Mr. Molaison's condition has not changed in any significant manner. He is currently a resident at Bickford Convalescent Home. He has been a recipient of Title 19 benefits as of 1992 and Bickford currently receives both his Social Security and State payment directly. He is given $30.00 per month on account for his personal needs. The records indicate that Thomas F. Mooney was appointed Conservator on November 17, 1992. I attempted to contact Mr. Mooney without success. The nursing home has indicated to me that they also have had no contact with Mr. Mooney. However, it appears that Mr. Molaison's financial and personal affairs are cared for by the administration at Bickford. There is certainly the need for a continued Conservatorship of Mr. Molaison. The question appears to be whether or not Mr. Mooney should continue in this capacity. I have no objection as the respondent's current situation is stable; that is both his personal and financial affairs are taken care of." After a brief hearing on the matter, a probate court judge agreed to Mooney remaining Henry's conservator, despite Mooney's lack of involvement in Henry's life.

In one important respect, however, Mooney did remain diligent: From the beginning of his conservatorship until the end of Henry's life, he never failed to provide consent to Suzanne Corkin for the continuation of "the MIT Research Project Known as the Amnesic Patient H.M." Whatever experiments Corkin and her colleagues wished to conduct, Mooney always signed the consents. Likewise, when the question arose of whether scientists could continue to experiment on Henry *after* his death, Mooney raised no objections. In fact, just one month after becoming Henry's conservator, Mooney returned to Henry's nursing home for another three-way meeting. This time, the third party was Suzanne Corkin. She explained to Mooney that now that he was Henry's conservator, he could legally authorize, in advance, the posthumous donation of Henry's brain to MIT and Massachusetts General Hospital. She'd brought along all the necessary paperwork, including an "Authorization for Brain Autopsy" form. At the bottom of the form were the following lines:

> I, Thomas F. Mooney, am the court-appointed guardian of the person of Henry G. Molaison. I also presently am Henry G. Molaison's closest living next-of-kin, and as such I am entitled by law to control Henry G. Molaison's remains upon his death. I hereby authorize the removal, retention and use of a whole brain specimen for diagnositic and/or research purposes by the Massachusetts General Hospital and the Massachusetts Institute of Technology's Department of Brain and Cognitive Sciences (or other agents), in the interest of determining Henry G. Molaison's cause of death and of advancing medical knowledge.

Mooney signed.

In 2002, Suzanne Corkin hosted a dinner party at her apartment in Boston's Charlestown neighborhood for the first of a series of meetings to determine precisely what should be done with Henry's brain after he died. Although Corkin was a neuropsychologist, she was not accustomed to working with actual brains. The wet work of neuroscience, the examination and analysis of those mysterious three-pound engines in our skulls, was not her area of expertise. The group she assembled that night was meant to address this deficiency. Present were neuropathologists, systems neuroscientists, and neuroanatomists, all more experienced than Corkin at working with gray matter itself rather than just analyzing the behavior the gray matter produced.

Some of the subjects the group debated were very specific. For example, it was clear that Henry's brain should be scanned again in an MRI machine after his death, but the precise number of scans, and the power of the machines used to scan it, was a matter of contention. The strength of the magnetic fields in MRI machines is measured by units known as teslas, and the machines available to Corkin ranged in power from one to seven teslas. While the higher-tesla machines were capable of producing more detailed imagery, they also produced greater amounts of heat in the tissues they were scanning. The team planned to scan Henry's brain for many hours to obtain the most accurate images possible, but they also wanted to make sure they didn't heat it so much that its physical integrity was damaged. At the same time, they wanted to scan it fast while it was still in Henry's skull, to be able to harvest the brain before any decay had begun to set in. There were clearly going to be trade-offs.

Other issues were more fundamental. Chief among them was who would be put in charge of the most important part of the project: the postmortem, post-harvesting, post-MRI processing and preservation and analysis of Henry's brain. Corkin and her group had debated the pros and cons of various approaches and the merits of various researchers. Once he died, she knew she would have to give up at least partial control, that she would have to relinquish the most valuable part of Henry into somebody else's hands. The committee needed to find somebody worthy of that precious cargo.

Jacopo Annese, who had flown in from California for that dinner party, eventually became the committee's first choice. Annese was relatively young, in his early forties, and in the early stages of a promising career as a neuroanatomist. He'd begun that career at the University of Florence in Italy, then continued it at NYU, McGill, and UCLA before he finally settled at the University of California, San Diego. While at UCSD, Annese honed his innovative techniques for preserving human brains in both histological and digital form, a technique that promised to give scientists the opportunity to continue their research with Patient H.M. long past Henry's death.

It was after the committee selected Annese that he asked Corkin if he might meet Henry while he was still alive, which led to that 2006 visit to his nursing home, where they ate their mostly silent lunch and then wheeled Henry back to his room. On the way out, Annese noticed a snapshot of Henry tacked to one of the bulletin boards near the entrance. Nobody was looking, and he had to resist the temptation to take it, to slip it into his pocket, to keep it as a sort of totem, something he could ponder in his off hours and use to help him imagine his way further into Henry's mind before the day he had to start digging into his brain.

TWENTY-EIGHT

PATIENT H.M. (1953–2008)

New employees at the Bickford Health Care Center always received a briefing on Henry and his special circumstances. For example, they were directed never to speak to anyone outside the center about Henry, as the fact that he resided there was a closely guarded secret. If a stranger called inquiring after Henry, the staff member receiving the call was supposed to give a noncommittal response, neither confirming nor denying his presence, and then immediately phone Henry's conservator, warning him about the snoop. The cloak of anonymity placed over Henry was effective: He had lived at the center for decades, and though he was the most famous patient in the history of neuroscience, no outsider ever found him.

New employees were also briefed on the special rules that applied specifically to Henry's dying and his death. Suzanne Corkin drafted these rules, and they were printed out and always attached to Henry's chart.

So on the morning and afternoon of December 2, 2008, as Henry began fading from respiratory failure at age eighty-two, Corkin, as per protocol, received periodic phone calls keeping her abreast of the situation. Corkin had last seen Henry a month prior, and by that time his dementia was profound, and he had become completely mute. He wouldn't answer any of Corkin's questions, just stared at her, blank and uncomprehending. His value as a useful living research subject had come to an end.

When Henry's heart finally stopped, another call was placed to Corkin, and then someone at the center rushed to the freezer and dug out the flexible Cryopaks that Corkin had ordered placed there in anticipation of this moment. By the time the hearse arrived, the ice blankets were wrapped securely around Henry's head, keeping his brain chilled to slow decomposition.

Everything went smoothly, according to plan, and a couple of hours and 106 miles later, the hearse pulled into the parking lot of building 149 in the Charlestown Navy Yard, the Athinoula A. Martinos Center for Biomedical Imaging in Boston, where Corkin was waiting. The body bag was unzipped and the ice blankets unwrapped. Corkin had known Henry for forty-six years, met him for the first time when she was still a graduate student in Brenda Milner's lab at McGill. Of course, her relationship with Henry, transactionally speaking, was that she'd known Henry, not the other way around. Forty-six years of meeting someone for the first time, introducing herself to an old friend.

And now this last meeting that only she would remember.

During the night that followed, Corkin watched as Henry underwent a series of high-resolution MRI scans. Then, the next morning, she attended the harvesting. She stood on a chair outside the autopsy room in the Mass General pathology department and peeked through a window as a neuropathologist named Matthew Frosch, assisted by Jacopo Annese, who had flown in on the red-eye from San Diego, sawed off the top of Henry's skull and, with the care of obstetricians delivering a baby, pulled his brain into the light. Corkin had spent the bulk of her career pondering the inner workings of Henry's brain. That morning, she finally got to see it. Henry, a man she had known for almost a half century, had died the day before, but there, gleaming under the bright lights, was the part of him that she'd always been most interested in. As she gazed at Henry's brain, only one word could describe the feelings that coursed through her. She was, she later wrote, "ecstatic."

After the harvesting, the brain sat for a while in a bucket that was inside a cooler, steeping in a preservative solution, hanging upside down, suspended by a piece of kitchen twine looped through its basilar artery. When the brain was firm enough to travel safely, Corkin rode to Logan International Airport with the cooler. She accompanied it to the gate of a JetBlue flight from Boston to San Diego. There were camera people following her. It was a self-consciously historic moment. Henry's death had been announced on the front page of *The New York Times,* which described him as "the most important patient in the history of brain science" and revealed Patient H.M.'s real name to the world. Corkin already had a book and movie deal. She put the cooler down near the gate, and Annese picked it up. She watched him walk down the ramp with it and disappear into the plane.

It was hard to let go.

Henry's brain sat on a small rectangle of formaldehyde-slick green marble, under a sheet of plastic wrap. Annese was wearing a medical smock, blue rubber gloves, and safety glasses. He plucked off the plastic wrap, picked up a scalpel, and began to peel away the pia mater, a thin, sticky membrane that covers the brain. He was alone, music on—it was a Beatles kind of night—and everything went perfectly. He removed the oxidized clips my grandfather had left behind fifty-five years before, set them aside. Then the membranes, the blood vessels, all the obstructing tissue, stripping everything away until he was left, finally, with Henry's naked brain. His peelings usually lasted three to four hours, but with Henry he took his time, made sure everything was just right. He peeled for five hours straight.

He was still on a bit of a high, still pinching himself. He was part of the group that Corkin had convened years before to decide what to do with Henry's brain postmortem, and so of course he had known for a while that the group had decided to give the brain to him as the cornerstone of his Brain Observatory. But it was still a shock to actually have it in his possession. This prize, this valuable artifact, this revolutionary brain. Despite all the planning, all the verbal agreements, he hadn't been absolutely sure he'd wind up with it until he boarded the flight with the cooler in hand. A part of him, the fatalistic Italian part, thought that something would happen at the last minute, that Corkin would change her mind, take Henry back.

But she hadn't.

He presented the gate agent at Logan with two tickets, boarded the plane with time to spare. He gave Henry the window seat. He'd penned words on the cooler with a black Sharpie, along with his phone numbers and email address: "Diagnostic specimen. Fragile. If found, please do not open. Contact Dr. Annese immediately." He landed in San Diego, and a UCSD official escorted him and Henry straight to his lab.

So far, everything seemed to be going just as smoothly as could be. After the peeling, he embedded Henry's naked brain in gelatin, froze it solid. The freezing had to be done quickly, using a liquid-cooled, custom-built device steaming with dry ice vapors, like a witch's cauldron. The embedding was, he's not too modest to say, a masterpiece.

But it was only the prelude to what came next.

For the fifty-five years following his operation, Henry had lived mostly in seclusion and anonymity. During most of that period, a select coterie of scientists had shielded him from the prying eyes of the outside world.

That was about to change.

I was sitting at a high table at Alpine Bakery, in Whitehorse, the capital city of Canada's Yukon Territory, eating a scone and drinking coffee. It was one of those cold, clear subarctic December days, a time of year when the sun barely nudges itself above the mountains, spraying weak light through the trees, giving everything a shadowy purplish tint. Outside, people with cold red faces hurried down Main Street, their bodies swaddled in expensive garments made from goose feathers and lamb hair. Inside it was warm and homey. I cradled my coffee in one hand, cracked open my laptop, and logged into my email.

I'd been living in the Yukon for more than three years. I moved up there from Atlanta shortly before my daughter, Anwyn, was born at Whitehorse General Hospital in September 2006. Long story. Her mom and I had known each other almost our entire lives, ever since we became playmates as expatriate toddlers together in Mexico City, but we'd been out of touch for most of our adult lives. We reconnected during a tumultuous trip to Ecuador in the fall of 2005, a trip that started out platonic and ended with us becoming trapped together for three days in a little oil town on the Colombian border called Lago Agrio (Bitter Lake). Anti-petroleum-industry protesters had taken over the town the day we walked into it, and they sealed off all access, dismantling bridges and barricading the airport. Eventually the army was called in to handle the situation, which made it much worse. Molotov cocktails, tear gas, the acrid smell of burning banks. A shared hotel room. Our friendship turned into something else.

We saw each other again a couple of months later, when she flew down to Mexico over Christmas break while I was on assignment there. A few weeks after she left, I called her from a pay phone in Catemaco, a small town near the city of Veracruz famed for both its lake snails and its history of *brujería,* witchcraft. I had a digital camera with me, and when she told me she was pregnant I snapped a picture of myself with the phone pressed to my ear, wanting to document a moment when I knew my life had just changed forever.

She was living in the Yukon. I was living in Atlanta.

One of us had to move. My job was portable.

We lived together for two years after Anwyn was born, and then we didn't. I moved into a small place of my own. I had a steady stream of assignments and commuted for my stories, usually a long way. I went to Antarctica and ran a marathon. I went to Laredo, Texas, and interviewed a teenage cartel hit man. I went to Wasilla, Alaska, and got frostbite snowmobiling with Todd Palin. That December afternoon, I'd recently returned from Jamaica, where I spent a week with Usain Bolt, the fastest man in history, who spent most of our time together eating junk food and playing *Call of Duty* and practicing his turntable skills, leaving the impression that everything I'd ever heard about how success is the result of hard work and dedication might simply be untrue.

I'd kept tabs on Henry's story over the years, reading the new papers that came out about him. When he died, and the veil was dropped from his name, I waded into the flood of published tributes. Eventually the tributes slowed to a trickle. I set up a Google Alert for "Henry Molaison," and in the months that followed his death it would only periodically ping me, advising me of some new Henry-related tidbit.

That afternoon in the coffee shop, however, Google had flooded my in-box with stories. Most of them contained a link to a UCSD-hosted website for the Brain Observatory. I clicked on it. It took a few moments to open. In the center of my browser, the throttled Yukon bandwidth was churning out a pixelated and choppy live stream video. The video was dark, so I amped the brightness on my screen. There was a white square in the center of the video, taking up most of the space. Around the square was a misty fog of billowing dry-ice smoke, like something you'd see during a magic show. In the center of the white square was a rough oval blob, and it was pink. A silver metal bar moved slowly across the square, from the bottom of the screen toward the top, and as it moved, the pink blob curled up ahead of it, crumpling and furrowing unpredictably, like the breaking edge of a wave, but in super-slow motion. When the bar approached the top of the block and had run the entire length of the pink blob, a human hand would appear, wearing stylish purple medical gloves and holding a paintbrush. The paintbrush dabbed at the pink stuff, disengaging it from the white block completely and then lifting it into an individually numbered tray filled with some sort of clear solution. Then the bar returned to the bottom of the block and started again.

Despite all my efforts years before, I had never met Henry while he was alive.

But now, with the click of a trackpad, I, along with hundreds of thousands of other people around the world, was getting a live view of his brain's dissection.

Just like that, I was back on the Henry beat.

I sat and drank coffee for a while longer, watching Henry's brain being slowly sliced apart. The video stream was quiet, but occasionally the purple-gloved hands would put a little yellow Post-it note on a stick held in front of the camera, giving some sort of tidbit or trivia or shout-out.

"Listening to the White Album now," one of the notes said. Behind it, the dissection continued.

When I first visited the Brain Observatory shortly after the cutting was completed, there was a slim book sitting on a shelf right next to Jacopo Annese's desk in his glass-walled office. Unlike a lot of the other books in this place—*A Study of Error, Serial Murder Syndrome, Man and Society in Calamity, Flesh in the Age of Reason, The Open and Closed Mind*—this one didn't have a very lively title. But *Localisation in the Cerebral Cortex,* by Korbinian Brodmann, is to Annese as vital a book as can be. It was originally published in 1909 and contains a series of meticulously hand-drawn maps of the human brain, divided into fifty-two so-called Brodmann areas, each unique in its neuronal organization and, consequently, its function. Brodmann gleaned the borders of his areas through a rough and painstaking combination of microscopy and histology, and he did a great job, all things considered. Out of an uncharted cerebral wilderness, Brodmann created an enduring Rand McNally road atlas of the mind, one that my grandfather used to direct his surgeries and that most neuroscientists and neurosurgeons still use today.

As a fellow anatomist, Annese admired Brodmann's work immensely and had even written a glowing tribute to him that appeared in the journal *Nature.* But he hoped to make Brodmann's old maps irrelevant.

That's what the Brain Observatory was all about.

If Korbinian Brodmann created the mind's Rand McNally, then Jacopo Annese was creating its Google Maps.

A short walk from Annese's office, past an imported espresso machine and through a secure, airtight door, was the wet lab. At the far end of the lab, a number of tall, glass-fronted refrigerators stood against a wall. Many of them contained plastic buckets, and though the plastic was murkier than the glass, it was still possible to see what was inside. Most of the brains were human, but there was one from a dolphin. The dolphin brain was huge, significantly bigger than any of the human ones, though Annese cautioned that it would be a mistake to read too much into size.

What's true of individual brains is true of brain collections as well. With his Brain Observatory, Annese was setting out to create not the world's largest but the world's most useful collection of brains. Each specimen would, through a proprietary process developed by Annese, be preserved in histological and digital form, at an unprecedented, neuronal level of resolution. Unlike Brodmann's hand-drawn sketches, Annese's maps would be three-dimensional and fully scalable, allowing neuroscientists to zoom in from an overhead view of the hundred-billion-neuron forest all the way down to whatever intriguing thicket they liked. And though each brain is by definition unique, the idea was that as more and more brains came online, the commonalities and differences between them would become increasingly apparent, allowing, Annese hoped, for the eventual synthesis of the holy grail of any neuroanatomist: a modern, multidimensional atlas of the human mind, one that conclusively maps form to function. For the first time, we'd be able to meaningfully compare large numbers of brains, perhaps finally understanding why one brain might be less empathetic or better at calculus or more likely to develop Alzheimer's than another. The Brain Observatory promised to revolutionize our understanding of how these three-pound hunks of tissue inside our skulls do what they do—which means, of course, that it promised to revolutionize our understanding of ourselves.

And what could be a better cornerstone for the Brain Observatory, a better volume for Annese's collection, than the brain of Patient H.M.? The boxes filled with the cryogenic vials containing the slices of Henry's brain sat in their own freezers, to the left of the others, under lock and key. Precious cargo. San Diego is earthquake-prone, but there were backup generators and sensors that would automatically dial Annese's home and cellphone in the event of an emergency, so that wherever he was, he could jump into his Porsche, rush over, protect Henry.

Henry, just by being Henry, was helping bring Annese's larger ambitions closer to reality. People who'd read newspaper articles about Annese's work with Henry's brain had already called him up and made arrangements to donate their own. One of them, Bette Ferguson, a feisty ninety-one-year-old who was one of the original flying monkeys in *The Wizard of Oz,* would be dropping by the Brain Observatory soon to get her second set of MRI scans. Annese knew the publicity would continue, hoped it would continue to inspire donation. He had wanted to get the brain of the guy *Rain Man* was based on, but that didn't work out. Eventually he wanted to get somebody really big, a household name, Bill Clinton, someone like that.

But Henry's brain was more than just an attention-garnering curiosity. And it was more than just a proof of concept, something Annese could use to demonstrate to the world the power of his methods.

It was an object—2,401 objects now—that contained enduring mysteries still waiting to be solved.

The cutting was just the beginning.

Within a matter of months, Annese was planning to release a three-dimensional surface model of Henry's brain, built from the 2,401 high-resolution "block-face" images taken during the slicing. Those images captured the view of Henry's frozen, embedded brain just prior to each pass of the blade. This model would be at least ten times as detailed as anything one could possibly produce with an MRI machine and had the additional benefit of being derived from images of the actual brain rather than a computerized interpretation of it. And then, bit by bit, he planned to supplement that model with imagery of even greater resolution: A custom-built microscope scanner would digitize each of the mounted, stained slides at such a level of magnification that single neurons would be clearly visible. All of this, the resulting petabyte or two, would be accessible for free online, to researchers worldwide. Over the past fifty-five years of his life, Henry was hidden away while a select coterie of scientists gathered more data about him, his abilities and his deficits, than about any human in history. Now, after his death, Annese was poised to release Henry's brain into the wilds of the Internet, and the whole world would be able to reillumine that unprecedented volume of clinical data in the light of an unprecedented neuroanatomical map.

One of Annese's assistants poked her head into his office, told him that some more slides were ready for staining. A few minutes later, Annese held one of the fresh-dipped seven-by-five-inch slides up to the light, letting a purplish dye drip off the glass. The dye had adhered to the slide's cross section of pale, almost invisible brain tissue, darkening it, developing it like a photograph.

A cross section of brain looks a lot like an inkblot, a Rorschach, and this one at first glance gave the impression of the head of a vaguely sinister goat. But then Annese started guiding me through it.

"You can see here," he said, indicating a spot where the tissue looked darker, the neurons more cramped, "where your grandfather pushed up his frontal lobes."

We looked at another slide, and he pointed to an area that would have sat a little below and back from Henry's frontal lobes, a portion of the slide where no dye stuck, since there was no tissue for it to adhere to. It was a part of the lesion itself, the little bit of nothing that spawned everything. Though Annese didn't want to go into too much detail—not before his findings were officially published—he told me, sotto voce, that he'd already discovered some surprising new things about what my grandfather destroyed in Henry's brain and what he spared. For years, memory researchers assumed that the hippocampal stump that remained in Henry's brain was completely atrophic and nonfunctioning. According to Annese, however, that didn't seem to be the case. The little that was left of Henry's hippocampus looked like it was in pretty good shape, actually.

According to Annese, this was the sort of revelation that could shake up the field of memory science yet again. In 1953, when my grandfather closed a door in Henry's mind, did he leave it open just a crack? Did this explain the surprising exceptions to Henry's amnesia? So much of our understanding of how memory works is based on our understanding of how Henry's memory *didn't* work. But have we been misunderstanding him, at least in part, all these years? These were the sorts of questions scientists would grapple with and argue over in the years to come as the Brain Observatory went online, as Henry's mind was preserved everywhere and nowhere at once, as his cells were counted and his final mysteries came to light.

Annese put the slide on a rack to dry, and I looked at it again, the blank spot near the middle, the hole you could see right through.

TWENTY-NINE

THE SMELL OF BONE DUST

One night, half past midnight, half-tipsy on pink champagne, I stood beside Jacopo Annese and watched him bring a drill down onto a man's exposed skull. The sound of high-rpm machinery filled my ears, the smell of bone dust wafted up to my nostrils, and with a little ecphoric jolt I was reminded of being in a dentist's office midfilling. I was wearing my best shoes, and I stepped back from the body to avoid the splatter. I hadn't expected to be there that night, doing what we were doing.

Earlier in the evening, I'd attended a fundraising cocktail party for Annese's new nonprofit, the Institute for Brain and Society. Annese had conceived of the institute as a sort of complement to the Brain Observatory, one that would engage the public through educational outreach programs and museum exhibits and maybe even a café and gift shop. The money generated by the nonprofit, he said, would then be cycled back into funding the research conducted at the observatory. The party took place at a fancy condominium near Balboa Park, and there were maybe a hundred people in attendance, mingling and munching on "brain-healthy" appetizers, lots of fish and folic acid. Just past the condo's entrance, in the front hallway, framed portraits of several of Jacopo's "donors"—the men and women who had agreed to give him their brains—hung on the wall. Some of the donors in the portraits had already passed away, but others, like the nonagenarian Bette Ferguson, were still alive. In fact, when I made my way past the portraits and into the central room of the party, I spotted her sitting in a corner, on a low bench by a window.

"Bette!"

She looked toward my voice, her eyes sort of unfocused.

"It's Luke," I said, "the writer."

"Luke!" She grabbed my hand. "Sit down, sit down."

I'd met Ferguson two years earlier, during my first reporting trip to San Diego. Annese brought me along to her apartment while he conducted one of his periodic interviews with her. A key aspect of Annese's work, and his collection, was the accumulation of as much premortem data about his donors as possible. This data ranged from childhood stories to IQ tests, from MRI scans to the chronicles of their marriages and divorces. So during that first visit I learned a variety of things about Ferguson's past and present. I learned that she collected angel figurines, that she'd been married five times, and that she'd grown up near Los Angeles, where her mother worked for a while as a waitress in a café on the Paramount lot. I learned that when she was fourteen, she auditioned for, and got, a part as one of the flying monkeys in *The Wizard of Oz.* I learned that she was proudly Irish, though she'd never been to Ireland. I learned that she wasn't religious, strictly speaking, but that she believed in something beyond life. When she spoke of death, she didn't use that word. "Graduation" is what she called it.

We'd seen each other a few other times since then, once in Annese's laboratory, where I attended one of her MRI scans, and twice more when she took me to her favorite BBQ joint, a place called Phil's. She was friends with Phil, and everyone there knew her and would whisk her right to a table no matter how long the line was outside.

The last time we'd spoken was on the telephone, a few weeks before the party. She asked how my daughter, Anwyn, was doing, and I told her that she'd just played a munchkin in her school's production of *The Wizard of Oz* and that she'd been excited when I bragged that I was acquainted with one of the real-life flying monkeys. Ferguson suggested I buy Anwyn a kids' book written by Rush Limbaugh called

Rush Revere and the Brave Pilgrims, and I told her I'd think about it. At the end of the call, I mentioned I might be coming to San Diego for Annese's party, and she said she was looking forward to being there herself, if she hadn't graduated beforehand.

At the party it was clear that her eyes had gotten bad and that people had become walking blurs. Her mind, though, seemed as sharp as ever. When a large guy stood nearby shoveling canapés into his mouth, Ferguson leaned toward me and said in a too-loud voice, "Now, there's a fat man right over there, right?" then laughed, aware she was being inappropriate and clearly not giving a damn.

While she and I caught up, sipping pink champagne, Annese was too busy to join us. He was making the rounds, talking to as many attendees as possible, trying to convince them to donate either their money or their brains. At about ten-thirty P.M., he came over and pulled me aside.

"Something's come up," he said, and explained that one of his donors had died, that the body was about to be delivered to his lab, and that he was going to have to do the harvesting that night. He didn't want to leave the party early, since he'd been planning it for months, but there wasn't any way around it: The brain needed to be removed, stat.

Then a thought occurred to him.

"Do you want to come along?"

When we arrived at the Brain Observatory that night, Jack was already there waiting for us. I suppose you could say someone else was waiting for us, too, though he was lying in a red-zippered bag on a stretcher in the back of Jack's minivan.

Jack was big and friendly, jeans and a Padres jersey, a warm smile and a lazy eye. This was his job, delivering bodies. To the coroner or the medical examiner or the funeral parlor or wherever. His minivan was beige and unremarkable, had no logo or anything else on the outside that would give you a hint as to its contents. He made sure the air-conditioning was always in tip-top shape. Annese opened a rear door to the lab and helped Jack negotiate a tight corner with the stretcher. They rolled the body down a hallway, then turned left into the MRI room. A clock on the wall said it was eleven P.M. Annese asked Jack when he'd picked up the body from the hospice, and Jack said he'd had him for about an hour. He'd died that morning at about ten A.M., so that meant he was in cold storage for about twelve hours before Jack picked him up, not frozen but close to it, and had only had an hour to thaw. If the body was too cold it could interfere with the MRI results. They unzipped the red canvas body bag. Inside, a man was swaddled from head to toe in a baby-blue sheet. They each gripped a handful of sheet and lifted him off of Jack's stretcher and onto a second stretcher that was built entirely out of plastic and nonferrous metals so it could be used inside the MRI machine. Annese placed a hand gently on the forehead through the sheet, like a parent taking a child's temperature, and decided that he was warm enough for the scanning.

The MRI room was actually two rooms, the first a sort of control center, with the computer used to operate the machine, and the second containing the machine itself. Annese wheeled the stretcher into the second room and prepared to insert it, body and all, into the hollow center of the machine. Jack said he had to leave but that he'd be back to collect the body later.

"Just let me know when," he said. "'Cause I know Luke wants to hit the last call. I'll get you guys a lap dance!"

Annese laughed.

"You have any other cases tonight?" he asked Jack.

"I've got two. Actually, I'm sorry, I've got three. One in Tri-City, up in Oceanside, one in Chula Vista, and one in San Diego. But like I said, whenever I know that I've got a case for you, I try to make sure I set X amount of time aside, so that way when you call I'm ready to go."

After Jack left, it took Annese another five minutes or so to finish prepping for the scan. He had to position the man's head as close as possible to the dead center of the machine, where the magnetic field was strongest. During the premortem scans, the man could adjust himself somewhat while lying in the machine, shifting up and down or left and right a bit if Annese needed him to. Now Annese had to make sure he was positioned exactly right from the start.

Eventually he came back into the first room, sat down in front of the computer, and began the scan. A loud, abrasive, rhythmic pulse began to sound. This was normal. An MRI machine works in two stages. First the magnet, which was about five thousand times more powerful than Earth's magnetic field, halted the normally chaotic spins of the hydrogen atoms that were the primary ingredient of the man's body, lining them up so that they were locked into place, their protons pointing either in the direction of his head or his toes. The second stage was what caused the loud pulsing sound: A precisely tuned radio signal was bombarding the atoms, knocking the protons briefly out of polarity. Each time the radio signal turned off and the pulse subsided, the protons would snap back into their magnetic formation, like springs that had just been released. This would create a tiny burst of energy, and it was this energy that was becoming visible to Annese now, producing a ghostly, Shroud of Turin–esque image on his computer screen.

The scan took about an hour to complete, the machine making its way slice by virtual slice through the man's skull, mapping its contents. Tissue tends to heat up after prolonged periods in an MRI machine, which means that prolonged periods can be challenging for living subjects, even if lying in a tight tube doesn't bring on claustrophobia, as it often does. For this final scan, however, Annese didn't need to worry about the man's comfort, so he didn't rush things. I sat on a couch and tried to ignore the ornery pulse of the machine.

When the scan finished, Annese prepared for the harvesting. He grabbed a roll of plastic sheeting from somewhere, and some duct tape, and we headed down a hallway, passing through the book-and-computer-filled front office of the Brain Observatory, through the airtight biosafety door, and into the laboratory itself. The lab was colder than the surrounding rooms and had a vague antiseptic, chemical odor. It was clearly an active workspace, somewhat cluttered, with a lived-in feel, and in a dim, dark corner of the lab I could hear the gentle whir of a machine that had been left to do its work overnight, slowly slicing a frozen brain into thousands of infinitesimally thin sections. On a granite tabletop by a sink there were a variety of items that had been left out to dry, mostly laboratory beakers and pipettes and test tubes but also a couple of martini glasses and one large glass container with a taped-on label that read, ORCA CEREBELLUM. Annese had an ongoing arrangement with SeaWorld.

Annese unfurled the plastic sheeting and taped it in place so that it covered an approximately twelve-by-eight-foot area between two rows of cabinets. Behind him, against the far wall of the lab, there was the long bank of glass-fronted refrigerators that contained most of his growing collection of brains, including the remaining slices of Henry's, the slices that hadn't yet been mounted on slides. There was a bit of a backlog, too many brains, too few histological technicians, too much mounting to be done.

A student assistant of Annese's arrived at about midnight, and we returned to the MRI room to retrieve the man inside. Annese and the assistant carefully lifted him from the MRI stretcher back onto the more basic one that he'd arrived on, then wheeled him off to the lab. They stopped in the middle of the plastic-protected area, and Annese took some of the same tape he'd used to affix the plastic sheeting and now used it to bind the wheels of the stretcher so that it would not move back and forth during the harvesting. The task before him was one that required an unusual combination of strength and finesse, in that the brain is a delicate organ, easy to damage, but one that is shielded by the body's most formidable fortress.

Annese armored up. He covered his shoes with stretchable booties, his torso with a hospital smock, his face with a rounded, clear plastic visor. On his hands, rubber gloves; up his nostrils, two wads of cotton. Brains contain so much: memories, ideas, emotions, perceptions, aspirations, desires. They also contain pathogens, dangerous proteins and viruses and bacteria, some found nowhere else in the body. It's important to protect yourself.

He rolled the light blue sheet that covered the man's face down a little ways, to a point where his bushy eyebrows were visible. He had a formidable crop of white hair, unthinned by time. It sprouted densely from his ears as well, in that familiar old-man way. The skin of his forehead and his temples looked waxen and yellowish, thicker and heavier than living skin.

Annese picked up a scalpel, then remembered something and put it down. He walked to a nearby computer, brought it out of sleep, and clicked and typed for a few moments. Mozart's Twenty-fifth Symphony announced itself with a delicate flurry of strings, tinkling out of some speakers positioned in various spots around the laboratory. He always works with music, and music was itself once his work: As a young student abroad, he would play covers of old pop songs on his acoustic guitar in London subway stations, busking for change. Back then, he'd aimed his songs at the audience, trying to read them, their preferences. He did the same thing now, trying to match his own mood with the sensibilities of the person lying before him. A few months before, one of Annese's best friends had died—Roberto, the owner of an Italian restaurant in San Diego. The two had known each other for only a few years, but they'd become close, bonding over their common heritage and their shared passion for food and wine. Roberto had signed up to be one of Annese's brain donors, and when he died Annese did the harvesting. It was a strangely intimate evening, those last hours with Roberto. He was all alone, hadn't wanted an assistant.

"It feels like a Mozart night," he said now, returning to the stretcher.

He picked up the scalpel and bent down, using his fingers to part some of the thick white hair, finding the skin below. He placed the point of the scalpel at a spot just where the top of the cartilage of the ear connects to the side of the head, then began to cut, moving upward slowly in an arc, pausing every few inches to part the hair again if he needed a better view. He moved the scalpel up one side of the head, crested the summit, then moved it down the other side, stopping at the top of the other ear. He bent low again, using one hand to pull the hair forward a bit, making the incision visible. There was very little blood. Annese inserted the scalpel underneath the incision, pointing it in the direction of the forehead, and carefully slid it back and forth, severing the binding tissue between the skull and his scalp. Then he put the scalpel down and used his thumbs and fingers to gently roll the scalp down off of his skull. When he finished, the top of the skull was completely exposed. The white towel that had covered all of the man's face had shifted lower and was now just concealing his mouth. His eyes were still invisible, though, since the front portion of his scalp had been rolled down over them.

"You doing all right?" Annese asked, turning to me.

"I'm fine," I said.

He picked up an electric drill, a customized device that cost nearly twenty thousand dollars. It was plugged into a nearby wall socket, and Annese kicked the cord out of the way before he got to work. When he activated the drill, it made a high whining sound. He told his assistant to grab a plastic sports water bottle he'd positioned nearby, one with a long narrow spout. He told him to use the water inside to keep the skull irrigated, to minimize the dust.

He used the drill to bore a small hole in the left temple. Imagine a mom giving her kid a simple bowl cut, snipping a level shelf around her kid's head, just above the ears. That's the basic trajectory taken by Annese's drill, cutting around the skull, up from his left ear, straight across his forehead, down the other side, around the back, then finally rising up again till it met up with the initial hole. When Annese's drill had completed its orbit, the entire top of the skull came loose. It didn't fall off, though. There were still things keeping it in place, an adhesive tangle of membranes and arteries. It wouldn't take much force for Annese to remove it completely, but before he did, there was something I wanted to do.

I found a blank sheet of paper in a printer. I drew two circles in the middle if it, each about the size of a silver dollar, with about two inches of space between them. Then I folded the paper in two so that the crease bisected each circle. I used scissors to cut along the lines, like a kindergartner making a simple mask, and opened the paper back up. It now had two silver-dollar-size holes in it.

"You got it?" Annese asked.

"Yeah," I said.

"Okay. Put it on."

I laid the paper across the man's forehead, positioning the holes so that each was about one inch north of the eye sockets. Then Annese began to gently pry off the top of the skull. His assistant had a flashlight, and he shone it into one of the holes. I leaned in and peered inside the cranium. I saw a jumble of pink and shadow glistening in the flashlight's shaky illumination, like when the dentist holds up a mirror and you catch a glimpse of the back of your own throat. My initial visceral impressions faded, and I started to get my bearings and recognize some of the structures I was seeing. The basic view I had was of the underside of the frontal lobes. I could see some of their convolutions, those famously intricate folds. From this angle, looking down at the frontal lobes from above, it was as though their underside were a cliff and I was peering over the edge at night. And there, in the shadows below, I could see a hint of some other structures. The temporal lobes were nearly invisible to me, just hazy forms in the depths.

"That's exactly the view your grandfather had at the beginning of Henry's operation," Annese said.

I tried to imagine myself in my grandfather's operating room, peering through the holes that he'd just made in Henry's skull, holes the size of the ones I'd cut in the piece of paper. I tried to imagine how the view would have shifted slightly when he levered Henry's frontal lobes up and out of the way, finally getting a clear look, in the light of his headlamp, at the hidden and mysterious structures he was targeting, nestled in their moist, crepuscular cave. I'd often wondered what my grandfather had seen, standing at the head of his table, in his operating room, peering through the trephined holes in Henry's skull, right before he'd made his devastating and enlightening cuts. Now there it was. I looked through the holes for as long as I could, trying to commit the sight to memory, and then Annese removed the blood and serum-stained paper and tossed it into a biohazard bin. It was time to proceed with the harvesting.

He worked the top of the skull completely off, then laid it aside, rim up, like a bowl. The brain was now completely exposed, and Annese placed one hand underneath it, then reached below with a scalpel and neatly severed the top of the spinal column. He put the scalpel aside and placed his other hand underneath the brain. He tugged gently, making sure it was completely detached, that there were no other arteries or membranes or bands of nervous fibers that he'd missed. Then he pulled the brain out into the light.

THIRTY

EVERY DAY IS ALONE IN ITSELF

On a hot July morning I drove an hour west of Washington, D.C., to the small town of Warrenton, Virginia, to meet with Karl Pribram. I'd first come across Pribram's name in the original 1957 Milner/Scoville paper that introduced Patient H.M. to the world. It was just a glancing mention, used to illustrate the severity of H.M.'s amnesia, and came during a description of the day in 1955 when H.M.'s formal testing began: "Just before coming into the examining room he had been talking to Dr. Karl Pribram, yet he had no recollection of this at all and denied that anyone had spoken to him." From that point on, Pribram seemed to pop up everywhere I looked in my research. He interned as a neurosurgeon under Paul Bucy in Chicago in the late 1930s, went down to Florida to work under Karl Lashley—the man who came up with the theory of equipotentiality, which Brenda Milner and Patient H.M. helped overthrow—in the early 1940s, then moved to John Fulton's laboratory at Yale in 1948, helping to oversee much of the psychosurgery-related work there, including the Connecticut Cooperative Lobotomy Study. Then, in 1949, he moved to Hartford and became the director of research at the Institute of Living, my grandmother's asylum, overseeing Mortimer Mishkin, the man who would eventually develop the first monkey models of H.M.-like amnesia. In the late 1950s, Pribram left Connecticut and moved to Palo Alto, California, where he headed up the neuropsychology department at Stanford. He remained a professor emeritus there and also held a professorship at Georgetown University.

I knocked on Pribram's door. His longtime partner, Katherine Neville, a writer, philanthropist, and former model three decades his junior, opened the door. She was on her cellphone and she smiled and waved me into the living room to wait. Like my grandfather's old house, Pribram and Neville's was stocked with all sorts of things that wouldn't look out of place in a museum, including, at the far end of the living room, an eight-foot-tall carved wooden sculpture from Thailand, of a mythological creature called a Naga. Neville was on the board of directors of the Smithsonian.

I waited on an overstuffed couch for a while, reviewing the questions in my notebook. There was an urgency to this meeting: When I'd spoken with Neville on the phone, she warned me that Pribram was very ill with colon cancer, that he'd almost died several months before and still hadn't recovered.

Eventually Neville came back into the living room. Karl was ready to meet me, she said, and led me back to the ground-floor room he'd been using since his latest round of chemo had made him too weak to climb the stairs. I'd seen pictures of Pribram online, and in those pictures he always looked like God as wrought by Michelangelo: long white locks, a broad face, a thick, well-trimmed beard, piercing eyes, a gaze brimming over with raw knowledge.

Now Pribram sat in a bamboo chair, ninety-three years old, his hair and beard still regal but everything else looking shrunken and fragile. We shook hands, and I was careful not to squeeze too hard. He had pushed himself forward a bit for the handshake, and when he settled back into the chair he gave a soft groan.

I sat on a footstool near the chair, leaned in toward him, and began asking him questions. Neville had warned me that Pribram had been slowed by his latest round of chemo. He was still all there, she told me, but I had to be patient and give him time to answer. This seemed good advice, not least because Pribram was irritable; several times he snapped at Neville or his caregiver if either interrupted him while he was speaking, or at me whenever he thought my questions were inane or unclear. Every time he got distracted, either by an interruption or a fearsome attack of thrushy coughs, I had to remind him what he'd been talking about. Still, the memories he was able to pull up seemed vivid and clear, and his vocabulary was formidable. I got the sense his mind was working, just at a much different pace.

In some ways, it seemed like his memories had been worn down until only the sharpest, most salient facts and anecdotes remained. The impressions he had left were the impressions that had hit him hardest.

"What are your recollections of John Fulton?" I asked.

"John Fulton?" he said, then paused, looking up at a point in the distance for a full twenty seconds, while I fought the urge to fill the silence.

"John Fulton," Pribram said finally. "Brilliant. Financially dependent on his wife. A drunk. Bitter that he never won the Nobel Prize. Ran one of the world's great laboratories."

Then he stopped and waited for my next question. Later, when I asked him about Fulton again, fishing for more details, he told me the exact same thing, no more, no less. That was Fulton for him: brilliant, great lab, wife-dependent, bitter, Nobel-coveting drunk. Those impressions, and those alone, seemed to be what remained of Fulton in Pribram's mind. The rest had slipped away.

And so I threw other names at him, waited for the gears to turn, and listened.

Paul Bucy. Karl Lashley. Brenda Milner. Charles Burlingame. Walter Freeman.

He had something to say about each of them, some little thumbnail sketch, accurate, minimalistic, often cutting.

Eventually I asked him about my grandfather.

"Do you remember Bill Scoville?" I asked.

"Yes," he said.

"What were your impressions of him?"

Pribram looked off into the distance. By now I was used to the pace and waited in silence for his memories to coalesce.

He shifted his eyes back down and met mine. He paused again. And then he told me.

The first story I ever sold to a publication was the one I'd first told to my grandmother in one of the many letters I sent home from Egypt. I remember the act of writing that letter, almost twenty years ago on my rented houseboat, tapping the words out on a clunky old laptop. Earlier that afternoon, while rowing a single scull on the Nile, I'd found a body floating facedown in the river. The letter was about that, and about what happened next, how I'd drifted alongside the corpse for a half hour, hollering in awful Arabic at strangers on the shore, trying to get help. I'd never seen a dead body before, and to this day the sight remains vivid in my mind: the wet jet-black hair of a teenage boy, the tops of his ears, his collared patterned shirt. The rest of him was invisible in the murk of the river. After I wrote the letter and sent it off, I decided I'd try to get it published. I didn't change much, basically just dropped the "Dear Bambam" at the beginning and gave it to the editor of a Cairo-based newspaper called *The Middle East Times*.

Many years later, I sat beside Bambam during Thanksgiving dinner and told her about a different story I'd begun working on, one about a man my grandfather had once operated on. Bambam was by then ninety-eight years old, a tiny woman in a blue dress. I asked her what she remembered about the saga of Patient H.M. She leaned toward me, inclining her hearing aid. I repeated the question, and she shook her head.

"I don't remember anything," she said.

I changed the subject. I'd recently spent the night in a monastery, reporting a story for *National Geographic*. I told her about the novelty of sleeping in a cell and waking early with the monks for dawn services.

She nodded. I couldn't tell if she had heard or understood much. Then she leaned toward me again and began to speak.

"There once was a monk from Siberia,
Whose life grew drearier and drearier.
He leapt from his cell with a terrible yell,
And eloped with a mother superior."

She finished the limerick, smiled, and settled back into her chair.

The next day I visited her at her assisted living home in Lexington, Massachusetts. We ate in the cafeteria, had a terrible lunch of watery scrod. She poured salt in her coffee, thought it was consommé. I'd brought along some old letters that her son, my uncle Barrett, had written to her. Barrett loved her deeply, and they corresponded a lot. She kept a bag of these letters by her bed, and visitors would read them to her, over and over. I'd grabbed a handful before we walked to the cafeteria. I opened one up, unfolded it on the table, read it aloud.

It was a letter Barrett had written shortly after my grandfather died, in 1983. He wrote about his grief, about the immensity of it. "If things go as are natural," he wrote, "you also will die before me, and then my heart will truly break."

Things did not go as are natural. Barrett died in a plane crash three years after he wrote that letter. My grandmother named Barrett after her younger brother, who also died too young.

I read a few more letters to her, and then before I left I reminded her of the monastery limerick she'd recited the night before and asked if she knew any others.

There was a long pause.

"There once was a man with a beard,
Who said it's just as I feared . . ."

She trailed off.

She didn't remember how it ended.

. . .

Karl Pribram settled his eyes back on mine. "Everybody liked Bill very much," he said. "He was a very liked person."

He paused.

"But," he said, and paused again.

I leaned in closer, waiting to hear what he would say next.

From the beginning of his career, Pribram always rubbed certain people the wrong way. In 1948, for example, while weighing whether to hire Pribram on at his Yale laboratory, John Fulton asked Paul Bucy for his opinion of Pribram, and Bucy responded with a deliciously sharp sketch. After noting that Pribram was the son of "an outstanding Jewish Austrian bacteriologist," he described Pribram's years as an intern under Bucy during medical school, where, according to Bucy, Pribram "was very immature in almost every way. His behavior was very irritating to many people. He was like a puppy that is always under foot." Bucy did note that Pribram "is not unintelligent and has had an unusually extensive training in neurology and neurological surgery," but stressed that "most people are very irritated by his personal manner" and that "the greatest difficulty, if there were any, would be his irritation of you and the other people in the laboratory."

Another prominent neurosurgeon who'd taken Pribram under his wing, Percival Bailey, wrote a similarly caustic letter to Fulton:

"I can tell you a great deal about Karl Pribram," he wrote. "He is, as you say, intensely eager to do investigative work. He is, in fact, so eager that he succeeds in making himself intolerable to whomever he works with. He comes, unfortunately, of a long line of rather distinguished physicians in Vienna and he feels that he is letting the family tradition down by becoming a mere practitioner. He will certainly work hard. What he will accomplish I do not know since he has never shown any evidence of originality. I should hate to have him drive you to drink on my recommendation."

Eventually, however, Karl Lashley came through with a glowing recommendation, declaring Pribram "one of the three or four most promising research men that I have had to work with me," describing "a quick, keen mind, a genuine scientific curiosity, and an enormous capacity for work."

Fulton took a chance, hired him on. And he never seemed to regret it, although he did write Pribram a pointed and fairly reproving memo six months into his tenure at Yale. The memo was titled "Thoughts on Secretaries," and in it Fulton told Pribram that although he had "come to admire your industry and enthusiasm," he had also "become aware of the fact that you have considerable difficulty in getting on with those who are serving you. I am convinced that you are unconscious of the basis of your difficulty." He listed a number of rules he wanted Pribram to follow from then on. "Whenever you ask any of the gals in the Department to do something for you, approach them as though you were asking a special favor," Fulton wrote. "Never give them work just because you have to keep them busy." Fulton stressed to Pribram that "you lose caste with everyone in the Department whenever you raise your voice with animal boys, secretaries, or anyone else in the group," and advised him that a new secretary would be joining the lab the following morning. "I feel that you, rather than she, is on trial," he wrote. He ended on a warm note, though, sending Pribram "every good wish for the success of the Lobotomy Project."

As it turned out, Pribram wouldn't be irritating secretaries or animal boys or anyone else at Fulton's lab for much longer. A few months after Fulton sent that letter, Pribram moved to Hartford, where he started his own laboratory, becoming the director of research at the Institute of Living. And among the people he would get to know while in Hartford was a fellow neurosurgeon who, unlike him, was a very liked person.

In June 1972, my grandfather invited a reporter from the *Hartford Courant* to his office and gave an extended interview, which formed the basis for a three-part series of front-page articles. Walter Freeman had died the month before, and with Freeman gone the country's most passionate lobotomy advocate, and most prolific lobotomist, was my grandfather. The interview was a calculated attempt to haul the reputation of the lobotomy out of the gutter. Long after the majority of neurosurgeons stopped performing lobotomies, my grandfather continued to do so on scores of patients, ranging from psychotics to neurotics, in asylums and in his private practice. Even as the 1950s turned into the 1960s and 1970s, he continued, though the numbers dwindled significantly.

In 1970, my grandfather founded the International Society for Psychiatric Surgery. Composed of several dozen like-minded lobotomists, the organization was a self-conscious attempt to bring respectability to the widely disparaged field. In a "Letter from the President" to the entire membership regarding an upcoming meeting of the society that would be taking place in London, my grandfather stressed the importance of seeking "basic scientists rather than surgeons for our guest speakers in order to convince the world, and ourselves, that we indeed are a Society of dedicated scientists rather than trigger-happy barber surgeons." His continued advocacy for psychosurgery didn't have any clear financial motive: Lobotomies by that time constituted only a very small percentage of his total surgical work, and his bread-and-butter jobs, as for many neurosurgeons, were spinal surgeries on slipped disks. His reputation by that point also rested on far more than his history as a lobotomist. He had been the president of the American Academy of Neurological Surgery and was the founder and honorary president of the World Federation of Neurological Societies. He was a professor of neurosurgery at Yale, remained the chief of neurosurgery at Hartford Hospital, and had accumulated a number of other honors, including a teaching chair in his name at the University of Connecticut. His reputation among fellow neurosurgeons was almost unparalleled and was helped by his constant travel and networking: One neurosurgeon I spoke with recalled visiting a prominent neurosurgeon in the Soviet Union during the height of the Cold War. Hanging behind the man's desk were three framed portraits: one of Vladimir Lenin, one of Ivan Pavlov, and one of William Beecher Scoville.

So he didn't need to invite that reporter to his office. But he wanted to. He had some things to get off his chest. Psychosurgery, he knew, had been getting a bad rap in the media.

Now, my grandfather told the *Courant* reporter, he wanted to "show the other side."

He began by conceding that Thorazine and other new antipsychotic medications had eliminated the need for lobotomies in some cases, and he'd also come to see that there were a variety of mental conditions beyond the reach of psychosurgery. For example, he thought it was useless to operate on "psychopaths genetically or constitutionally born without a sense of moral obligation to society, and totally lacking in feelings of guilt or sympathy toward their fellow creatures." He also hesitated to operate on "spoilt children" and "social rebels." Regarding these latter types, he said, "they never warrant psychosurgery because they are not mentally ill but rather are reacting against an unfortunate environment."

Nevertheless he believed there was still an important place for the procedure in the medical armamentarium. "I am more impressed with psychosurgery every year," he said. "I wonder why more people don't have it done." He told the reporter that he performed about eight lobotomies a year and that his preferred method remained the orbital undercutting procedure he'd developed decades earlier. Three months prior to the interview, a woman died of a cerebral hemorrhage while he was performing one, but he boasted that this was the only lobotomy-related fatality he'd ever caused. In general, he believed the procedure to be safe and almost harmless, though he conceded that it did somewhat blunt a person's faculties and ambitions. Patients of his might go on to become anything they chose, "even a brain surgeon," though my grandfather considered it doubtful that these hypothetical lobotomized neurosurgeons could ever become leaders in their field.

Then he trotted out one of his patients, a woman named Mary, for the reporter to meet. She was forty-nine years old, and the reporter described her as "plumpish, grey-haired," and "bubbly." She'd spent most of her career working as a secretary. A few years prior, she began to get extremely anxious. "I used to get panicky with worry over the simplest things, over everything," she said. She described a panic attack triggered by seeing an open window in a nearby apartment: "Knowing a child lived in that apartment, I'd get sick with worry that he'd fall out." Eventually, her husband sent her to the Institute of Living, where she was institutionalized for ten months. While there, she received intensive psychoanalytic therapy as well as electroshock therapy, among other treatments.

"The psychiatrists gave me hope and helped somewhat," Mary said, "but it just wasn't enough."

Institute staff told Mary's husband that a lobotomy was her best option. Her husband, incidentally, was present during the meeting with the *Courant* reporter, and one of the resulting articles described how Mary "furrowed her brow in concentration while being questioned and at times asked her husband for help in answering."

Three years had passed since Mary's lobotomy.

"My old feelings gradually left me," Mary said, "and now I feel very happy—all the time." She mentioned that she'd been able to resume her job as a secretary for a while, "but now I have a longing to stay home, so I quit." All in all, Mary gave her lobotomy, and my grandfather, a glowing review. "The operation made me realize how wonderful life is," she said. "Dr. Scoville is the best doctor in the world!"

It is unclear how much of what my grandfather said during this interview can be taken at face value. Certain of the things he said were not true. For example, at one point he said that he'd performed exactly twenty lobotomies in the operating room at the Institute of Living. This is incorrect: I have access to records documenting at least forty-six, and those records are likely incomplete. Whether his untruths were the result of deceit or carelessness is impossible to say.

Maybe some rearview blurriness is just the inevitable by-product of a life of great drive, one spent always moving forward at great speed, rarely pausing to reflect on where you've been. Maybe that, too, explains why the reporter described my grandfather as being "contemptuous" of slower psychoanalytic approaches to the treatment of mental illness. "No man is worth that much of another man's time," my grandfather told him. "I like fast results."

The following year, on March 6, 1973, Senator Ted Kennedy presided over a subcommittee hearing meant to explore the current state of psychosurgery in America.

Kennedy called the meeting to order with a brief overview of the issues at hand: "The nature and functioning of the human mind has fascinated scientists for centuries," he said. "In recent years they have begun to understand that this is the basis of behavior and have developed tools and techniques to modify and control it. There are those who say the new behavioral research will enable us to realize our full potential as a nation and as a people. There are others who believe that the new technology is a threat to our most cherished freedoms. . . . Few areas of biomedical research have been as controversial as the behavioral research we are to hear about today. Some federal scientists recently circulated a petition urging the National Institutes of Health and the National Institute of Mental Health to refuse to sponsor research into psychosurgery. It is our hope that today's hearing will air both sides of the controversy and help us as a society come to understand and master this new technology so as not to become the victims of it."

Over the course of the afternoon, several witnesses testified before the subcommittee, and most had views about psychosurgery ranging from neutral to positive. When Kennedy asked the director of the National Institute of Mental Health whether psychosurgery was an effective therapy, the director responded like this:

"Do I think it is a valid technique for behavioral disorders? My answer is a crisp 'maybe.'"

And when a neurosurgeon from the University of Mississippi was called to testify, he began with a series of five lobotomy success stories, such as one about a twenty-four-year-old man who "had attacks of nervousness and aggression since childhood. . . . Psychosurgery was performed twelve years ago. He no longer has outbursts of aggression, is happily married, and supervises five other workers at his place of employment."

Others stressed that the psychosurgery of the 1970s was very different from the psychosurgery practiced decades prior, that the modern procedures were much less damaging, much more precise, and could hardly be called lobotomies at all.

Then Peter Breggin took the stand. Breggin was a Washington, D.C.–based psychiatrist who had published a number of books and articles critical of psychosurgery. After Kennedy introduced him, Breggin said, "The psychosurgeons represent the greatest future threat we are going to face for our traditional American values, as promoted in the Declaration of the Independence and the Bill of Rights. This totalitarianism asks for social control of the individual, at the expense of life, liberty, and the pursuit of happiness. It undermines Jefferson's self-evident truths. These men, I believe, are doing nothing more than giving us a new form of totalitarianism. . . . It creates for themselves an elitist power over human mind and spirit. If America ever falls to totalitarianism, the dictator will be a behavioral scientist and the secret police will be armed with lobotomy and psychosurgery."

After Breggin finished his opening statement, Kennedy began questioning him.

Kennedy: Do you think all psychosurgery ought to be made illegal?

Breggin: Yes. It is not, in my opinion, a medical procedure any more than the mutilation of an arm as punishment of a crime is a medical procedure. The mere fact that a physician performs the mutilation does not make it a medical procedure. That was established at Nuremberg.

Kennedy: It is your position that the government ought to prohibit psychosurgery?

Breggin: Yes. Very definitely yes. I think it falls into the class of atrocities, as defined in Nuremberg. Let me get to the specifics on what is going on at the present time, in regard to lobotomy and psychosurgery. . . . William Scoville, president of the Association of

> Psychosurgery, is a lobotomist. Do not believe what you have been told today, Senator, about the demise of the lobotomy. There is a great deal of lobotomy going on in this country right now.

Earlier, in a written report submitted to Congress, Breggin had described my grandfather as having "replaced the deceased Walter Freeman as the nation's spokesman for lobotomy and psychosurgery." Now, on the Senate floor, Breggin painted a bleak picture of the procedures my grandfather was a spokesman for, describing a "permanently mutilating operation" that destroyed "spiritual and emotional responsiveness." Breggin highlighted Hartford, and Hartford Hospital, as important centers for psychosurgery, though he warned that nearly every city in the country had at least one active lobotomist. After Breggin spoke for several minutes, Kennedy appeared to grow impatient.

> Kennedy: Doctor, did you ever think that they might be right and you might be wrong?
>
> Breggin: Senator, all I can do . . .
>
> Kennedy: I am sure you are familiar with other examples from medical history: Dr. [Ignaz] Semmelweis saw childbirth defects caused by bacteria transmitted by surgeons; and he was ostracized, and he turned out to be right. Dr. Morton was ostracized because of his beliefs about ether. He was right. Copernicus thought the earth was not the center of the universe. And he was given a very hard time. Now, why do you think you're right, and they are wrong?

Breggin floundered, thrown off-balance by Kennedy's barbed questions. He told Kennedy that he considered *himself* to be much more like those medical mavericks Kennedy mentioned than the lobotomists were. The lobotomists, he pointed out, were almost all high-ranking members of the medical establishment.

"The anti-psychosurgeons are the ones likely to be burned at the stake," Breggin said, "not the psychosurgeons."

Breggin ended his testimony then. After the hearings, Kennedy co-sponsored a bill that created a national commission intended to examine, among other topics, whether psychosurgery should be permitted to continue in the United States.

On February 25, 1984, my grandfather and his second wife, Helene, set out from their home in Farmington, Connecticut, to the birthday party of his brother-in-law in New Hope, Pennsylvania.

He was seventy-seven years old and still a practicing neurosurgeon, maybe the oldest in the country. He had no intentions of ever stopping—he'd once written that he was "not intending to retire until stricken by God or man"—despite the fact that age had caused his skills to slip. Over the previous few years, there had been a great deal of tension between him and the management at Hartford Hospital. They worried about the mistakes he might make, and their consequences. Not long before, while in the operating room, he'd been working on a patient's spine, leaning over the incision, using his custom tools to move away the layers of fat and muscle. Then he paused and glanced at one of the nurses.

"Which way is the head?" he asked. Nobody was sure whether he was joking.

The director of the neurosurgery department was now a younger neurosurgeon named James Collias, whom my grandfather had hired two decades before. The hospital put Collias in the awkward position of attempting to control the damage his mentor and former boss might do, increasing the amount of supervision he received and limiting the number and types of procedures he performed. Collias and my grandfather had a series of meetings, and then Collias sent my grandfather a letter on Hartford Hospital letterhead laying out the new rules:

1. Surgical privileges restricted to disc and carpal tunnel procedures, without exception.

> 2. Allowed to schedule no more than one elective operative case per day, and only during regularly scheduled surgery hours.
> 3. Not to start any scheduled elective surgery after 5:00 p.m.
> 4. Never to perform any surgery without neurosurgical house staff or attending assistance at all times.
> 5. To refer all stat and emergency cases (including complications of scheduled surgery) requiring night (after 5:00 p.m.), weekend (Saturday and Sunday), or holiday surgery to neurosurgical attending on call, or of choice.
> 6. Not allowed to schedule elective Saturday surgery.
>
> These restrictions will be reviewed periodically by the Department Director and additions, deletions or modifications made at his discretion at any time. Any breach of the above restrictions by William B. Scoville, M.D., will result in the immediate revocation of all surgical privileges.

My grandfather submitted to all the restrictions, though eventually Collias discovered that he had secretly obtained full neurosurgical privileges at another nearby hospital to be able to continue to perform brain operations.

It's unclear whether he was still performing lobotomies in 1984. The International Society for Psychiatric Surgery had disbanded the year before, but in many ways that was because there was no longer a need for its boosterism. In 1977, the National Commission for the Protection of Human Subjects of Biomedical and Behavioral Research, which Kennedy had launched after his 1973 Senate hearings, completed its report. Rather than declare that psychosurgery should be banned, as Breggin and other activists had hoped, the commission instead "determined unanimously that there are circumstances under which psychosurgical procedures may be appropriately performed" and that "psychosurgery should not be prohibited." The single most important item leading the commission to this conclusion was a federally funded investigation spearheaded by MIT, which looked at hundreds of lobotomy patients in an attempt to determine "the possible side effects of these operations." This was the largest survey of its kind ever conducted, and it had been accelerated at the request of the commission, which received advance notice of its findings. Those findings were, to many, a shock. After taking a close look at "the neurologic and psychologic sequelae" of the lobotomy patients, the MIT researchers failed to find "any obvious 'costs' of the intervention."

One of the lead authors of that MIT paper, incidentally, was Suzanne Corkin.

What exactly that future of psychosurgery would look like was still in doubt, though it seemed clear that new generations of psychosurgeons would focus on ever more selective lesioning combined with miniaturized brain-stimulating electrical implants. This was in many ways a realization of a prediction my grandfather made in his 1953 paper "The Limbic Lobe in Man," in which he first mentioned Patient H.M. At the end of that paper, he speculated that "who knows but that in future years neurosurgeons may apply directive selective shock therapy to the hypothalamus, thereby relegating psychoanalysis to that scientific limbo where perhaps it belongs?"

Although my grandfather rarely indulged in introspection, during the twilight of his career he must have felt some satisfaction in knowing that the subspecialty he had devoted so much of his life to was going to endure, in one form or another. Psychosurgery of the sort Walter Freeman and my grandfather pioneered was nearly extinct, but surgery of the mind would continue to evolve.

His own life, on the other hand, was in many ways the same as it had been for decades. He still loved to operate and still spent as much time doing so as he could, even as his skills faded and restrictions piled up. He was always good at defying rules and never stopped enjoying living on the edge. His children with my grandmother had all grown up and begun their adult lives, but he'd started a new family with his second wife, and in a recent update he'd sent to his Yale class biography, he wrote that his current interests included "my new children, younger than my grandchildren." He still loved fast cars, too. In fact, he'd sent his son Barrett a letter just two weeks earlier asking for his help in acquiring a new European sports car. Barrett lived in Frankfurt, where he worked for a pharmaceutical company.

"I am extremely anxious for you to check the car situation for me," he wrote. "My interests have increased considerably since our last conversation and I must get it off my mind. I have spent all my time dreaming and dreaming of wonderful cars." He listed the cars he was most interested in, which included a BMW M635CSi, a Porsche Carrera (with the Targa top), and a Porsche 944 with low-profile tires. He gave Barrett the names of some connected German friends who might be able to speed up the exportation process, and ended with an urgent plea: "Will you please put this matter in your highest priorities, for I dream of having one more good sport car before I die and these should last until then."

He never got any of those last dream cars of his. On the way to his brother-in-law's birthday party, he missed his exit, stopped the car he'd chosen for this trip—a relatively sedate Honda Prelude—and started reversing up the New Jersey Turnpike, back toward the turnoff.

The car was hit by another vehicle. His wife was not seriously injured, but he was killed instantly.

In 1968, Brenda Milner and Suzanne Corkin co-authored a paper called "Further Analysis of the Hippocampal Amnesic Syndrome: 14-Year Follow-Up Study of H.M." The paper detailed the minutiae of Henry's performance on a variety of new tests, like his experience with something called the Gollin Incomplete Pictures Test, which required him to try to identify drawings of an object even when the drawings were incomplete. For example, he was shown a drawing of an airplane that bore only a sketchy hint of wings and fuselage and tail. If he failed to identify it, he'd be shown a version of the same picture, only this time it would be slightly more filled in. There were five versions of each picture, with the final one complete. Henry, as expected, performed about as well as nonamnesic control subjects when first given the test. "This finding constitutes further proof that H.M.'s perceptual abilities are largely intact," Milner and Corkin wrote. Also unsurprising was that when they gave him the test an hour later, his performance improved somewhat but not nearly as much as people with normal memories. "On the first exposure to the task," they concluded, "H.M.'s performance was almost indistinguishable from that of the control subjects, but on retest he shows considerably less improvement than they do."

This was one of the last papers Milner contributed to about H.M. By that point, she had already moved on, focusing on other cases and other neuropsychological puzzles. Years later, during one of my conversations with Milner, she described her relationship with Corkin as follows: "I've had all the recognition I need. I'm getting prizes and the rest of it, and I've done a lot more in my life than study H.M. And so, you know, I wish her well. . . . There's no denying that I'm the person that did the basic work with H.M. and that she was my student." Milner added that in her opinion, Corkin was a very competent, careful, and hardworking scientist but was "not very creative. I think she would probably admit that. I've had other students who were more creative, but I've not had students that were more dedicated."

The 1968 paper wasn't particularly revelatory, scientifically speaking. Like most of the work Corkin oversaw with Henry over the following decades, it refined the groundbreaking discoveries Milner had made early on rather than breaking new ground itself. The paper did, however, contain one of the single most heartbreaking descriptions of how Henry actually experienced the world.

"Every day is alone in itself," he's quoted as saying. "Whatever enjoyment I've had, and whatever sorrow I've had."

Days.

Weeks.

Months.

Years.

They leave traces, some faint, some strong.

They pile up.

They order themselves into sequences in our minds, chains of causes and effects.

They become stories.

In the end, this is the difference between Henry and us: Henry could no longer hold on to the present, could no longer make new memories, which meant that he could no longer tell or even understand stories, at least ones that lasted more than a few moments.

We can.

And we can do more than that, too.

We can alter stories.

Sometimes the things you discover in the present change, irrevocably, your understanding of the past, adding new perspective, calling into question old interpretations.

These shifts might occur in the archives of an asylum, in the office of a scientist, in the basement of a library.

Or they might happen while you're sitting on a footstool in an old man's home, listening to him dredge up some impressions of your grandfather.

Sometimes just a few words can change everything you thought you knew about the story you thought you were telling.

"Everybody liked Bill very much," Karl Pribram said. "He was a very liked person. But."

He paused.

"I felt that anyone who did psychosurgery on his wife was sort of suspect. So I had my reservations about that."

He told me what he knew, or at least what he remembered.

He thought the operation had taken place around 1950 and that it was an orbital undercutting procedure, the more precise, less blunting lobotomy that my grandfather had invented. He thought that the operation was performed at Hartford Hospital, not the Institute of Living, in part because he, Pribram, as head of research at the asylum, had expressed ethical reservations about the plan.

Katherine Neville and a full-time caretaker named Marlene were both sitting in on my interview, and each told me that they remembered Karl telling them the story about the psychosurgeon colleague of his who'd operated on his own wife.

"I didn't know it was your grandfather," Neville said.

I sat there, pen frozen, mind racing. Was it possible? I thought back on the history I'd been steeping in. My grandfather had operated on hundreds of women with symptoms just like his wife's. Why wouldn't he have done the same to her? If he believed in what he was doing, if he believed he could help.

But could he have kept it a secret?

I remember Thanksgiving dinners when I was a kid, my grandfather holding court at the head of the table, his second wife to one side, my grandmother sitting mostly silent a few seats away. I'd hardly known anything about my grandmother's illness back then. It wasn't a subject the family would ever discuss directly. By the time I was in my late teens I'd picked up a few scraps of information: I knew Bambam had once had some sort of breakdown and that she still took Thorazine to deal with her symptoms. It wasn't until I began researching this book, though, that I learned any real, hard details. Most of what I uncovered was a surprise not just to me but to my mother, whose knowledge about my grandmother's troubles and treatments had been nearly as vague as my own. My grandfather's instinct to keep secret the details of his wife's illness had been present from the start: The first letter he sent to his parents after the breakdown ended with a request: "Please do not tell this to a living soul but yourselves—not even Aunt Alice. (At first unless you are sure she will not talk to the Cheneys.)" The Cheneys were my grandmother's side of the family.

But could that secrecy have run deeper than I'd ever imagined?

I knew there were precedents. Madness has always carried a stigma, and many institutionalizations are covered up. Some families are very good at keeping entire lives in the shadows: When Senator Ted Kennedy presided over those psychosurgery hearings in 1973, it was still not public knowledge that his own younger sister, Rosemary, had been lobotomized by Walter Freeman and James Watts three decades earlier, in 1941. Rosemary's operation was botched, and she remained in an institution till her death, deeply damaged.

I also knew that lobotomy patients often had no recollection of the operation and would deny having received one if asked. If my grandfather in fact operated on my grandmother, she may not have known it. Her brain may have carried a wound her mind was unaware of.

I looked at a Miró print on Pribram's wall. The print was blue and white and black, had overlapping shapes, sharp colors. I looked back at Pribram. He peered at me with rheumy eyes for a few moments before speaking again.

"It was a different era," he said. "And he did what at the time he thought was okay: He lobotomized his wife. And she became much more tractable. And so he succeeded in getting what he wanted: a tractable wife."

Over the next several months I would try to confirm what Pribram had told me. I looked for proof it had happened, or proof that it hadn't. I came up empty. Her medical records were gone, her body was cremated. The people I asked either didn't know or wouldn't tell me. I stared at old family photos, trying to spot faint scars. I told my mother, and it shook her, of course. She said that if he'd done it, he'd surely done it with the best intentions. I told Dennis Spencer, the former resident of my grandfather's who went on to become the head of the Yale department of neurosurgery.

"I never heard that," he said. "But it doesn't surprise me. It sounds like something Bill would do."

Here's what I do know for sure: My grandmother was not a tractable wife. Pribram was wrong about that, at least.

Despite everything. Despite whatever happened before her breakdown. Despite whatever happened after. The electroshock treatment. The hydrotherapy. The fever room. The Thorazine.

The operating room?

Whatever they did to my grandmother at the asylums, however bad it got, whatever they took, whatever *he* took: What remained was strong.

In 1957, the same year my grandfather and Brenda Milner published their paper about Henry, my grandmother took a trip to Reno, Nevada. She walked into one of the local shops that specialized in quick, no-contest divorces. She filled out each of the forms they gave her. The man working there took advantage of her, told her that it was a good-luck tradition to leave her wedding ring behind, to just drop it in the cardboard box behind the counter, and she did so, even though she could have made good money selling it. But maybe it was worth it, just dropping it in the box and walking away.

She left my grandfather, moved to New York City on her own.

By that point, her children had all gone off to college.

She started over.

She got a small apartment on the Upper East Side, landed a job as an assistant at an advertising agency. Eventually she pursued a master's degree in remedial reading, and taught at a school for children with special needs.

Her children grew older, had children of their own. She was a good grandmother. I remember how, during one visit to New York when I was twelve years old, I begged her to take me to the new Arnold Schwarzenegger movie, *Commando,* because it was R-rated and I couldn't go alone. She didn't enjoy it, but she went. I remember her leaning over and asking me if his muscles were real.

During her free time, she volunteered at an organization called the LightHouse, reading books aloud to blind people, men and women whose losses were clear for everyone else to see.

Some of our losses are more subtle, more hidden, more secret.

Some of our strengths are like that, too.

THIRTY-ONE

POSTMORTEM

Suzanne Corkin's office was on the fifth floor of MIT's Brain and Cognitive Sciences Complex, a $175 million facility that opened in 2006. The buildings were part of a recent MIT construction boom, right across the street from the $300 million Frank Gehry–designed Stata Center, which was packed with AI researchers. The Brain and Cognitive Sciences Complex had the feel of a modern corporate headquarters, perhaps one belonging to an Internet firm, its gleaming metal and polished marble leavened with touches of playfulness like the foosball table that sat on a fifth-floor balcony overlooking the cavernous central courtyard. High-definition TVs hung everywhere, advertising daily lectures and symposiums. Several architectural firms were involved in designing the individual buildings that made up the complex, and its website boasted it was an "exemplar of collaborative design, and is designed to inspire future collaboration." Many of its laboratories had glass walls facing the hallways, and anybody walking nearby could watch the researchers inside at work, huddled over their pipettes or their laptops, all pursuing their own private mysteries.

Corkin's office sat at the end of a long red hallway. It was actually more of a suite: There was a front office with a desk for a secretary, a coatrack, and a bookshelf stacked with the latest issues of various scientific journals. I walked through the front office and into the larger one beyond it. I sat at a small round table, and Corkin took a seat across from me and offered me a French chocolate from a glass bowl.

She was fighting a serious illness, and appeared physically frail. Her dog, Trooper, got up from a dog bed and wandered back out to the first room. Corkin asked if I could go check that the door to the red hallway was closed. She told me a story about how Trooper had once gotten out of the office, descended five stories, and walked out the front entrance and onto the street, which presumably she could have done only if somebody held the lobby door open for her.

"It's MIT and you'd think people here would be smarter than that," Corkin said.

This was our second interview. The first had taken place the week before. She agreed to that interview after many denied requests, and I was looking forward to finally getting the chance to ask all the questions that had been building up. That first interview, Corkin had shown up a few minutes late for our one P.M. appointment, and at about one-fifty P.M. somebody poked his head into the office and reminded Corkin that they had a two o'clock meeting scheduled. I'd been expecting much more time, and when that first interview came to a close we'd barely had a chance to edge past Corkin's childhood in Hartford and into the early years of her career at McGill and MIT. None of those first questions had been particularly tough, but her answers were generally curt, offering up the bare minimum of information in a clinical and dispassionate way, even when they dealt with major milestones in her past. When I asked if she could describe for me any details of the first time she met Henry, to help me set the scene of that initial encounter with the man who would go on to define her life's work, this was her response:

"No, but that's not surprising. Because what you're asking for is an episodic memory, and episodic memories typically don't last that long, no matter what the situation is. Now, I'm sure there are exceptions, when, you know, say, somebody's being *raped,* and she remembers every little detail of that event. But what probably happens in cases like that, that are very emotional, is that they were repeated many times after. They were rehearsed, mentally, and became semanticized. So no, I don't remember what it was like to first shake hands with Henry, but if I did it would probably be fiction rather than fact."

Unsure I'd get another chance to talk with her, I told Corkin that there were a number of documents she presumably had in her Patient H.M. files—including some of Henry's specific testing data, his brain donation form, certain unpublished transcripts of interviews with him, and my grandfather's original operative report—that I hoped she'd let me take a look at. She responded with a list of reasons that might not be possible, saying that she wasn't sure where all those documents were, and that she'd have to check with the MIT lawyers and her literary agent before sharing anything. Also, the idea of providing unfiltered research data to a layperson such as myself made her uncomfortable.

"I mean, you're not a psychologist," she said. "You're not trained to administer or interpret these tests. So there's always the danger that you will misinterpret things."

I decided to conduct this second interview somewhat differently. Knowing that I might not have much time, I skipped straight to my most pressing questions. Some of those had to do with Jacopo Annese. I'd learned of a surprising conflict that had flared up between Corkin and Annese, and I wanted to hear Corkin's side of it. When I brought up Annese's name, Corkin's expression, which tended to be fairly flat, visibly soured, as though she'd just tasted something unpleasant.

"He was technically good," she said. "Good with his hands. He's a high-level technician." Then, referring to the period leading up to and during Annese's slicing of Henry's brain, Corkin said, "At this point I still trusted him. We were friends. I thought this was a legitimate collaboration all in good faith. . . . He prepared in advance for the brain and got all sorts of fancy equipment, and refrigerators and backup freezers and alarm systems."

I asked the obvious follow-up question: What had happened to make her stop trusting him?

"I'm not going into this at all with you," she said.

"Not at all?" I said. "Because I'd like to . . ."

Corkin cut me off.

"You would and so would everybody else." She laughed. "*The New York Times* would *love* to have this story."

Funny thing is, *The New York Times* did have the story. They just didn't know it. The first crack in the relationship between Corkin and Annese, at least as best as Annese can tell, was literally chronicled in the pages of the *Times* in the form of a tiny, unheralded correction to one of its articles.

Here's what happened. On December 22, 2009, the *Times* published an article on the cover of its science section. The article was called "Building a Search Engine of the Brain, Slice by Slice," and it was about the project to preserve and archive Henry's brain, digitally and histologically, at the Brain Observatory at UCSD. A reporter had traveled to San Diego for the slicing, and his piece was solidly reported and evocatively written. At one point he described how, as the blade cut ever deeper into the block of milky-white frozen gelatin that encased Henry's brain, revealing more and more of it, the visible part of the brain appeared to be "growing with every slice like spilled rosé on a cream carpet."

Annese and his laboratory were clearly at the center of the piece, but Corkin had come to San Diego to witness the slicing, and the piece included a few quotes from her. In the sixth paragraph was the following line: "The dissection marked a culmination, for one thing, of H.M.'s remarkable life, and of more than a year of preparation for just this moment, orchestrated by Suzanne Corkin, a memory researcher at the Massachusetts Institute of Technology who had worked with Mr. Molaison for the last five decades of his life."

Annese liked the article, but that particular line puzzled him. Truth be told, it got under his skin, as he was not a man without ego. It was true, Annese knew, that Corkin had played a critical role in making that day happen. Without Corkin, Annese would never have had Henry's brain to work with in the first place. But to say that over the past year she had "orchestrated" the work that culminated in Henry's historic forty-eight-hour live-streamed dissection—a dissection that certainly represented, among other things, the most prominent single moment in Annese's career—well, that just wasn't accurate. A year before, at that JetBlue gate at Logan Airport in Boston, in front of those television cameras, Corkin had handed off Henry's brain to Annese, and ever since then she'd actually *kept* her hands off. Annese received the occasional email from her during the year that followed, but he was basically left to his own devices, assembling his team, applying for grants, acquiring equipment. The fact was, Annese worked harder than he ever had in his life, to make sure the slicing went off without a hitch and to lay the groundwork for the critical tasks that still lay in the future, and Suzanne Corkin had very little to do with any of it.

Annese wrote the *Times* reporter an email. He laid out, in diplomatic terms, his problems with that particular line of the article. He didn't want to minimize Suzanne Corkin's role in the project; he just felt that the article had overstated it. "I think of it this way," Annese wrote. "Suzanne is writing the personal and scientific biography of H.M.; we at the Brain Observatory are writing the anatomy. At some point in 2010 we'll compare notes and knock memory researchers off their chairs!" Annese didn't want to feel like he was doing anything behind Corkin's back, so before he pressed send, he cc'd her on the email. The following day, *The New York Times* made a small change to the online version of the article.

Here's how the new line read: "The dissection marked a culmination, for one thing, of H.M.'s remarkable life, which was documented by Suzanne Corkin, a memory researcher at the Massachusetts Institute of Technology who had worked with Mr. Molaison for the last five decades of his life."

It was such a subtle change that you might have to read the two lines one right after another to notice that the clause where Corkin was said to have "orchestrated" the dissection no longer existed. Annese was happy with the change at the time, but looking back, he believes that tiny tweak may have been one of the sparks for what would eventually become a secret custody battle waged between two major universities—MIT and UCSD—over one singularly famous human brain. Like all custody battles, this one would have victors and casualties, and among those casualties would be Annese's career, or at least his career as he knew it.

How to chart the death of a relationship?

In the case of Suzanne Corkin and Jacopo Annese, what may have begun as a minor unvoiced grievance slowly festered into something much worse.

Annese kept in touch with Corkin after the slicing, sending her updates addressed "Cara Sue" and receiving back smiley emoticons pecked out on her BlackBerry. On the surface, at least, things seemed as collegial as ever. Then, five months after the slicing, in April 2010, Corkin asked Annese to send her all of the block-face images he'd taken of Henry's brain. These were the high-resolution pictures made during the slicing, when a camera mounted above the microtome snapped a shot just before each pass of the blade. Corkin explained that Jean Augustinack, a young postdoc at the Martinos Imaging Center in Boston, where Henry had received his MRI scans, needed the images because she was planning to write a paper analyzing the anatomy of Henry's lesion.

Annese wasn't sure what to do. The block-face data set was one of the first products of his ongoing work with Henry's brain, which so far represented thousands of hours of toil, not to mention almost $750,000 in grant money. His plan had always been to make the block-face images public, accessible to anybody, but he intended to first write his own paper based on the data. In science, as in journalism, being scooped is a real fear, and the idea of giving up his data before he had a chance to publish made Annese's stomach turn.

After exchanging several emails, Annese phoned Corkin to discuss the matter and try to plan a way forward, and in his recollection she informed him, ominously, that *all* of the data he'd produced in San Diego ultimately belonged to MIT and Massachusetts General Hospital, not UCSD, and that his hesitancy to promptly turn over any of the block-face imagery she'd requested, along with portions of the tissue needed for a neuropathology exam, might end up damaging his career. Corkin also let him know that, in her view, he didn't have the "neuroanatomical competence" to do an effective analysis of Henry's lesion, and she mentioned that in certain press reports about his work with the brain, he had given short shrift to Corkin's role.

Up to that point, Annese had felt that his laboratory and Corkin's group were equal partners, bound by mutual respect. It suddenly was very clear the feeling hadn't been mutual.

After the phone call, Annese wrote Corkin an email defending his competence and stressing that it was his "responsibility to define the most opportune use of the data produced in my laboratory and the timing of its release. . . . We have the expertise to explore this complex data and I plan to share the results of the analysis as soon as the work is complete." He added that he would love for her to collaborate on any resulting papers, and that he hoped they could "always reach agreement about content, interpretation, and author listings." As for how reporters had portrayed Annese in the media, he told Corkin that he had always praised her work with H.M., and her foresight, and that if she was "concerned with instances in which I took unmerited credit for ideas, methods, or any science related to the project, I ask that you share these with me openly and freely." He ended with a plea for reconciliation: "I hope this clearly expresses my scientific and academic principles and what you should expect from me and my team. It would be a shame if misunderstanding polarized a potentially very productive collaboration; so I hope this can be avoided. As must be obvious, my enthusiasm and that of my lab for this project is second to none."

Corkin and Annese were both going to be attending the upcoming meeting of the Society for Neuroscience, at the San Diego Convention Center, in late November 2010, and at the end of October Annese sent Corkin some images she'd requested for use during her keynote presentation. Annese also provided a USB drive containing block-face images from Henry's medial temporal region. He did not, however, send the complete block-face data set or any of the tissue for the neuropathological exam. In the meantime, Annese and his lab mates continued the Henry-related work of the Brain Observatory, processing slides, building a three-dimensional digital model of the brain, and developing the software for the Web-based brain atlas. Annese knew there were unresolved tensions between him and Corkin, but he chose to put them out of his mind, hoping he could figure out a way to smooth things over down the line.

The relationship was already over, though. Annese just didn't know it yet.

In July 2011, William Bradley, the chairman of Annese's department at UCSD, received an email from Bruce Rosen, a colleague of Corkin's who ran the Martinos Bioimaging Center at Massachusetts General Hospital. The email was cc'd to Corkin and her colleagues at MIT/MGH but not to Annese.

"I wanted to establish a friendly contact with you to solicit your help in resolving a potentially challenging issue between Jacopo Annese and the team of investigators here at the MGH and MIT, led by Sue Corkin, who have been involved in studying H.M. for nearly five decades," Rosen began. He complained not only that Annese had been denying them access to "critical data that we need for our further analysis," but that "several press articles have come out about this brain, which fail to mention who the brain was donated to, and who was involved in this research. This certainly further inflamed the situation, even beyond the scientific issues. . . . Folks here are pretty frustrated, to say the least, and have even talked about legal action to get the brain back." He asked Bradley to tell Annese that "just running off with the brain is not really the right thing to do." Then he made a not-so-subtle threat: "I'm worried," he wrote, "that we may have some more fundamental psychopathology to deal with, but I hope that I'm wrong, and that he is just a misguided young scientist without appropriate mentoring to understand just how damaging to his scientific career this sort of behavior will be."

Bradley responded: "I suspect that what we have here is a failure to communicate rather than Jacopo running off with the goods. I haven't seen any evidence of that since he has been here at UCSD. But I should let him respond. I suspect this will all work out."

Rosen wrote back to Bradley, reiterating that Corkin and her group had "been trying for over a year to get him to return data, and the brain" and that they "were quite explicit about what they were expecting." He also reminded Bradley that "the group here was pushing for legal action, since the brain was donated to MGH legally, but I convinced them that you would help prod him to do the right thing. I think all sides would be greatly embarrassed if such a famous historical figure as H.M. became the object of a legal squabble."

In March 2013, a summit meeting was held in New York City to negotiate the conflict that had arisen over the control of Henry's brain. The meeting was held in a boardroom on the eighth floor of the headquarters of the Simons Foundation, on Fifth Avenue, and was chaired by Gerald Fischbach, director of life sciences at the Simons Foundation. The foundation may have been chosen to preside over the conflict in part because it was a neutral entity, one of the few major private or public scientific foundations that had not at one time or another funded research on Henry. Annese and Corkin were present, as was the president of the Dana Foundation, a $240 million agency that had funded some of the postmortem work with H.M.'s brain. A number of other researchers and administrators from UCSD and MIT and MGH were there, too, though the focus was clearly on resolving the discord between Annese and Corkin. Things had not improved since Rosen's email a year prior. In fact, communication had broken down entirely. Annese and his colleagues at the Brain Observatory had proceeded with their work, completing a 3-D digital model of Henry's brain based on the block-face imagery and performing the first postmortem measurement of the lesion. Annese had also written and submitted a paper based on that work to a journal called *Nature Communications*. Whether or not the withholding of the requested block-face imagery and the tissue samples was warranted—settling the ownership of that data was one of the issues on the agenda of the meeting—Annese's act of writing and submitting that paper without consulting with Corkin and her colleagues might be viewed at the very least as a violation of academic etiquette. The paper was sent to four peer reviewers. Three of the reviewers were very positive. The fourth reviewer was Suzanne Corkin. The meeting in New York City was called soon after she received it.

A week before the meeting, MIT had sent administration officials at UCSD a formal demand that they turn over all 2,401 slices of Henry's brain to a representative of MIT, who would shepherd it to a brain bank called the MIND Institute at the University of California, Davis. The MIND Institute was run by David Amaral, a former colleague of Corkin's. Amaral was present at the meeting in New York, as was Matthew Frosch, the neuropathologist who had, along with Annese, performed Henry's brain extraction, and Jean Augustinack, the MGH researcher who was working on her own paper based on postmortem neuroanatomical analysis of Henry's brain.

After Fischbach called the meeting to order, Annese began by telling the group that he hoped they could quickly draft a retroactively applicable materials transfer agreement to settle the dispute. A materials transfer agreement, or MTA, is a legal contract that typically accompanies any important biological specimen when it is moved from one institution to another. It sets guidelines as to whether the transfer is permanent or temporary, and defines ownership. An MTA had not been drafted when Corkin gave Henry's brain to Annese: Instead she just passed him the cooler containing the brain at the JetBlue gate four years before, with no paperwork whatsoever. Annese now told the group that he didn't really care where Henry's brain ultimately resided, though he wanted to make sure that whatever happened his laboratory would receive fair credit and recognition for all the time and money and labor it had invested and that the conflict they were in the midst of wouldn't somehow damage his reputation.

"I never felt like I owned the brain," Annese said. It appeared, at that moment, that the question of where Henry's brain would ultimately be housed—whether it would remain with Jacopo Annese at UCSD or be moved at MIT's request to David Amaral at UC Davis—might be settled quickly.

"And how would you like to set this up," Gerald Fischbach, the meeting chairman, asked, "if it doesn't really matter where the tissue resides? Do you want to decide that now? Or decide it by committee?"

Corkin interjected.

"It's not a decision by committee," she said. "Because it's a decision by the people who own the tissue. And that's MGH and MIT. And our decision is that we would like it to go to David's lab."

"In what sense do you own the tissue?" the chair asked.

"We have a brain donation form," Corkin said.

She then passed around a copy of the form, the one signed by Tom Mooney on December 19, 1992, a month after he became Henry's conservator.

The meeting lasted two and a half hours and became increasingly contentious. One of the items on the agenda was a discussion of the *Nature Communications* paper, which some at the table did not want to be published. Annese made what he hoped would be taken as a conciliatory gesture, telling Corkin that the journal's editor had agreed to add her name as a co-author.

Corkin said that adding "names to that paper wouldn't make it publishable. It needs to be rewritten. As a serious scientific document."

"It's not a serious scientific document?" Annese asked.

"It's not sophisticated scientific writing," Corkin said.

"That's not what the other three reviewers said."

Jean Augustinack, the researcher who planned to collaborate with Corkin on her own anatomical paper based on the MRI imagery of Henry's brain, jumped in: "Maybe the point is that [Annese's paper] is not the *definitive* anatomical paper."

"Of *course* it's not," Corkin said. "Of course it's not."

The chairman intervened.

"Somebody just read it and reviewed it," he told Corkin, "and they already determined it was publishable." Then, referring to Annese's and Agustinack's similarly themed papers, he continued: "We've got the knotty subject of two publications here. And I'm uncomfortable criticizing a publication that I have not been a part of. If Jacopo feels that, in his scientific judgement, it merits publication, and the reviewers have accepted it, well . . . It may not turn out to be *the* definitive paper. And that may be left to you, Sue, and your colleagues, or to you, Jacopo."

"It's just melodrama," Corkin said, referring to Annese's paper.

"The reviewers found it very sound," Annese said.

"You call him *Henry*," Corkin said. "I mean, it's so chatty! 'During the operation, Henry . . . ' "

"Please, Sue, send me your comments," Annese said. "One of the reviewers said the data was stellar."

The bickering continued, and the meeting ended a half hour later. In a follow-up memo that was sent to all the participants, Dana Foundation adviser Guy McKhann summarized what he saw as the major decisions made. He wrote that Corkin and her colleague Matthew Frosch would both be added as co-authors on Annese's paper and that they would provide their comments and edits to Annese by April 15, after which Annese would resubmit to the journal. McKhann also wrote that while the "ultimate location" of Henry's brain would someday be the MIND Institute at UC Davis, it would be up to Annese and Amaral to determine what tissue would be transferred and when.

Annese and Corkin both quickly sent the memo back to McKhann, each having heavily edited it. Regarding the ultimate location of the brain, Annese wrote the following: "Dr. Corkin *requested* that the ultimate location of the tissue will be with Dr. Amaral at UC Davis." As for Corkin, she struck through the line that said Amaral and Annese would collaborate on the timing and details of the transfer. Instead, she wrote, UCSD would simply "transfer all of H.M.'s brain tissue, including the tissue that has already been mounted," to an agent of MIT. At that point, "MIT and MGH will write a materials transfer agreement, which will allow UC Davis certain rights to use and distribute the tissue owned by MIT and MGH."

As far as Corkin was concerned, she and her colleagues owned the brain, period, and Annese had no say in the matter whatsoever.

Despite the ongoing and unsettled question of where Henry's brain would reside, the New York meeting did accomplish at least one thing: Now that Corkin and Frosch had been retroactively added as authors on Annese's *Nature Communications* paper, Corkin and Annese *had* to communicate to complete the revisions process, breaking the long and chilly silence between them. What this new correspondence revealed was that despite what she'd said during the meeting, Corkin's central problem with the paper, the one she pushed back on hardest, didn't have much to do with Annese's chatty writing style. Instead she was concerned with something Annese had discovered in Henry's brain.

Specifically, Annese's analysis of the block-face images and the tissue itself had revealed a previously unreported lesion in Henry's frontal lobe. The lesion was in the left hemisphere and appeared to have been caused by a man-made object. In his draft of the paper, Annese speculated that my grandfather had created the lesion during the operation while he was levering up Henry's frontal lobes to access the medial temporal lobes. This was a significant finding. As one of the paper's anonymous peer reviewers pointed out, "much of the neuropsychological literature on H.M. has made the case that so-called frontal function was intact." This stood in stark contrast to the new paper, where, the reviewer stressed, "Annese and colleagues unequivocally demonstrate that frontal white matter was affected, likely by the surgical approach. Their future investigations that will examine detailed histology in this region will be of great importance, because if there *is* significant white matter pathology within frontal systems, that may have implications for a retrospective reinterpretation of H.M.'s neuropsychological findings." In other words, for the previous six decades neuropsychologists such as Corkin had interpreted their experimental results with Henry under the working assumption that his lesions were restricted to the medial temporal lobes. The discovery of this new lesion in a different part of Henry's brain might call some of their conclusions about the functions of the medial temporal lobes into question and require a reexamination of all that old data.

When Corkin sent Annese her revisions of his paper, she deleted all references to the frontal lesion. In a note to Annese, she explained that "the frontal lobe lesion does not appear on either the in situ scans [the MRI scans made while the brain was still in Henry's skull] or the fresh brain photos" and that "any consideration of it would be highly misleading." She followed up with an email stating that she and her colleagues at MIT and MGH "believe that there is a good chance that the alleged orbitofrontal lesion is a handling lesion," meaning that it was caused after death, during the extraction and subsequent handling of the brain. She added that "there is no intent by the Martinos [MGH] group and me to hide evidence."

Annese responded with a series of images from in situ MRI scans that, contrary to Corkin's assertions, appeared to give clear views of the lesion, demonstrating that it could not have resulted from the postautopsy handling of the brain. Annese also sent imagery of some of the slides bearing that portion of Henry's frontal lobes, and these slides seemed to confirm that the lesion resulted from the use of a man-made object, such as a flat brain spatula. The lesion, Annese wrote to Corkin, "was previously unreported (we ascertained it was present even in the 1992–93 MRI scans) and together with other data represents new evidence in the case. I really don't understand the reluctance; this is not image data reconstructed from K-space; this is real, flesh and blood.

There's a lesion outside the medial temporal lobes, it is conspicuous and it should be reported. Remember, the goal of this paper and the archive is to catalyze new investigations as well as new debates, like the one we have been having."

The arguments over the lesion and other aspects of the paper continued for months and devolved into acrimony. An outside mediator hosted a conference call, and eventually a compromise was reached. The frontal lesion would stay in the paper, though it wouldn't be featured as prominently as it had been in earlier drafts.

On January 28, 2014, *Nature Communications* published the article, "Postmortem Examination of Patient H.M.'s Brain Based on Histological Sectioning and Digital 3D Reconstruction." Annese was the first author, but Corkin and Matthew Frosch were authors on it as well. The paper's published findings included the discovery of a "circumscribed lesion in the orbitofrontal cortex," which was described as "new evidence which may help elucidate the consequences of H.M.'s operation in the context of the brain's overall pathology."

The publication of the paper felt to Annese like a victory, but it may have been a Pyrrhic one.

Shortly after the New York City meeting, lawyers from MIT began negotiating with lawyers from UC San Diego, finally drafting a materials transfer agreement that would retroactively settle questions left hazy during the brain's move from Boston to San Diego. In the first draft of the contract, UCSD proposed an eighty-twenty split: That is, they would hold on to approximately 20 percent of the brain, including most of the slides processed by Annese's lab. The rest of the tissue would be given back to MIT, which planned to loan it to David Amaral's brain bank at UC Davis. MIT rejected that proposal, demanding that *all* of the material in question be transferred away from UCSD. "Materials" were defined as "the Brain, Unmodified Derivatives, and any other substances created through the use of Materials." That would include the slides Annese had made.

On September 18, the chairman of Annese's department sent Annese an email:

"As you know," he wrote, UCSD had "been negotiating with MIT's lawyers to try to let you keep at least 20% of H.M.'s material . . . unfortunately, they are adamant that they want it all back now, and will likely sue the university if they don't get it." He added that UCSD's chief counsel had reviewed the matter and did not believe UCSD would prevail against MIT if such a suit were filed.

Six weeks later, the head of UCSD's department of research wrote MIT's lead counsel a letter in which she stated that UCSD agreed "to transfer all brain tissue to an MIT representative" and that she was "hopeful that this brings the matter to a mutually agreeable close."

Almost five years after watching Jacopo Annese board that JetBlue flight in Boston, Suzanne Corkin had regained control of Patient H.M.

On February 5, 2015, Annese sat on the floor of his condominium and sorted through the slides of Henry's brain. The raw, unprocessed tissue samples were still in their refrigerators at the lab, but all the slides were there at home with him. They filled three boxes and fifty trays, 214 slides in all. He removed each from its tray, one by one, and inspected it carefully. He was looking for cracks, for decomposition, for any problems that he might have to alert the administration about. In the lower right-hand corner of most slides was a little laser-etched logo of his own design. THE BRAIN OBSERVATORY, it read in cursive script. He hoped that Corkin wouldn't order these etchings scratched off, but he knew that was going to be out of his control.

One week prior, on January 29, Annese had submitted his resignation to UCSD. "I believe that, regretfully, this is the only way to provide a dignifying narrative to the changes in course of the H.M. project," he wrote to the vice chancellor. He had decided he could no longer fight. He'd felt the university shift its support, felt that they weren't willing to fight for him. He could have continued fighting, on his own, and a part of him had wanted to. A part of him was screaming for him to. He'd even told the university's HR manager at one point that he'd rather go to jail than give up Henry's brain. The whole thing just wasn't right, he felt. Corkin had given him the brain, and he had poured years of his life into it. But he knew it was a fight he wouldn't win. It was a fight that would take lots of money, lots of lawyers. He couldn't do it.

He was giving up the brain. He had to let it go.

He'd worried about whether he could do so, when the time came. He'd worried about whether inspecting the slides would be too hard emotionally. He thought that looking at them, handling them, would be like pouring alcohol on an open wound. The truth was, MIT and UCSD were both worried about the same thing. After Annese resigned, he sent a letter to Maria Zuber, MIT's vice president for research, informing her of his resignation and telling her that he hoped they could "coordinate future steps involving the remaining tissue from Patient H.M.'s brain" and that his Brain Observatory might somehow play a continuing role. This was interpreted as a veiled threat to hold on to or destroy the parts of the brain that remained in Annese's possession and provoked a flurry of emails between high-level administrators and attorneys at the two institutions. "I am writing to memorialize MIT's grave concerns about the safety and security of all of the tissue of the brain of H.M.," one MIT lawyer wrote to a UCSD counterpart. "Dr. Annese's email to Professor Zuber below indicates he may be taking unilateral steps to plan for the future of the tissue." UCSD's emails back to MIT were placid on the surface but increasingly unsettled behind the scenes. "Sounds like you have had another crazy day of H.M. brain chaos," one UCSD official wrote to another. "I think we all need a happy hour after this."

Finally, Annese's former chairman reached out to him, then sent a reassuring note to his UCSD colleagues.

"I just got off the phone with Jacopo who was quite reasonable," he wrote. "The only H.M. material that was transferred out of the RIL [Radiology Imaging Laboratory] were the slides which do not need to be refrigerated or anything. All the other material is still at the RIL, properly protected as always. He will meet you at the RIL with the slides at 4pm."

Four P.M. was approaching fast, but Annese was taking his time packing the slides. Each one, to him, was a little work of art, the culmination of a career spent honing his skills, the product of so much accumulated work. He thought about the journey he'd been on over the past several years, from the crowded autopsy room on that winter's morning in Massachusetts to the odd cross-country JetBlue flight with the cooler on the seat beside him; from the cutting that he'd livestreamed to the world to the countless private hours spent in his laboratory at night, Ennio Morricone in his ears, teasing apart the little pink slices of tissue that floated in front of him, each finally coming to rest on a bed of glass.

He knew that it was his ego that made it so difficult to let go, that recoiled at the thought of Amaral and Corkin taking his slides, his data. He'd grown to believe that for Corkin, this fight was about control, and recognition, and that for Amaral it was about acquiring a very important collection that he didn't make. If he was being honest with himself, he'd admit that a part of him *had* considered holding on to the brain, or at least some of those beautiful slides he made with it. Annese's ego gave him a sense of entitlement and attachment. A sense of outrage, too. Now he tried to tamp it down. He'd taken up yoga recently. In his better moments, he could conceive of science as being a public trust. It wasn't easy, though. Everybody liked to talk about "open data," but when it came down to it how many people could just walk away from what they'd spent years of their lives creating?

He had to try.

It was nearing Valentine's Day, and he'd been streaming an Internet radio station that had assembled a collection of anti–Valentine's Day songs, with titles like "Fuck You," "Evil Woman," and "Over You." He inspected his slides, one by one, the bitter songs ringing in his ears. And then a funny thing happened: He realized he didn't feel as bad as he'd thought he would. He realized that somehow, somewhere, at some point during this long, strange odyssey, he'd already let go. Looking at the slides, he began imagining a moment five or more decades in the future, tried to imagine some eager and brilliant young student somewhere holding one of those pieces of histological art up to the light, appreciating it for the historic and gorgeous and mysterious object it would always remain, not giving a damn who owned or controlled it.

Later that afternoon, Annese put the slides in the back of a rented van, drove them to the former site of the Brain Observatory, and handed them over to the vice chancellor, who was waiting there with two campus police officers to make sure everything was accounted for.

Back in Corkin's office, I pressed for her side of the story, her perspective on the fight over the brain.

"I don't want to talk about it," she said. "You know that we have a brain donation form, right? That says it all."

"I'm curious about the paperwork. What was the—"

She cut me off.

"I can't talk about the paperwork," she said.

I asked her about Tom Mooney, the man she'd arranged to become Henry's conservator, who had signed the brain donation form. On the form, Mooney had described himself as Henry's closest living relative. I asked whether she could tell me how, precisely, Mooney was related to Henry.

I had tried to ask Mooney the same question. I made numerous phone calls and even showed up on his doorstep once. He always made excuses for not being able to meet with me.

I asked Corkin whether she was aware that when Mooney became Henry's conservator and signed over Henry's brain, he was not in fact Henry's next of kin, that there had been a number of first cousins of Henry's living nearby at the time. I mentioned one of them, Frank S. Molaison.

"I was not aware of his existence," she said.

I asked whether she had ever done any genealogical research at all into the man she had studied for almost a half-century.

"No," she said.

"So," I said, "you were not aware that Mr. Mooney was not his next of kin?"

"No," she said.

I asked why she had started the process of finding a conservator for Henry in the first place.

"I just wanted another level of security. Another person who was not amnesic, and who had Henry's best interests at heart."

I asked what she meant by "security." Security from what?

"For Henry," she said. "For MIT."

And what were MIT's vulnerabilities?

"I don't know," she said. "I'd have to ask our lawyers that."

After the eighty-twenty deal collapsed, the MIT and UCSD lawyers continued to negotiate over a final point of contention: the high-resolution block-face images that Annese and his colleagues had acquired during the slicing. UCSD proposed that this data be kept on its servers, accessible to all but still the acknowledged property of the institution that had created it. MIT turned down this proposal. David Amaral sent an email to UCSD explaining why he also believed the digital data along with the physical materials should all be kept at his lab, under one roof.

"My major concern," he wrote, was "that posterity judges us as good custodians of this precious resource." He added that he was happy to acknowledge Annese's contributions to the analysis of H.M., but that he would "deeply regret" if portions of H.M.'s "data legacy were not put into a permanent, accessible resource because of politics or legal issues. We both grew up with H.M. in our textbooks and we need to work together to insure that the final chapter of his life and legacy is treated with respect and the highest scientific standards."

Amaral's email was forwarded to UCSD's assistant vice chancellor, Marianne Generales, who forwarded it to a colleague along with a one-line comment.

"What a crock!" she wrote.

But in the end, UCSD did agree to that final demand: The digital data, too, would go.

It struck me that this was not necessarily a bad thing, and that Generales's comment may have been unfair. However you interpreted the conflict leading up to the removal of Henry's brain from San Diego, once that transfer happened, it seemed reasonable to want everything consolidated: all the material, all the data. This would make it easier for scientists to continue their analysis of Henry's brain, mining it for whatever last revelations it contained. Amaral's email, from that perspective, made perfect sense: Who wouldn't want H.M.'s "data legacy" to be put into a "permanent, accessible resource"?

Ideally, of course, that trove would include the data that had been collected from Henry while he was alive, all the experimental and observational information that scientists had extracted from him after he left my grandfather's operating room.

So toward the end of my interview with Corkin, I asked what *she* intended to do with her Henry files, the raw, experimental data she'd spent her career gathering. The whole idea behind preserving Henry's brain, after all, was to be able to compare and correlate his neuroanatomical data with the unprecedented amount of behavioral data that already existed in his case, most of it presumably stored away in Corkin's files.

Me: Are you aiming to give his files to an archive?
Corkin: Not his files, but I'm giving his memorabilia to my department. And they will be on display on the third floor.

By memorabilia, she meant his personal effects—his bible, his journal, his crucifix—all of which she owned. She also claimed copyright to every known family photo of Henry and his parents.

Me: Right. And what's going to happen to the files themselves?

She paused for several seconds.

Corkin: *Shredded.*
Me: Shredded? Why would they be shredded?
Corkin: Nobody's gonna look at them.
Me: Really? I can't imagine shredding the files of the most important research subject in history. Why would you do that?
Corkin: Well, you can't just take one test on one day and draw conclusions about it. That's a very dangerous thing to do.
Me: Yeah, but your files would be comprehensive. They would span decades.
Corkin: Yeah, well, the tests are gone. The test data. The data sheets are gone. Because the stuff is published. Most of it is published. Or a lot of it is published.
Me: But not all of it.
Corkin: Well, the things that aren't published are, you know, experiments that just didn't . . . [another long pause] go right. Didn't. You know, there was a problem. He had a seizure or something like that.
Me: But you know, even what's published . . . As you know, if you look at the papers, in some sense each paper is just the tip of the iceberg of the work that was done, and the work that was done—all that data floating underneath—it seems to me that so much of that would be valuable to preserve. That people really may want to go back and review . . .
Corkin: There's no place to preserve it.
Me: There's no place to preserve it? Not at MIT? How many files are we talking about, roughly? Are we talking about a storeroom like this, full of boxes of papers?
Corkin: No, not that much.
Me: Are they mostly at your home now?
Corkin: Some of it was. No, not now. It isn't. No.
Me: It's just in storage somewhere?

Corkin: Most of it has gone, is in the trash, was shredded.
Me: Most of it was already shredded? Just recently?
Corkin: Yeah. When I moved.

She had moved to a new home not long before.

Me: When you moved you *shredded* it?
Corkin: Mm-hmm.
Me: And what is left, most of it you're planning to shred?
Corkin: Probably.

Elements of her story seemed to be shifting and flexing in real time. Whatever the details, though—whatever Corkin had or hadn't yet shredded—the whole idea of willfully shredding *any* of Henry's data struck me as horrific. What she was telling me appeared to violate not just basic research ethics and scientific accountability but also something harder to articulate.

Me: Not to sound too high-flown here, but I could see future generations being disappointed that the primary-source documents for the work that was done on Patient H.M. had been destroyed.
Corkin: Well, I mean, there are other famous amnesic patients, and their data aren't available to the public.
Me: But why would . . . It seems to me, and I think it gets back to this: He's somebody that has been so fundamentally important to our understanding of ourselves. And it seems to me that the data that was used to provide this understanding of ourselves is almost a common heritage.
Corkin: Yeah, but it's not peer-reviewed, for one thing. That's important. The stuff that's published is good stuff. Peer-reviewed. You can believe it. Things that, you know, experiments that might not have been good experiments, there might have been inadequate control groups. . . . There are all sorts of things that can go wrong with experiments. Not every experiment is publishable.

Me: But they can still be interpreted by other people. Maybe as we continue to understand how the brain works, and how memory works, some of this existing data of H.M.'s could be reilluminated by new theories, by new ideas, by new . . . It just seems a shame to destroy it. And it also seems—and this would be the darker interpretation of it—it locks in stone your own telling of H.M.'s story.
Corkin: Well, it's not just me. It's me and over a hundred colleagues.
Me: I know. But again, you're the principal investigator for the last many decades. And that is the story, then: When you destroy the data, that becomes the inalterable and sort of inviolate story of Patient H.M. And if you do destroy it, I can imagine people saying, well, there certainly could be a self-serving motive there.
Corkin: I don't think scientists would say that.
Me: Okay.
Corkin: I think people like you might say that.

She was wrong about that. It wasn't just people like me. Later I told a number of scientists what Corkin had told me about her shredding of Henry's data. Almost all of them reacted with horror. In addition, some told me it fit a pattern in Corkin's behavior. Donald MacKay, the UCLA psychologist who had documented Henry's language deficits, said that the shredding was "not the first outrageous data protection strategy she's used." In 2006, after Corkin had co-authored a paper about Henry's crossword puzzle performance, describing Henry as having "erudite puzzle-solving abilities," MacKay wrote to Corkin requesting to see the actual puzzles she had based the paper on, since MacKay thought it was highly unlikely that Henry was in fact capable of high-level puzzle solving. Corkin refused to provide MacKay with the puzzles, and MacKay told me he'd sent a complaint to the American Psychological Association, since he considered such data hoarding to be a violation of research ethics.

Even as a nonscientist, I couldn't help notice that some of the unpublished data I'd come across while reporting this book went against the grain of the established narrative of Patient H.M. Take psychologist Liselotte Fischer's assessment of Henry, the one she conducted the day before the operation. Fischer's report is the only documentation of Henry's preoperative cognitive abilities. The report is three pages long, and over the years, small portions of it have been published, but most of it has not. When I saw it, some of the unpublished parts immediately leapt out. In describing Henry's attempts to grapple with arithmetic problems, Fischer wrote that, "in solving, or rather attempting to solve, simple arithmetic examples, he has immediate difficulties in recalling the tasks, then usually goes off on a wrong computation track, then recognizes this, but has forgotten the actual task." Later Fischer concluded that, "while most areas of intellectual functioning, including Gestalt perception and abstraction, are intact, number concept and learning of new material are severely impaired, so that his functioning is grossly uneven." In other words, Fischer's report provides evidence that on the day *before* the operation that transformed Henry Molaison into the amnesic Patient H.M., Henry's memory was "severely impaired." This fact has, at best, only been hinted at over the years—the original 1957 paper about Henry, for example, mentioned that Henry had shown an "improvement in arithmetic" after the operation when measured against Fischer's preoperative testing, though the paper failed to mention that his preoperative struggles with arithmetic stemmed from an inability to recall the task at hand. More recently, Corkin wrote that although Fischer's testing of Henry had revealed a below-average ability to remember a random string of numbers, Henry's "deficits before the big event were likely a combined result of seizures and nerves."

The causes and significance of Henry's preoperative memory deficits can be debated—were they the result of his anxiety, or his seizures, or some preexisting neurological condition?—but it seems hard to justify omitting Fischer's full report from the scientific record of the most important research subject in the history of memory science, which is what has been done. I wondered what other surprises might be found in a full accounting of Henry's unpublished data, at least the data that hadn't already made its way to Corkin's shredder.

You didn't need to be a scientist to grasp the essence of what was happening: My grandfather had cut a hole into Henry's memory, and now one of the many people who had profited from that act was cutting another hole, this one into our memory of Henry.

It occurred to me to ask Corkin another question I'd been wondering about. According to probate court documents, the Bickford Health Care Center had received all of Henry's Social Security and Medicaid payments, from which they disbursed to Henry exactly thirty dollars a month—less than a dollar a day—for his "personal needs." "The MIT Research Project Known as Patient H.M.," meanwhile, generated incalculable personal, professional, and financial benefits for Corkin and her colleagues. I had always thought that most people who participate as research subjects, at MIT or other universities, received payments for their time, and Henry had, by all accounts, spent more time as a research subject than anyone in history.

Me: Did Henry ever receive compensation for his work as a research subject?
Corkin: We had . . .

She paused again.

Corkin: We were studying a lot of Alzheimer patients then, who were also amnesic. And also Parkinson's patients, who were demented. Some of whom were demented. And we got a waiver from the IRB [Institutional Review Board] not to compensate these people. Because they got benefits from being tested. And from physical exams. Like, Henry had a basal cell carcinoma removed from his ear, at no cost. And also, we just gave him things whenever we thought he needed something. Like clothes. Or a video player. Stuff.

In other words: No.

. . .

On the way out of Corkin's office, I noticed a framed photograph of Henry's brain hanging on a wall. The photo was beautifully lit, professionally shot. It had been taken in Annese's lab, right after the night Annese had peeled the brain, removing the various membranes that had cloaked it, leaving it fully exposed. The brain was, in its own way, beautiful, even if you divorced it from context, even if you didn't know who it belonged to or what it had taught us. Even if you didn't know anything about its story.

The photo showed the brain in profile, close up. It was pink, the pink of a ballerina's slippers, though a complex network of dark purplish veins crisscrossed its surface. It had a squat but aerodynamic profile: The frontal lobe angled down toward the surface like the nose of a sports car. A gentle curve sloped back from there, up and over the top of the cortex before dropping down again toward the rear, flaring inward at the base of the occipital lobe, then flaring outward around the thick foundation of the cerebellum. The surface of the brain, the portion visible in profile, was a labyrinth: ropey tubes of gyri folded over each other in complex, compact ways, sulcal trenches dividing them. Within the tangle of sulci and gyri there were other, broader and more subtle demarcations, including the one between the temporal lobe and the frontal lobe. You might, with an educated eye, notice that the cerebellum was smaller than most, withered away by years of Dilantin, but the real damage, the lesions that transformed Henry Molaison into Patient H.M., was not visible in the photo. The outer portion of Henry's temporal lobes was intact; my grandfather had only destroyed their innermost medial structures. To see that destruction, that vacuum, you'd have to flop the brain over onto its side, exposing its underbelly. In the photograph that hung on Corkin's wall, Henry's brain looked like yours, or hers, or mine, or anyone's.

There was something aquatic about it, like a creature you might encounter while diving too deep in a dark underwater cave. Staring at it, I remembered something Annese had told me once. He was complaining about the current trend of depicting the brain, with its myriad neural networks, as though it were some sort of electric metropolis. You could hardly glance at a newsstand's magazine rack, he pointed out, without seeing a CGI cover illustration of the brain looking like a coruscating fiber-optic fantasia, as though we were all walking around with Times Square blazing in our heads. Annese didn't like that all-electric metaphor. He saw the brain as more organic than that. Earthier. Not like a lightbulb; more like an oyster.

That didn't strike me as right either, though.

It's not a pneumatic pump, a telephone switchboard, or a tape recorder.

Maybe the human brain is an object beyond the reach of metaphor, for the simple reason that it is the only object capable of creating metaphors to describe itself. There really is nothing else like it. The human brain creates the human mind, and then the human mind tries to understand the human brain, however long it takes and whatever the cost.

I took one more look at the picture on the wall, at those ripples like molten copper, and tried to commit the moment to memory.

EPILOGUE

On a cold, windy morning in Connecticut—the car radio told me the city was experiencing twenty-year lows—I visited Hartford Hospital. I'd made an appointment to meet the hospital's archivist, and he'd prepared for the meeting by rounding up a bunch of my grandfather's old neurosurgical tools. He'd found them in storage. He brought them into a conference room, in a cardboard box, and placed them on the table. I pulled them out one by one.

There was a brace that looked almost like orthodontic headgear and could be adjusted with butterfly screws.

There were a handful of Scoville clips, used for pinching off aneurysms, made of an iron alloy called austenite.

There were oddly bent scissors, designed to deal with the tricky angles neurosurgeons navigate when cutting through veins or membranes.

There were scythe-shaped tools, probably hand-soldered by my grandfather, and some long, piercing tools like awls.

There were a few trephines, those circular bone drills with their raw, serrated edges.

Everything had been sitting untouched for years, if not decades, and many of the tools were flecked with bits of reddish rust.

At the bottom of the box I found a strange pair of glasses. They looked like 1950s-style horn-rims, but they had long magnifying lenses attached to the tops of the regular lenses: surgical loupes. It occurred to me that the last person to wear them had been my grandfather. I tried them on.

They were too small for me, and the sides clamped hard against my head. When I first put them on, my eyes naturally settled their gaze through the main lenses, which seemed to be just plain glass, no prescription. Then I shifted my eyes up, looking through the loupes. It took a few moments to adjust. I held my hand out in front of me. At two or more feet away, it was just a pinkish blob, like looking at a shell underwater without a diving mask. I moved it closer, and at about eighteen inches my hand began to materialize into a recognizable form. A foot away was the sweet spot. At that distance, my hand looked gigantic but incredibly sharp. I could only see a small patch of it at a time, though, and slowly scanned over my skin's alien topography. Looking through the loupes, I found that my field of vision was magnified but extremely narrow. I could make out the individual ridges on my fingertips, but everything else was a blur.

I made that visit to the hospital a long time ago, during my first reporting trip to Hartford, back when I was just beginning to look into the story of Patient H.M. A decade has passed. But still: I can imagine it. I can conjure up little scraps of the scene, flashes of how it felt looking through my grandfather's loupes in that conference room, on that chilly morning. I'm here and I'm there all at once.

And here, now, looking back at the beginning of this story from the very end of it, I'm thinking about my grandfather, about the people he helped and the people he hurt. I'm thinking how, when he put on those strange glasses and leaned in, all context, all sense of place, had to have just melted away. All he could see was what was right in front of his eyes.

I'm thinking about Henry. About what he sacrificed for us, and what we gave him in return. I'm thinking how, for all that separated my grandfather and him, that tight, constricted field of view was something they had in common. The holes my grandfather made in Henry's brain marooned Henry on a small island where all he had, all he could see, was a sliver of the present, a place where the past and the future were nothing but indistinct blurs. Unlike my grandfather, though, Henry had no choice but to see the world that way.

I'm thinking about some of the people who studied Henry. Brenda Milner, Suzanne Corkin, Jacopo Annese. How each took something from Henry's loss, and eventually each, in turn, lost Henry.

I'm thinking about John Fulton, Charles Burlingame, Walter Freeman.

I'm thinking about monkeys, and I'm thinking about men.

I'm thinking about women: Patient A.R., Patient R.B., Patient G.D. Hundreds more. Thousands.

I'm thinking about my grandmother, Bambam.

I'm thinking about my mother, my sisters, my daughter.

I'm thinking about how some stories are just the opposite of how things looked through my grandfather's loupes. Some stories, the longer you look at them, the wider their lenses become and the more they start pulling other parts of the world into view. Other people, other places, other times, all swirling together before they slip away, like when you're just waking from a dream.

Ninety-three-year-old Karl Pribram moved in his chair and groaned again and looked like he was trying to get up.

Marlene, his caregiver, asked if he needed to go to the bathroom or if he wanted lunch.

"Not particularly," he said, and settled back down. He looked at me. "A lot of different things are happening, and I'm not sure what I'm supposed to be doing. Am I supposed to be answering your questions?"

After he'd told me what he remembered my grandfather doing to my grandmother, Pribram had begun telling me about the day he first met Henry. We'd been talking for a while by this point, and he was getting foggier. People who undergo cancer treatment often develop memory problems, a result of the damage chemotherapy drugs do to the hippocampus. Pribram was trying to tell me his story of meeting the world's most famous amnesic, but he was struggling under the weight of his own failing memory, and his sentences were becoming a jumbled pile of fragments. As best as I could tell from what he'd said so far, sometime not long after Henry's operation Pribram had walked the short distance from his monkey lab at the Institute of Living to my grandfather's office at Hartford Hospital. He had arranged to meet Henry, and he was eager to do so. After all, this was a human being who had just received the same sorts of lesions that Pribram back then spent his days making in the brains of macaques. My grandfather introduced the two men, then excused himself, since he had been called to attend to another patient. Pribram and Henry were left alone. They began chatting. They talked for a while and eventually discovered they had something in common. They each, it turned out, dreamed of one day making an overland trip from Cairo to Cape Town.

"You were telling me about the day you met H.M.," I reminded Pribram. He nodded, then began to speak. His words came slowly. Short sentences, long pauses between each one.

I'll tell you exactly what happened.

H.M. and I were talking. And Bill was called out.

I guess Bill was there, too.

Bill was called out of the room. And we were left together. And I was.

Bill had to go out, was called out on a case. H.M. and I were left together. And we discussed our Cape Town–to-Cairo things that we were both so much interested in.

And Bill had to go do something else.

H.M. and I were left alone. And kept talking about Cape Town to Cairo.

He was called out. Scoville was called out for something.

And I was left with H.M.

So we kept talking about the whole Cape Town–to-Cairo business.

And then Scoville was called out and H.M. and I were left alone.

And we talked about our mutual interest in Cape Town to Cairo.

And then Scoville was called out of the room.

And H.M. and I were left alone.
And I remember vividly.
What happened was.
Something like.
Whatever was going on.
Called out of the room.
Came back in.
And H.M. said, "Have you been in here before?"

ACKNOWLEDGMENTS

I've been working on this book for six years, and hundreds of people have helped me along the way, providing editorial, emotional, and financial support. There's no way I can list everyone who deserves to be thanked, but here are a few.

Andy Ward is my editor at Random House. Andy is ridiculously talented, enormously patient, and contagiously enthusiastic. Are there better qualities for a book editor, or a human being? (If Andy were editing the preceding two sentences, he'd remind me that I have a bad habit of overusing adverbs *and* rhetorical questions, but I know for a fact that he avoids reading nice things about himself, so I think I'm safe.) Others at Random House who've been a big help include Juliet Brooke, Kaela Myers, Beth Pearson, Richard Elman, Jennifer Prior, Evan Stone, Kathryn Jones, Maralee Youngs, Lawrence Krauser, Greg Kubie, Carolyn Foley, and Caitlin McCaskey. Also, if the words in this book aren't any good that's my fault, but at the very least they're coming to you in a striking package. That's thanks to the Random House art department, including my cousin, art director Paolo Pepe, the cover designer, Evan Gaffney, and the interior designer, Simon Sullivan.

Tyler Cabot is my editor at *Esquire*. We teamed up on an article about Patient H.M. in 2010, and that article led directly to this book project. Tyler is a brilliant editor and a great guy, and one of the hardest things about the last few years has been having to cut way down on doing magazine stories with him. I hope to team up with him again soon. Others at *Esquire* who've helped me through this process in one way or another include Mike Sager, Tom Junod, Kevin McDonnell, Ryan D'Agostino, Chris Jones, Bob Scheffler, Peter Griffin, and David Granger. Also, my former editor at *Esquire,* Terry Noland, assigned me my first, aborted Patient H.M. story, way back in 2005, which got this whole ball rolling.

Sloan Harris is my agent at ICM, and he's had my back throughout, while also providing fantastic editorial notes. His passion for this project, and his occasional whip-cracking, have both been essential. Heather Karpas at ICM has also been great.

Archivists and librarians tend to be amazing people, possessing all the investigative skills of good journalists, but none of the ego. Among the ones who've helped me dig up material for this book are Steve Lytle at Hartford Hospital, Lori James at the Institute of Living, Melissa Grafe at the Yale School of Medicine, Lily Szcygiel at McGill University, Jessica Murphy at the Harvard Medical School's Center for the History of Medicine, Allen Ramsey and Mel Smith at the Connecticut State Library, and Jennifer Kinniff at George Washington University.

In order for me to tell this story, dozens of people have had to tell me their own. Some of those people are major characters in the book, and I imagine that their contributions are obvious to the reader. Jacopo Annese, Brenda Milner, Mortimer Mishkin, Karl Pribram, Suzanne Corkin, Dennis Spencer, Howard Eichenbaum: Each was invaluable in helping me understand one or another aspect of the larger story. (Sadly, two of them, Pribram and Corkin, passed away while I was completing work on this book.) Other people were featured less prominently, or didn't appear at all, but still provided crucial insights. These included Justin Feinstein, Karim Nader, Daniel Nijensohn, Goldie Nijensohn, Nir Patel, Ruth Klaming, Natalie Schenker-Ahmed, Colleen Sheh, Paul Maechler, Oliver Hardt, Alice Cronin-Golomb, Marilyn Jones-Gotman, Nancy Hebben, Duncan Hunter, Eileen Mucha, David Mumford, Donald MacKay, Lori Johnson, Phil Hilts, Jenni

Ogden, Linda DeLisle, Sandra Parlee, Myra Crowley, Oliver Hardt, Bette Ferguson, and more. Many, many more. Thank you all.

Then, of course, there's Henry. I never met him, and wish I had. Sometimes, thinking of Henry, I'm reminded of what Kurt Vonnegut wrote about the carpet bombing of Dresden, the inspiration for his novel *Slaughterhouse-Five*. Vonnegut, writing of the bombing, stated that "only one person on the entire planet got any benefit from it. I am that person. I wrote this book." I can't claim to be the only person to benefit from Henry's misfortune, but having written this book, I am now one of the many profiteers. I can't ever repay the debt I owe Henry. None of us can.

Apart from Andy and myself, my mother, Lisa, and my sister, Laska, have probably had more contact with this book than anyone. I've read drafts of most of the chapters aloud to them, leaning on their ears, their brains, and their hearts. I simply couldn't have done this without them. I know some parts of this story were difficult for them to hear, particularly the ones about my grandmother, Bambam. She died a few years ago, just before I dove into the most disturbing aspects of her own personal history. She was a quiet, private, much loved woman, and I have mixed feelings about dragging her darkest moments into the light.

Lots of good friends and family members have also listened to big chunks of this book, or just listened to me ramble on about it, and provided helpful feedback. These include Vance Jacobs, Matt Moyer, Amy Toensing, Sara Tillett, Angie Conte, Katherine McCallum, Genesee Keevil, Mich Gignac, June Keevil, Sarah Dohle, Kat Roberts, Bil Roberts, Gail Roberts, Darby Newnham, Ian Stewart, Moira Sauer, Carrie Watters, Andrea Learned, Olda Dittrich, Didi Pershouse, Edward Pershouse, Paige Williams, and Justin Heckert. Two eagle-eyed and sharp-brained neuroscientists at Stanford, Aaron Andalman and Matthew Lovett-Barron, read through the entire manuscript and gave me extremely useful notes, though any remaining scientific errors are on me.

Finally, my daughter, Anwyn. In the years I've been working on this book, I've been away too much, distracted too often. That sucks, and I want to change that. Partly for Anwyn, but also for selfish reasons, since a lot of my favorite days on this planet have been spent with her, building snow forts in the Yukon, exploring caves in Greece, or climbing mountains in New York. Love you, Anwyn. Can't wait for our next adventure.

INDEX

Academia Municipal Taurina, 234–35
Ace Electric Motors, 21–22, 135
Adams, Rebecca, 150–51
Adirondacks, 105–8
adrenal glands, 94, 152
adrenaline, 72
air pressure experiments, 110–12
Alabama, 118
Albright, Fuller, 91–94, 97
alcoholism, 228
Alexandria, 116
Alexandria Quartet (Durrell), 33
Alzheimer patients, 404
Amaral, David, 388–91, 393, 396, 398–99
ambiguity, 305–7
American Academy of Neurological Surgery, 364
American Association of Neurological Surgeons, 266
American Medical Association, 82, 117
American Neurological Association, 92, 184–87, 189
American Physiological Society, 127
American Psychiatric Association, 147
American Psychological Association, 402
amnesia, 199, 220–23, 241, 253–58
  gluttony and, 42
  of Henry, *see* Molaison, Henry Gustave, as Patient H.M., amnesia of
  infantile, 108–9
  of Jason, 252–53
  in monkeys, 242, 244–45
  in movies, 220
  pain tolerance and, 42–43
  short-term, 68, 172
amygdala, 128, 156, 173, 195, 211, 213, 224, 244, 318, 328
anesthesia, 15, 102, 117, 169, 191–94, 198, 210
aneurysms, 153
animal experiments, 80–81, 111, 114, 127–28, 129, 148, 173, 195–96, 238–45
  ethics or morals of, 243
  human experimentation compared with, 130, 133, 176, 205, 238, 277
  at Institute of Living, 238–40
  memory and, 185, 187, 240–41, 242, 244–45
  uncus in, 192
animal magnetism, 186
animals, petting of, xii
ankh, Egyptian, 18–19
Annese, Jacopo, 334–35, 337–39, 342–56, 409
  Brain Observatory and, 338–39, 342–56, 381–98
  Corkin's conflict with, 381–98
  fundraising party of, 347–48
  harvesting and, 337, 351–56
  Henry visited by, 322–24, 335

Annese, Jacopo *(cont'd):*
lab work of, xi–xii, 338, 347, 351–56, 386
writings of, 388–93
anomia, 248
anosmic patients, 173
anterograde amnesia, 298
anxiety, 85, 99, 198, 227, 319, 321, 366, 403
aphasia, 160, 176
apnea, 192
Apollo, 179
appetite, 42
archival research, 123–24
Armstrong, Neil, 300–301
Army, U.S., 51, 52
Asclepius, 4, 19
Associated Press, 124
Assyrian surgeons, 15
astronauts, 300–301
asylums:
use of word, 60
see also Connecticut State Hospital; Institute of Living; mental asylums
Athinoula A. Martino Center for Biomedical Imaging, 337, 384, 386, 392
Atlanta, 280–81
Atlanta, Ga., 285, 289
Atlantic, 142
Atlantic City, N.J., 182–87, 189
Auckland, University of, 276, 302
auditory cortex, 175, 180
Augustinack, Jean, 384, 388, 390
Aurora (monkey), 128
Auschwitz, 115
"Authorization for Brain Autopsy" form, 333
autopsies, of brain, 82, 95
aviation, 110–11
axons, 7, 30, 36

baboons, 238–41
Bacon, Francis, 205
bacteria, 92
Bailey, Percival, 362
bank clerk, epileptic, 176
Barnum, P. T., 90
Beatles, 299–300
becerros, 234–38
Beck Depression Inventory, 319
Becky (chimpanzee), 80–81, 204
"Bedlam 1946" (*Life* article), 99–101
Beecher, Henry Ward, 137–39, 142–43
behavior, 367
neural circuitry and, 163
behavioral and personality changes, 107, 127, 131, 146–47, 149, 196–97, 207, 215, 318–19
belladonna, 165
Bellevue Hospital, 51, 91, 92–93
Belsen, 100
Bergamini, Mary T., 332
Bickford Health Care Center, xii, 321, 326–27, 329, 332, 336–37, 404
Annese's visit to, 322–24, 335
briefings on Henry at, 336
bicycle accident:
of Henry Scoville, 140
Henry's fall and, 3, 5–7, 21–22, 39, 136, 140
bilateral medial temporal lobotomies, 229, 232, 318
Bill of Rights, U.S., 368
birth defects, 221, 369
blacks, medical experiments on, 117, 118
block-face images, 384–86, 388, 391, 398
Block Number Five, 113
Bolt, Usain, 340–41
bone dust, smell of, 347
bone flap, 169
Boston, Mass., 89–93, 162, 163, 268–69, 301–2, 337–38, 393
Athinoula A. Martino Center for Biomedical Imaging in, 337, 384, 386, 392
Corkin's dinner party in, 334–35
Logan Airport in, 338, 339, 383
Boston City Hospital, 93
Bradley, William, 386–87
Brady, William, 41

brain:
autopsies of, 82, 95
behavior and, 163
of bull, 248
bullet to, 87
complexity of, 36
connection making of, 30–31, 140
cortex of, *see* cortex
CSF and, 141
of dolphin, 343
in Edwin Smith Papyrus, 18
electrodes in stimulation of, 166, 167, 168, 172, 174, 180–87, 189, 211
epilepsy and, 165
frontal lobes of, *see* frontal lobes
harvesting and, 337, 351–56
Hippocrates's view of, 4–5
histological and digital preservation of, 334, 343, 345, 382–83, 388
"homunculi" illustrations of, 167–69, 183
infection of, 252
length of axons in, 36
maps of, 342, 345
memory and, xiv, 185–88
of monkey, 127
pain tolerance and, 43
of primate, 241
of Tan, 165–66
tumors in, 82, 140, 142, 143, 153, 166, 211
two memory systems in, 255–56
as viewed until nineteenth century, 94–95
white matter of, 81
young, 108
see also Molaison, Henry Gustave, brain of
brain damage, 221
accidental, to midbrain, 193, 194
epilepsy and, 166
reliance on for history of modern science of, 94–96
brain-imaging technologies, 140–42, 209, 211, 212, 222
brain injuries, 36
treatment of, 15–18
Brain Observatory, 338–39, 341–56, 405
Corkin's conflict with, 381–98
donors to, 344, 347–56
mission of, 343
brain stem, 29, 143, 264, 326
brain surgery, 52, 75–79
of Patient P.B., 160
see also leucotomy; lobotomy; psychosurgery
Brain Unmodified Derivatives, 393
Breasted, James Henry, 17
Breggin, Peter, 368–71
bridges:
climbing of, 27–31, 35–37
connections and, 35
sensory, 31
Broca, Paul, 15, 165, 172–73
Broca, Pierre, 95
Broca's area, 95, 96
Brodmann, Korbinian, 342, 343
Brodmann areas, 342
Brodmann's areas 9 and 10, 123
Brooklyn Heights, N.Y., 137–38
brucella, 92
Bucy, Paul, vii, 126–33, 173, 195, 206–7, 238–40, 318
on Pribram, 362
Pribram's work with, 357, 362
Buddhism, 320
Buffalo, N.Y., 331
"Building a Search Engine of the Brain" (*Times* article), 382
buildings, internal template for, 31
bullet:
to brain, 87
experiments with, 112
bullet story, 12
bulls, bullfighting, 234–38, 245–48
Burckhardt, Gottlieb, 79–80
burglar story, 12
Burlingame, C. Charles, 60, 68, 69, 71–74, 120–26, 238, 359
fame of, 152

Burlingame, C. Charles *(cont'd):*
Fulton's close relationship with, 126
as industrial psychiatrist, 73–74
Joint Committee of State Mental Hospitals meeting and, 101
Lewis's talk and, 119
personality change and, 146–47, 149
presidency lost by, 147
Simon's tensions with, 124–26

Cairo, 34, 360
California, University of, 124
at Davis, MIND Institute at, 388–91, 393, 398–99
California, University of, at San Diego (UCSD), 334, 339, 341
in custody battle for Henry's brain, 384–99
department of research at, 394
lawyers at, 393, 395, 398
see also Brain Observatory
Cambridge, Mass., 274, 275, 310, 317
see also MIT
Cambridge University, 162, 171, 220, 282
camping trips, 139
Canada, 9, 144, 179, 254, 339–42
see also Montreal Neurological Institute
capote, 235–36, 245–46
carotid arteries, 15
carotid sinus nerve, 128
cars and driving, 257, 260–61, 268, 370, 372–73
cartoons, 311–12
castration, 164–65
Catemaco, 340
Celsus, 116
Central Park (Manhattan), 117
central sulcus, 167
cerebellum, 170, 326
orca, 352
Cerebral Cortex of Man, The (Penfield), 167
cerebral hemorrhage, 365
cerebrospinal fluid (CSF), 141, 142, 209
cerebrum, 211
Charlestown Navy Yard, 337
Chatterbox, 60–61, 71
Cheney Brothers Silk Manufacturing Company, 64–65, 73–74
chewing motions, compulsive, 127–28
Chicago, Ill., 229–32, 280–81, 357
Chicago, University of, 126, 127–28, 173
Chicopee Falls, 310
childbirth, complications of, 117
children, 365
medical experiments on, 117
misbehavior in, 86
scraped skulls of, 15
Chile, 90
chimpanzees, experiments on, 80–81, 129, 148, 173, 176, 204
China, Chinese, 64, 124
Christianity, 188, 191
Christmas memories, 279
circus, freak show in, 90
City and Regional Magazine Association, 280–81
Civil War, U.S., 137
Cleopatra, 116
climbing:
bridge, 27–31, 35–37
pyramid, 31–35
Clinical Research Center (CRC), 274, 293–97, 299–308, 313–15, 317–20, 329
Clinton, Bill, 344
clips:
inside Henry, 324–25, 338
Scoville, 143, 263, 407
Cobb, Stanley, 91–93
cocaine, 169
Codman & Shurtleff, 145, 324–25
cognitive tests, on chimpanzees, 80–81
Cold War, 364
Collias, James, 370–71
Collins, Bill, 260–62, 265
Colt, Samuel, 3
Colt Park (Hartford), 3, 21–22
coma, 72, 108, 198
Commando (movie), 378

concentration camps:
medical experiments at, 110–16
state mental hospitals compared with, 100
confidentiality agreement, 284–85
confusion, 84
Congress, U.S., 100, 367–71, 376
Connecticut, 55–74, 139
Joint Committee of State Mental Hospitals of, 100–101, 103–4
Milner's visits to, 223–29, 241, 253–56
see also specific places
Connecticut, University of, 205, 260, 364
Connecticut Colony for Epileptics, 204
Connecticut Cooperative Lobotomy Study, 101–4, 125, 131, 148, 153–54, 194, 357
Connecticut General Assembly, 56, 60
Connecticut State Hospital, 98–104, 148
annual report of (1948), 154
in-house newsletter of, 103
Milner's visit to, 227–28
OR 2200 at, 102–3, 104, 191–99, 227
Shew Hall at, 98–99, 100
Connecticut State Library, 286–87
Connecticut State Medical Society, 56
Connecticut Training School for the Feebleminded, 204
connections:
brain's making of, 30–31, 140
neuronal, 36
storytelling and, 35
consciousness, 94
brain surgery and, 77–79, 84, 169, 198, 211
Cooke, Jerry, 99–100
Copasso, Marie, 144–45, 147–49
Copernicus, Nicolaus, 369
Corkin, Suzanne, 267–85, 322–30, 372, 379–405
Annese's conflict with, 381–98
author's contacts with, 282–85, 289
author's interviews with, 379–82, 397–402, 404
background of, 267–69, 271, 380
dinner party of, 333–34
ecstasy of, 337
education of, 268, 269, 273, 337, 380
Henry and, 269, 271–73, 275–77, 305–8, 318–20, 332–39, 380–84, 409
Henry files of, 399–404
Henry inherited by, 275–76
Henry's death and, 333–39
as Henry's gatekeeper, 276–77, 281–85, 289, 322–23
Henry's MRIs and, 324–26, 334, 337
Lisa Dittrich's friendship with, 267–69, 281, 282
Milner's co-authored paper with, 373–74
at the Neuro, 269–73
in San Diego, 382, 385–86
Scoville's relationship with, 268, 274
Cornell University, 51
Correll, Robert, 241
cortex, 142, 318
auditory, 175, 180
entorhinal, 156, 211, 213
hippocampal, 198
memory, 183
motor, 167–68, 174, 180
olfactory, *see* limbic lobe
primary auditory, 96
somatosensory, 168–69, 174, 183
cowpox, 117
cows, 92, 99, 179
craniotomy, 143
CRC, *see* Clinical Research Center
criminals, 116, 130
Cronin-Golomb, Alice, 310
Crowley, Myra, 331
CT scans, 324
curiosity, 171, 249, 275
Cushing, Harvey, 184

Dachau concentration camp, Experimental Station at, 110–16
Dana Foundation, 387, 390
Dandy, Walter, 51, 52

D'Aulaires' Book of Greek Myths, 179–80, 188
Davidoff, Leo, 205
death penalty, 281
death row, 281
debauchery, 164
Declaration of Independence, U.S., 368
delayed-nonmatching-to-sample task, 244–45, 246
delayed-response studies, 240, 241, 244–45, 246
Delgado, José, 248
delusions, 186, 194, 198
  electrode-provoked, 181–82
dementia, 404
  of Elizabeth Molaison, 316, 321, 327
  of Henry, 323, 326
dendrites, 7
depression, 81, 87
Descartes, René, 186
desire, capacity for, 44
Dickens, Charles, 57–59
Dictaphone, 122, 199
Dilantin, 140, 142, 203, 323
disease, 116–17, 118, 130
disorientation, 84
dissection, human, 116
Dittrich, Lisa Scoville, 22, 52–55, 106, 280, 376, 409
  Corkin's friendship with, 267–69, 281, 282
  family background of, 190–91
  father's relationship with, 190, 199–200
  memories of, 108–9, 190–91
Dittrich, Luke (author):
  at *Atlanta,* 280–81
  Brain Observatory visited by, 342–56
  bullfighting of, 236–38, 245–47
  contract sent to, 284–85
  early jobs of, 33–34
  education of, 33
  Esquire pitch of, 281–82
  in Greece, 179–80
  in Mexico, 234–38, 340
  photo of, 340
  pyramid climb of, 31–35
  sledding memory of, xiv
  at Sunset Hill, 105–7
  writing start of, 33–34, 360
  in Yukon, 339–42
Dittrich-Tillett, Anwyn, 179–80, 188, 340, 348–49, 409
diving, from George Washington Bridge, 27–29
docility, 196
doctors:
  different natures of, 80
  see also specific people and topics
Doctors Trial, 113–15, 118–19
dolorimeter, 42–43
dolphin brain, 343
dreams, 25–26, 46
drugs:
  antianxiety, 321
  antipsychotic, 365, 376, 377
  Dilantin, 140, 142, 203, 323
  for epilepsy, 140, 142, 165, 177, 203, 208, 210, 323, 326
  mind-altering, 127–28
Dudin, Florence, 144
Duke University, 308–9
dural ridge, 191
Durrell, Lawrence, 33
dwarfs, 94, 152

eating habits, 85, 87, 91, 239
Ecuador, 340
Edwin Smith Papyrus, 16–19, 170
ego (the self), 86, 129
Egypt, 360
  author's move to, 33
  Giza Plateau and pyramids in, 31–35
  Mohammedan invasion of, 31–32
Egypt, ancient:
  human experimentation in, 116, 130
  medicine in, 16–19
Egyptian Supreme Council of Antiquities, 32

Egypt Today, 34
Eichenbaum, Howard, 277, 318–19
Eisenhower, Dwight D., 305
electric drill, 354
electric shock treatment, 69–73, 82, 107, 120, 166, 231, 366, 377
electroencephalography, electroencephalographs (EEG), 172, 174, 209, 211, 212
electrophysiologists, 192
electropyrexia cabinet, 66–67
emotion, 94
  of Aurora, 128
  frontal lobes and, 86–87
  lack of, 87–88
  of white Americans vs. Asians, 124
emotional pain, 44
endocrinology, 91, 92–93
England, 137, 162
Ensisheim grave site, 14
entorhinal cortex, 156, 211, 213
environment:
  change in, 87
  high-altitude, 110–11
epilepsy, 4–5, 79, 163–66, 269–70
  causes of, 164, 165–66
  of Henry, 7–8, 21–22, 135–36, 140, 142, 201–4, 208–15, 223–24
  psychomotor, 159–60, 171–72, 176
  spasmodic mouth movements in, 128
  surgery for, 160, 163, 164–65, 171–72, 174–77, 180–82, 196, 207, 222, 223, 269
  see also seizures
epileptic focus, 209, 211, 212, 213
episodic memory, 255–56, 272, 278–80, 380–81
equipotentiality, theory of, 357
Erasistratus, 116
Esquire, 281–82, 285
eugenics, 92
euthanasia, 92
evolution, 86, 120, 187, 248
experimental neurosis, 81
Experimental Station; Experimenting on Living Humans for the Benefit of Mankind, 110–16
  air pressure and, 110–12
  Doctors Trial and, 113–15, 118–19
  freezing experiments, 112–15
  hypothermia and, 112–15
"Experiments for Rewarming of Intensely Chilled Human Beings by Animal Warmth" (Rascher), 114–15
eyes, 85

facial nerves, 128
facial paralysis, 176
Fairfield State Hospital, 100–101, 104, 148
Farmington, Conn., 56, 370
farms, 99, 116–17
Farrell, Bill, 41
Far Side, 311–12
fear, 71–73, 85, 87, 128
"Fear-Free Psychiatry," 71
feminine traits, 196
Ferguson, Bette, 344, 347–48
Ferrari, Enzo, 260
Ferrari factory, 260
fetuses, male vs. female, 116
fever therapy (pyretotherapy), 66–67, 68, 87, 108, 120, 121, 377
"first, do no harm" (primum non nocere), 80
First Experimental Station of the Luftwaffe, *see* Experimental Station
Fischbach, Gerald, 387–89
Fischer, Liselotte, 209–10, 403
fissure of Sylvius, 181
flat brain spatula, 147–48, 191, 195, 211
Florida, 185, 204–8, 357
Fort Sumter, 137
Foster, James, 261–62
fractional lobotomy (orbital undercutting), 148–49, 152–55, 265, 365, 375
France, 14, 95

Francis A. Countway Library, 89–91, 96–97
Frankfurt, 372
Frankie (Henry's playmate), 10
Freeman, Walter, 77–79, 82–88, 129, 359, 369, 372
  asylum touring of, 155
  celebrity of, 83
  death of, 363–64
  family background of, 82
  illegal operations of, 154
  lobotomy work of, 83–88, 148–55, 377
  at Psychosurgical Conference, 144–45, 149–55
  sloppiness of, 148
  tools of, 149, 150–51, 155
freezing experiments, 112–15, 121
French, 163, 172–73
Freud, Sigmund, 86, 184
Freudians, 171
frontal lobes, 76–77, 80–81, 238, 275
  function of, 86–87, 91, 94, 204
  of Gage, 90, 91
  of Henry, 345, 355, 391–92
  importance of, 86
  inferior, 95
  lobotomy and, 102, 132, 147–50, 152, 155, 191, 195, 211
Frosch, Matthew, 337, 390, 391, 393
Fulton, John, 80–81, 122–24, 126, 148–49
  Bucy's correspondence with, 126, 129–33, 206–7
  Freeman's letter from, 155
  at Harvey Cushing Society, 204
  lobotomy spread by, 129–30
  optimism of, 132
  Pribram's work with, 238, 357, 359, 362–63
Furman, William, 281
"Further Analysis of the Hippocampal Amnesic Syndrome" (Milner and Corkin), 373–74

Gage, Phineas, 90–91, 95, 220
Gahm, Norman, 263
Galesburg State Research Hospital, 230
galvanic shocks, paranoia and, 186
gangrene, 112
garages, as inspiration for surgical tools, 145, 149
Generales, Marianne, 391
genitals, 168
George Washington Bridge, 27–31, 35–37
George Washington University, 77–79, 150
Germany, Nazi, 110–16, 118–21, 298
  corruption of medical science in, 114
  Lewis's talk on, 119–20
  moral degradation of, 115, 118, 120
  paper trail in, 114
Giza Plateau, 31–35
Giza pyramids, 16, 31–35
glial cells, 6
gluttony, 42
goats, 91, 92
Goddard, Harvey, 135–36
Gollin Incomplete Pictures Test, 373–74
Google, 288, 341
Goose Bay, Labrador, 52
GPS, 257
grand mal seizures, 21, 22, 159
Grant, Ulysses S., 137
gray matter, 94, 333
Great Britain, 162
Great Depression, 99
Great Pyramid, 31–35
Greece, 179–80, 188
Greek myth, 4, 19, 179–80, 188–89
grooming, personal, 85
Guadalajara, 234–35
Gunner (Scoville patient), 264–65
guns, 135, 203
gynecology, 117

Hall, Marshall, 164
Hallissey, Arline (later Mrs. Pierce), 39–43, 47
hallucinations, 182, 191, 194, 197
hallucinogens, 127–28
hammer test of emotion, 124

handwashing, obsessive-compulsive, 86
Hartford, Conn., 5–7, 20–21, 38–42, 51, 52, 224–27, 271, 274, 275, 313, 320, 357, 363, 369, 380
author's visit to, 285–89, 407–8
Colt Park in, 3, 21–22, 39
Joint Committee of State Mental Hospitals meeting in, 100–101
mental asylums in, 56–74; *see also* Institute of Living
nursing homes in and near, 288–89
St. Peter's in, 39–42, 271
University Club in, 153
veterans' hospital nursing center in, 264–65
Hartford Bureau of Vital Records, 286
Hartford Courant, 101–2, 363–67
Hartford Hospital, 51, 52, 93, 102, 259–64, 271, 364, 369
author's visit to, 407–8
Emily Scoville's psychosurgery at, 375
Henry at, 140–42, 224–27, 241, 271, 410
macaque lab at, 241
Milner at, 224–27
new rules at, 370–71
Zehnder at, 122, 124
Hartford Regional Center for the Mentally Retarded, 317, 321
Hartford Retreat for the Insane, 56–60
see also Institute of Living
Harvard Medical School, Francis A. Countway Library at, 89–91, 96–97
Harvey Cushing Society, 204–8, 215
head wounds, 14, 17–18, 52, 262, 273
Hearst, William Randolph, 79, 83
Hebb, Donald, 163, 170, 172
Hebben, Nancy, 328–29
hematoma, 143
hemispheres, 175, 209, 211–14
left, 213, 221, 326, 391
motor cortex and, 167
psychomotor seizures and, 172
right, 166, 213, 326
temporal cortex of, 183
Hepburn, Katharine, 107
Hermes, 179, 188
Herophilus, 116
Herrick, Lillian, 316–17, 321, 327, 330
Hesiod, 188
hieratic script, 16
hieroglyphs, 16, 170
high-altitude physiology, 110–11
Himmler, Heinrich, 114
hippocampal gyrus, 224
hippocampus (Ammon's horn), 128, 156, 173, 192, 195, 198, 206, 211, 213, 221–22, 229, 232, 244, 318
function of, 205
of Henry, 298, 326, 345–46
Hippocrates Asclepiades, 4–5, 19, 165
Hippocratic oath, 19
Hollywood, Fla., 204–8
Hollywood Beach Hotel, 204–8
Holmes, Oliver Wendell, 137
Homer, 188
Homo sapiens, evolution of, 86
homosexuality, homosexual behavior, 128, 146–47, 191, 194, 210
homunculi illustrations, 167–69, 183
Hopkinton, Mass., 108–9
hormones, 91
Horowitz, Vladimir, 184
Horrax, Gilbert, 205
Horsley, Victor, 169
housewives, 86
Hudson River, diving into, 27–29
human experimentation, 132–33, 332–33
animal experiments compared with, 130, 133, 176, 205, 238, 277
drawbacks of, 130–31
in historical perspective, 116–18
informed consent and, 327, 329
by Nazis, 110–13, 118–21, 327
humor, sense of, 78–79, 311–12, 330
hydraulic theory of the action of the brain, 186, 187
hydrotherapy, 63–64, 68, 108, 377
hygiene, 146
hyperglycemic coma, 72, 108
hyperthermia, 198

hypnopompia, 25–26
hypothalamus, 372
hypothermia, 63–64, 112–15, 120, 121
"Hypothermia" (Marin-Foucher), 121
hysterics, 90

ice packs, 196
ice pick, 150–51, 155
id, 86
Illinois, 229–32
immunology, modern, creation of, 117
"Impressions of the Psychological Factors in Nazi Ideology" (Lewis), 119–20
Inca:
  skulls, 14
  surgical instruments of, 13
industrial psychiatry, 73–74
infantile amnesia, 108–9
infection, deaths from, 117
informed consent, 327–29
insane, use of word, 60
insanity, *see* mentally ill, mental illness
Institute for Brain and Society, 347–48
Institute of Living, 59–74, 119–24, 259, 316
  admonition to guests at, 60
  annual report of (1949), 238
  arrival at, 59–60
  Burlingame Building at, 145–46
  Butler One at, 69
  clinical notes of, 64, 67–68, 70–71, 105, 108–9
  Connecticut State Hospital compared with, 98, 103
  consulting staff of, 74, 101
  description of, 59, 60–61
  employment manual at, 69
  for-staff-only newsletter at, 71–72, 103
  hydrotherapy at, 63–64, 68, 108, 120
  in-house scientific journal of, 121
  Joint Committee of State Mental Hospitals meeting at, 101
  Lewis's talk at, 119–20
  lobotomy at, 101, 120–21, 123, 144–51, 366, 375
  monkey research lab at, 238–40
  naming of, 60
  OR at, 144–51, 238
  patient fear at, 71–73
  Pomander Walk at, 61
  Pribram and, 224, 238–42, 357, 363, 375, 410
  pyretotherapy at, 66–67, 68, 108, 120, 121
  shock treatment at, 69–73, 107–8, 120, 121, 366
  South One at, 62, 63–64
Institutional Review Board (IRB), 404
insulin shock treatment, 72, 73, 82, 108, 231
intellect, 94, 95, 131
intelligence quotient (IQ), 160, 190, 210, 226–27, 231, 271, 328, 348
International Society for Psychiatric Surgery, 364, 371
Italy, 260
Ithaca, N.Y., 51
"it is better to do something than nothing" (melius anceps remedium quam nullum), 80

Jacobsen, Carlyle, 80–81
Jamaica, 340–41
James, Lori, 307
Japan, Japanese, 124, 293
Jason (San Diego man), 251–53, 256–58
Jasper, Herbert, 174
Jenner, Edward, 116–17
Jews, 110
Jimmie (Henry's playmate), 10
Johns Hopkins, 93
jokes, 312
Journal of Neurology, Neurosurgery & Psychiatry, 232–33, 269
Journal of Neurosurgery, 215–16
Journal of the American Medical Association, 91–92

Keen, William, 82
Keene Valley, 105–8
Kendrick, John F., 229–30
Kennedy, Foster, 92
Kennedy, Jacqueline, 297
Kennedy, John F., 296–97, 298, 303, 312
Kennedy, John J., 329–30
Kennedy, Robert, 296, 303
Kennedy, Rosemary, 376–77
Kennedy, Ted, 367–71, 376–77
keyhole approach, 260
kindling effect, 8
Klaming, Ruth, 250–51, 253, 257
Klüver, Heinrich, 127–28, 173, 238–40, 318
knowledge, advancement of, 118
Kubie, Lawrence, 184–85

Lago Agrio (Bitter Lake), 340
language:
  comprehension of, 96
  creation of, 188
  Henry's use of and problems with, 305–10, 402
  infantile amnesia and, 108–9
Lashley, Karl, 185–87, 189, 357, 359, 363
laughter, 31
Learned, Mrs., 108
left frontoparietal area, 14
left-handedness, 39
left hemisphere, 213, 221, 326, 391
Lenin, Vladimir, 364
lesion method, 96, 127
"Letter from the President" (Scoville), 364
leuco, 81
leucotome, 81, 83
leucotomy, 81–83, 88, 146, 148–49
Lewis, Nolan, 119–21
Lexington, Mass., 361
Liberson, W. T., 192, 211
Life, 99–101
LightHouse, 378
light-housekeeping rooms, 21, 38
light show, on Giza Plateau, 32, 33
Lima, Almeida, 81
Limbaugh, Rush, 348–49
limbic lobe, 172–73, 205, 206–7, 214, 215, 326
"Limbic Lobe in Man, The" (Scoville), 372
limbique, 172–73
Lincoln, Abraham, 137, 138
living experiments, 112
lobectomy, 172
lobo, 83
Lobotomobile, 155
lobotomy, 83–88, 101–4, 129–33, 204, 318–19, 364–72
  at Connecticut State Hospital, 102–3, 104, 191–93
  decline in use of, 265, 364, 365, 369
  evaluation of costs and benefits of, 130–31
  fractional (orbital undercutting), 148–49, 152–55, 265, 365, 375
  Fulton's role in spreading of, 129–30
  ghoulish aspect of, 132
  incidence of, 154, 155
  at Institute of Living, 101, 120–21, 123, 144–51
  lack of control in experiments of, 125, 130
  popularity of, 197
  Scoville's performing of, 88, 101, 102–3, 104, 120–21, 123, 131, 143–49, 152–56, 191–99, 206–16, 227, 229–31, 240–41, 259–60, 265–66, 324–26, 328, 338, 342, 345, 355–56, 364–67, 375–78, 391–92, 408–9
  social acceptability and, 85–88
  traditional, drawbacks of, 154
  transorbital, 150–51, 153, 154–55
  use of term, 83
  Zehnder's research on, 123, 125
Lobotomy Quartet, 132
Localisation in the Cerebral Cortex (Brodmann), 342
Logan International Airport, 338, 339, 383

London, neurology conference in, 80–81
"Loss of Recent Memory After Bilateral Hippocampal Lesions" (Scoville and Milner), 232–33, 269, 357, 378, 403
Louisiana, 8, 279, 286, 287
love, love affairs:
  falling in, 279
  unfortunate, 147
Lucy (chimpanzee), 80–81
Luftwaffe, 110–14
lumbar puncture, 141
lungs, ruptured, 111–12
Luxor, 16

macaque monkeys, 127–28, 133, 195–96, 318
  at Hartford Hospital, 241
  at NIMH, 244–45
McCaw General Hospital, 51–55, 135
McGill Film Society, 220
McGill University, 163, 219–20, 238, 243, 275, 334
  Corkin at, 269, 273, 337
McGuire, Edward, 326–27, 330–31
MacKay, Donald, 307–10, 402
McKenzie, Kenneth George, 144
McKhann, Guy, 390–91
MacLean, Paul, 204–6
madness, *see* mentally ill, mental illness
magic, 17
magnetic resonance imaging, *see* MRI
malaria, 67, 112
malpractice suits, 143
Manchester, Conn., 9, 10–11, 64–65
Manchester, England, 162
manic depressives, 227–28
Mansfield Training School, 204
Manteno State Hospital, 229–30
Maranello, 260
Marin-Foucher, M., 121
Marlene (Pribram's caretaker), 376, 409
marriage, epilepsy and, 165
Marslen-Wilson, William, 20, 22, 293–97, 299–302, 313–15
Mary (Scoville's patient), 365–66
Massachusetts General Hospital, 51, 93, 325
  Henry's brain and, 385, 391, 392
  Henry's death and, 333, 337
  Martinos Bioimaging Center at, 384, 386, 392
Massachusetts Institute of Technology, *see* MIT
masturbation, 44, 86, 128, 194
maze test, 270, 271–72
measles, 176
medial, defined, 172
medial temporal lobes, 155–56, 192, 211–15, 275
  brain infection in, 252
  electrode stimulation of, 180–82, 211
  epilepsy and, 172, 213
  function of, 197, 213, 298
  monkey research and, 238–42, 244
  surgery on, 160, 180, 195–98, 206–7, 209, 212–15, 221–24, 227, 229, 232, 241, 318
  use of term, 172
medical arts, symbol for, 19
Medical Corps, 52
Memento (movie), 220
memory, xiv–xv, 36, 96, 179–89, 220–33, 244–49, 252–56, 277–80
  animal experimentation and, 185, 187, 240–41, 242, 244–45
  Corkin's work with, 270, 271–72
  dreams and, 26
  of Emily Scoville, 107
  episodic, 255–56, 272, 278–80, 380–81
  Henry's earliest, 9–11
  impairment of, 84, 128, 206, 215–16, 224–25, 228–29, 231, 232; *see also* amnesia
  institutional, 123–24
  of Lisa Dittrich, 108–9
  long-term, 298
  maze test and, 270, 271–72
  of Patient A.Z., 199
  of Patient P.B., 159–61

Penfield's theory of, 182–89
of postoperative monkeys, 240–41
short-term, 220
Socrates's view of, 189
after surgery for epilepsy, 176–78, 269–70
two systems of, 255–56, 272, 275
Wechsler Memory Scale and, 225–26, 228
Memory's Ghost (Hilts), 283
memory supplements, 257
memory traces, 298
meningioma, 143
mental asylums, 56–74, 131, 197, 364
crisis and exposé of, 99–101
Freeman's touring of, 155
Hartford Retreat for the Insane, 56–60
Milner's visits to, 227–31
in Washington, D.C., 82
see also Connecticut State Hospital; Institute of Living
mental hospital, use of word, 60
mentally ill, mental illness, 151
air pressure experiments and, 112
epidemic of, 99
first neurosurgical attacks on, 79–80
homosexuality as, 146
roadside cages for, 56
search for physical cause for, 82
use of word, 60
see also specific topics
mental retardation, 92
Mercedes Gullwing, 260
Mesantoin, 203
mescaline, 127–28, 173, 195
Mesmer, 186
Metrazol shock treatment, 72, 73, 82, 108
Mexico, 234–38, 340
Mexico City, 340
mice, in experiments, 129
midbrain, 193, 194–95
Middle East Times, 34, 360
Middletown, Conn., 98–104
Milner, Brenda, 159–67, 219–33, 248–49, 269–72, 357, 359, 409
author's first interview with, 219–20
background and education of, 161–62, 220
Connecticut visits of, 223–29, 241, 253–56
Corkin and, 269–72, 274, 337
Corkin's co-authored paper with, 373–74
curiosity and inquisitiveness of, 171, 275
discoveries about Henry of, 255–56, 272, 275, 311
Henry research ceded by, 275
Henry tested by, 224–27, 231, 254, 271, 272, 373–74
at the Neuro, 159–61, 163–66, 170–78, 269–72
Patient P.B. and, 159–61, 221, 223, 225
Penfield's relationship with, 171
poverty of, 171
Scoville's co-authored paper with, 232–33, 269, 357, 378
unilateral operations and, 175
Milner, Peter, 162, 163
Minnesota, University of, 185, 187
Miró, Joan, 377
mirror tracing test, 250–51, 253, 271, 272
Mishkin, Mortimer, 238–46, 248–49, 357
Mississippi, University of, 367–68
MIT (Massachusetts Institute of Technology), 9–11, 20–22, 39, 42–46, 201–3, 316, 328
Behavioral Neuroscience Laboratory at, 276
Brain and Cognitive Sciences Complex of, 379–80
Clinical Research Center at, 274, 293–97, 299–308, 313–15, 317–20, 329
Corkin at, 273–76, 283–85, 370–402
in custody battle for Henry's brain, 384–99
exhibit of personal belongings of Patient H.M. at, 38, 399–400
Henry's death and, 333

MIT (Massachusetts Institute of Technology) *(cont'd):*
  lawyers at, 283–85, 289, 327, 380, 394–95, 398
  lobotomy investigation headed by, 371–72
  sleep laboratory at, 25
MIT Research Project Known as the Amnesic Patient H.M., 329, 332, 404
Mnemosyne, 180, 188–89
"Modern-Day Patient H.M., A" (working title for paper), 257
Molaison, Elizabeth (Lizzie), 21, 135–36, 203, 204, 208–9, 224, 279, 287
  aging of, 320
  death of, 44, 327
  dementia of, 316, 321, 327
  Henry's relationship with, 316–17, 320
  in Montreal, 270–71
Molaison, Frank S., 331, 397–98
Molaison, Gustave, 3, 8, 135–36, 203, 204, 208–9, 294
  background of, 279, 286, 287
  death of, 10, 44, 45–46, 327
  as electrician, 3, 21, 136
  height of, 10
Molaison, Henry Gustave, vii, xii–xiii, 208–16, 252–57, 270–89, 293–346, 372–75, 408–11
  appearance of, 41–42, 322
  background of, 8–9, 271, 286
  behavior and personality of, 316–21
  birth certificate of, 38
  brain surgeon ambition of, 201, 202, 215
  childhood ambitions of, 26
  at Clinical Research Center, 274, 293–97, 299–308, 313–15, 317–20
  conservator of, 283, 327–33, 397–98
  crush of, 40–43, 47
  death of, 233, 275, 333–39
  dementia of, 323, 326
  earliest memory of, 9–11
  education of, 20–22, 38–42, 135, 331
  electroencephalograph of, 209
  epilepsy of, 7–8, 21–22, 135–36, 140, 142, 201–4, 208–15, 223–24
  fall of, 3–11, 39, 136, 140, 331
  finances of, 332, 404
  hygiene of, 316, 317
  hypnopompia of, 25–26
  informed consent and, 327–29
  Jason compared with, 252, 255, 256
  jobs of, 21–22, 135, 203, 317, 321
  language use and problems of, 304–10, 402
  loss of desire in, 44
  many childhood moves of, 3, 20–21, 331
  meds taken by, xiii, 140, 142, 203, 208, 210, 323, 326
  Milner's ceding of control of, 275
  Milner's discoveries about, 255–56, 272, 275, 311
  Milner's feeling of friendship for, 233
  at MIT labs, 9–11, 20–22, 25–26, 39, 42–46, 201–3, 274, 276–77, 283, 293–97, 299–308, 313–15, 317–20, 329
  movie inspired by, 220
  MRIs of, 324–26, 334, 337, 384, 392
  mysteries of, 313–15
  nearsightedness and glasses of, 26, 201, 202
  the Neuro visited by, 270–73
  non-epilepsy health problems of, 323, 326
  obituary of, 216
  pain endurance of, 42–43
  passivity of, 318–19, 328
  photos of, 38, 40, 135, 276, 335
  placidity and patience of, 44–45
  pleasures of, xii–xiii
  pneumoencephalogram of, 140–42, 209
  questionnaires given to, 319–20
  sense of humor of, 311–12, 330
  violent streak and outbursts of, 316–18, 320, 321, 328
  weight gain of, 42

Molaison, Henry Gustave, brain of, 323–26, 328, 333–35, 337–39, 343–46, 355–56
block-face images of, 384–86, 388, 391, 398
conflict and custody battle over, 381–99
eighty-twenty deal and, 393–94, 398
fall and, 5–7
Nature Communications article and, 388–93
New York summit over, 387–91
photos of, 392, 405–6
slides of, 394–97
surgical transformation of, xiii, 44, 46–47, 215–16, 241, 298, 318–19, 328, 329, 391–92
Molaison, Henry Gustave, as Patient H.M., 240–42, 244, 245, 270–89, 293–346, 372–75
amnesia of, 244, 252, 253–55, 272, 275, 278–79, 298, 323, 326–31, 346, 357, 373–74, 403–4, 408–9
author's first attempt to meet, 282–85
author's first knowledge of, 280
author's keeping tabs on, 341
author's pitch about, 281–82
author's search for, 285–89
birth of, 216, 271
compensation of, 404
Corkin as gatekeeper of, 276–77, 281–85, 289, 322–23
Corkin as lead investigator for, 275–76
desperation of, 230
exhibit of personal belongings of, 38, 399–400
fame of, 271, 273, 309, 311, 336, 401
Henry's transformation into, 8, 47, 403, 405
interviews of, 134–35, 279–80, 283, 293–315, 329
Milner's testing of, 224–27, 231, 254, 271, 272, 373–74
name of, 282–83, 285, 287–88, 338
from 1953–2008, 336–46
as pure, 273–74
Scoville and, 134–36, 140–43, 202–3, 208–15, 241, 245, 269, 279, 408, 411
Scoville-Milner article about, 232–33, 269, 357, 378, 403
secrecy about, 8
unpublished data about, 399–404
Molaison (grandfather), 9–10
Moniz, Egas, 80–83, 88, 132, 148–49, 197, 204
monkeys, 114, 127–28, 129, 192, 195–96, 357, 409
Institute of Living research on, 238–40
storytelling about, 199–200
Montreal, 163, 224, 254, 269–73, 275
Montreal, University of, 171
Montreal Neurological Institute (the Neuro), 159–61, 163–66, 169–78, 208
architecture of, 170
Corkin at, 269–73
funding of, 169
Henry's visit to, 270–73
interdisciplinary approach of, 169
operating room 2 at, 174–75
Penfield's ambitions for, 170
Penfield's dominance of, 170–71
photo for newsletter of, 219–20
Mooney, Thomas F., 326–27, 329–33, 389, 397–98
moon walk, 300–301
moral treatment, 59
morgues, as inspiration for surgical tools, 149, 150
mosquitoes, malarial, 112
motor cortex, 167–68, 174, 180
motorcycle accident, 261–62
motor winding, 21–22
MRI (magnetic resonance imaging), 324–26, 334, 337, 344, 348, 350–51, 384, 392
two stages of, 351
Mrs. Frisby and the Rats of NIMH (O'Brien), 242
murder, attempted, 230
Murray, Elisabeth, 247
muses, Greek, 180, 188

music, 64, 65, 106, 181, 182, 299–305
brain surgery and, 75–79
Mystic, Conn., 45
myth, Greek, 4, 19, 179–80, 188–89

nakedness, 85, 99, 113
National Commission for the Protection of Human Subjects of Biomedical and Behavioral Research, 371–72
National Geographic, 360–61
National Institute of Mental Health (NIMH), 242–45, 367
National Mental Health Act (1946), 100
Nature, 342
Nature (goddess), 170
Nature Communications, 388–93
Nazi Germany, *see* Germany, Nazi
nervous system:
Bucy's experiments with, 127–28
CSF and, 141, 142
neurasthenia, 91–92
neuroanatomy, neuroanatomists, xi–xii, 322–24, 333–35, 337–39, 342–46
neurologists, 144, 169
neurone pathway, 183
neurons, 6–7, 30, 87, 209
connections and, 36
sensory, 31
neuropathologists, 333–34
Neuro-Psychiatric Institute of the Hartford Retreat, 60
see also Institute of Living
neuroscience, neuroscientists, 333–34, 342
neurosis, neurotics, 87, 364
neurosurgery, neurosurgeons, 51–53, 93, 166–78, 180–84, 248, 342, 364
as adventure, 261
brain surgery and, 76–79, 123, 125, 131, 149, 151–55, 197
early history of, 14–15
first attacks on mental illness in, 79–80
at McCaw General Hospital, 51
as neural cartographers, 166
overcrowding problem and, 101
overreaching of, 130
at Psychosurgical Conference, 144, 152, 155
two frightening qualities of, 13
see also specific people
neurotransmitters, 7
Neville, Katherine, 358–59, 376
New Biographers, 142
New Hope, Pa., 370
New Jersey, 184–87, 189, 259
New Mexico, neurosurgical conference in (1954), 222–23
New Psychology, 142
New York, N.Y., 51, 91, 92–93, 378
Central Park in, 117
summit over Henry's brain in, 387–91
New York (American liner), 161, 177–78
New York Academy of Medicine, 117
New York *Daily News*, 28, 29
New Yorker, 311–12
New-York Historical Society, 17
New York Neurological Society, 164
New York Times, xiv, 29–30, 36–37, 142, 382–83
author's reading aloud of, 106
Beecher in, 138
Freeman in, 83
Henry's death in, 338
New Zealand, 276
Niagara Falls, 331
Nixon, Pat, 295–98, 312
Nixon, Richard, 295–96, 298
Nobel Prize, 197, 359
North Atlantic, 112
Norway, 91–92
Norwich State Hospital, 100–101, 104, 148
novel reading, 257–58
novocaine, 75, 169
numbness, 182
nuns, 39
Nuremberg Code, 118–19
Nuremberg trials, 327
nursing homes, 288–89, 321
see also Bickford Health Care Center

Obama, Barack, 242–43
obsessions, 131
obsessive-compulsives, 86
Ogden, Jenni, 276, 302–5
olfactory bulbs, 31
olfactory cortex, *see* limbic lobe
Olmsted, Frederick Law, 59, 98
"On the Sacred Disease" (Hippocrates), 4–5
Orange Park, Fla., 185
orbital undercutting, *see* fractional lobotomy
Organization of Behavior, The (Hebb), 163
Oswald, Lee Harvey, 297, 301–2
ovariectomies, 164–65
overcrowding, of state asylums, 99–102
Oxford School, 268
oxygen, 111, 141, 142

Pacholegg, Anton, 110–14
pain:
  emotional, 44
  physical, 42–43, 102–3, 116
  tolerance of, 42–43
Palin, Todd, 340
Palo Alto, Calif., 357–58
panic, 72, 366
Papyrus, Edwin Smith, 16–19, 170
parachutes, parachuting, 65, 112
paralysis, 174, 176, 201, 202
paranoia, 87, 186, 193, 198, 230
paranoid schizophrenics, 228
Paris, museum of anatomical curiosities in, 95
Parkinson's patients, 404
passivity, 196, 318–19, 328
Patient 52, 84
Patient 53, 85
Patient A.R., 232, 409
Patient A.Z., 197–99, 232
Patient B.P., 194
Patient D.C., 230–32
Patient D.F., 181
Patient D.M., 191–93
Patient E.M., 193–94
Patient E.S., 196
Patient F.C., 177–78, 221, 223, 225
Patient G.M., 196–97
Patient H.M., *see* Molaison, Henry Gustave, as Patient H.M.
Patient I.S., 193, 194–95, 198
Patient M.B., 227–28, 232
Patient M.D., 194
Patient P.B., 159–61, 177, 221, 223, 225
Patient S.B., 180–81
Patient V.M., 196
Pavlov, Ivan, 364
Pearl Harbor, Japanese attack on, 293–94
Penfield, Wilder, 166–75, 220–23, 233
  caution of, 212, 222
  as Christian, 188
  dominance of, 170–71
  epilepsy surgery of, 163, 171–72, 174–77, 180–82, 208, 222, 223, 269
  "homunculi" illustrations of, 167–69, 183
  Lashley's critique of, 185–87
  memory theory of, 182–89
  the Neuro established by, 169
  Scoville's correspondence with, 223–24, 230, 241
  Scoville's encounter with, 222–23
penicillin, 118
Pennsylvania, University of, 25
personality:
  phrenology and, 95
  see also behavioral and personality changes
Personews, 71–72, 103
Peru, Inca skulls in, 14
petit mal seizures, 22, 203, 208, 210, 299–300
PET scans, 324
phenobarbital, 203
Philadelphia, Pa., 25
Philadelphia Inquirer, 142
Phil's (BBQ joint), 348
Phipps, James, 117

photographs, 287, 377
of author, 340
in "Bedlam 1946," 99–100
of Emily Scoville, 107
of Henry, 38, 40, 135, 276, 335
of Henry's brain, 392, 405–6
for the Neuro newsletter, 219–20
preoperative vs. postoperative, 85
of Scoville, 24–25, 92, 247–48, 364
phrenology, 95
physiologists, 131, 152
Pierce, Mrs., *see* Hallissey, Arline
Pindar, 188–89
Ping Pong, 106
pituitary hypofunction, 94
Plato, 189
Plymouth Church, 137
pneumoencephalograms, 140–42, 209
Pomander Walk, 61
Poppen, James, 51, 52
"Possibilities of Surgery for the Treatment of Certain Psychoses, The" (Moniz), 82
"Postmortem Examination of Patient H.M's Brain Based on Histological Sectioning and Digital 3D Reconstruction" (Annese, Corkin, and Frosch), 393
Precentral Motor Cortex, The (Bucy), 126
precision method, 83
Prefargier, 79–80
pregnancies, dissection of, 116
Presley, Elvis, 303, 312
Pribram, Karl, 238–42, 248, 357–60
Henry and, 224–25, 409–11
people rubbed the wrong way by, 362–63
on Scoville, 359–60, 362, 375–77, 409, 411
Price, Joseph, 164–65
priests, 41
procedural memory, 255–56, 272
procreation, epilepsy and, 165
pronunciation, correction of, 106
psychiatric research, federal funds for, 100
psychiatrists, 144, 151, 152, 169, 259, 366
psychic blindness, 128
psychoanalysis, 184, 366–67, 372
Psychological Corporation, 225
psychological tests, 171, 175–78, 209–10, 224–28, 231–32, 373–74, 381
aptitude, 162–63
psychology, psychologists, 159–63, 169–78, 209–10, 381
psychomotor seizures, 159–60, 171–72, 176
psychopaths, 365
psychoses, psychotics, 68, 199, 206, 207, 223, 232, 364
Milner's visits to, 227–28
psychosurgery, 83, 129, 131–32, 146–47, 151–55, 206–7, 238, 364–72
of Emily Scoville, 375–78
mainstream acceptance of, 151
nobility of, 197
Penfield's view of, 222
Scoville's advocacy for, 364–67
Senate subcommittee hearings on, 367–71
see also lobotomy
Psychosurgery (Freeman), 83–84
Psychosurgical Conference, 144–56
psychotherapy, 87
Public Health Service, U.S., 118
pure, use of word, 273–74
pyramid climbing, 31–35
pyretotherapy (fever therapy), 66–67, 68, 87, 108, 120, 121, 377

questionnaires, 319–20

radar, paranoia and, 186
radio, 135, 186, 187, 203
rages, of Gage, 90
Rain Man (film), 344
Ramsdorf, Marjorie, 331
Rascher, Sigmund, 114–15
rats, in experiments, 111, 129, 176, 185, 187
rattlesnake photo, 25

Ravensbrück concentration camp, 114
Reich, Peter, 329
"Relative Value of Life and Learning, The" (Slosson), 118
religious delusions, 194
REM sleep, 25
Reno, Nev., 378
"Report of the Committee Respecting an Asylum for the Insane," 56–57
Retreat Gazette, 57
retrograde amnesia, 298
rewarming tactics, 113, 114–15
rhesus macaques, 238–41
right-handedness, 14, 39
right hemisphere, 166, 213, 326
roadside cages, 56
Roberto (Italian restaurant owner), 353
Rockefeller Foundation, 169
Rolling Stones, 299
Roosevelt, Franklin Delano, 312
Rorschach test, 210
Rosen, Bruce, 386–87
Rosetta Stone, 16
Ruby, Jack, 301–2
Rush Revere and the Brave Pilgrims (Limbaugh), 349
Rutland and Burlington Railroad, 90
Rylander, Gosta, 131

sadness, 86–87
St. Peter's Church, 286
St. Peter's school, 39–42, 47, 271
San Diego, Calif., 250–53, 256–58, 337–39, 342–56
  as earthquake-prone, 344
  Phil's in, 348
  see also Brain Observatory; California, University of, at San Diego
Santa Fe International Corporation, 33–34
scalpels, use of, 83, 86, 87, 145, 147, 155, 191, 211, 222, 353–54
schizophrenia, 191, 193–94, 243
scientific method, 125, 130
Scoville, Barrett, 52–55, 107, 190–91, 268, 361, 372
Scoville, Emily (Emmie; Bambam), 61–74, 153, 190–91, 261, 268, 360–61, 372, 409
  childbirths of, 53, 93
  clinical notes of, 64, 67–68, 70–71, 105, 108–9
  death of, 106, 361
  divorce of, 378
  education of, 64, 65
  fear and dread of, 68, 71
  household managed by, 52–53
  husband's psychosurgery on, 375–78, 409
  hydrotherapy of, 63–64, 68, 377
  at Institute of Living, 59–74, 105, 107–8, 357
  marital problems of, 55, 67–68
  marriage of, 51
  mental breakdown of, 51–55, 67, 68, 106–7, 376
  moving of, 51, 53
  musical interests of, 64, 65, 106
  as nonbeliever, 191
  outgoing personality of, 107
  privileged upbringing of, 64–65
  pyretotherapy of, 66–67, 68, 377
  reinstitutionalization of, 108
  shock treatment of, 69–73, 107–8, 377
  suicide attempt of, 54, 106
  at Sunset Hill, 105–8
  voices heard by, 51, 53, 54
  weight loss of, 53
  withdrawn personality of, 107
Scoville, Gurdon, 23, 139–40
Scoville, Helene, 370, 372, 373, 376
Scoville, Henry Ward Beecher, 139–40
Scoville, Lisa, *see* Dittrich, Lisa Scoville
Scoville, Peter, 52–55, 93, 107, 190–91, 268
  father's relationship with, 153, 190
Scoville, Samuel, Jr., 23, 25, 67, 106–7, 139–40
  Beecher defended by, 142–43

Scoville, William Beecher (Bill), xiii–xv, 51–56, 136–43, 190–200, 202–16, 229–32, 359–76
as advocate for psychosurgery, 364–67
ambitions and goals of, 91
artifacts of, 12–13
as asylum's consulting staff, 74, 101
at-first-mysterious illness of, 91–92
author compared with, 219–20, 247–48
birthday of, 231–32
bridge climbing of, 30–31, 35–37
bullet story of, 12
bullfighting of, 247–48
Burlingame's meeting with, 74
career choice of, 25, 26, 93
cars and driving of, 260–61, 268, 370, 372–73
confidence of, 52, 104
Corkin's relationship with, 269, 274
darkness of, 143
death of, xiii, 233, 269, 361, 373
deaths caused by, 143, 365
education of, 23–24, 35, 36, 52, 92
family background of, 137–40, 142–43
as father, 153, 190–91, 199–200, 268, 372
first great loss of, 136–40
Fulton's correspondence with, 132
guilt of, 230–32
Hartford Courant interview of, 363–67
Harvey Cushing Society speech of, 204–8, 215
Henry and, 134–36, 140–43, 202–3, 208–15, 241, 245, 269, 279, 408, 411
Henry's brain and, 44, 47, 324, 338, 345, 355–56, 391–92
infidelity of, 68
introspection of, 214, 372
letters of, 22–24, 67, 91–94, 97, 106–7, 223–24, 230, 241
lobotomies performed by, *see* lobotomy, Scoville's performing of
marriage of, 51, 67–68
medical specialty selected by, 52
motivations of, 91, 152–53
motorcycle accident and busted spleen of, 261–63
at New Mexico conference, 222–23
photos of, 24–25, 92, 247–48, 364
precision of, 148
Pribram's views on, 359–60, 362, 375–77, 409, 410
as researcher vs. practitioner, 152–53, 213–14
residencies and appointments of, 51–52, 92–93
risk taking of, 212–14
scrubbing up of, 264
second wife of, 370, 372, 373
secrecy of, 375–77
Spencer's stories about, 260–66
surgery accidents of, 193, 194
surgical glasses of, 214–15, 407–8
talents of, 24, 35
tools of, *see* tools, of Scoville
wife's breakdown and, 67, 106–7
wife's meeting of, 65
as Wild Bill, 260–66
writings of, 215–16, 232–33, 269, 357, 378, 403
Zehnder's resignation letter to, 124–25
Scoville clips, 143, 263, 407
Scoville retractor, 143, 263
Scribe, 103
Sea World, 352
Second International Congress of Neurology (1935), 80–81
seizures, 175, 192, 196, 203, 204, 207, 212, 403
grand mal, 21, 22, 159
petit mal, 22, 203, 208, 210, 299–300
psychomotor, 159–60, 171–72, 176
speech and, 166
"Selective Orbital Undercutting as a Means of Modifying and Studying Frontal Lobe Function in Man" (Scoville), 152
semantic memory, 278–80
Semmelweis, Ignaz, 369
Senate, U.S., 367–71, 376

senses, 94, 96
  see also specific sense
sensory conflict, 162
sensory cortex, 183
sexual activity, 114–15, 128, 171, 198
sexual reorientation, 46–47
Shakespeare, William, 205
shock treatment, 69–73, 82, 87, 107–8, 120, 193, 196, 198, 372
shrapnel injuries, treatment for, 112
shyness, 87
Sifnos, 179–80, 188
sight, sense of, 128, 220
silk industry, 64–65
Simon, Benjamin, 102–3, 124–26
Simons Foundation, 387
Sims, J. Marion, 117
Sinatra, Frank, 301–2, 305
singing woman, brain surgery of, 75–79, 84
skiing, 190–91, 268
skin tests, 92
skulls, 150–51, 269, 355–56
  of Copasso, 144–45
  of Gage, 90, 91
  of Henry, 337, 355–56, 391–92
  man-made holes in, 14–15, 76–77, 79, 81, 87, 128, 149, 169, 355–56, 391–92, 408
  of monkeys, 128
  phrenology and, 95
  postoperative growth in, 15
  scraped, 15
slavery, slaves, 98, 118
sledding and sleigh ride memories, xiv, 10–11
"'Sleep'—Not 'Shock,'" 72
slipped disks, 364
Slosson, E. E., 118
smallpox, 116–17
smartphones, 257
smell, sense of, 31, 173, 175, 182, 192–93, 326, 347
Smith, Edwin, 16–17
Smith, Mel, 287
Smith College, 268
smoking, 256–58
snake-twined staff, 19, 25
social acceptability, 85–88
social misfits, Freeman's view of, 85–88
social pathological phenomenon, 120
social rebels, 365
Society for Neuroscience, 385–86
Society of Neurological Surgeons, 144
Socrates, 189
somatosensory cortex, 168–69, 174, 183
"Somesthetic Function After Focal Cerebral Damage" (Corkin), 273
"son of a bitch, son of a bitch, son of a bitch" story, 259, 265–66
sound, 96, 181–82, 220
South America, 13, 90, 165
South Coventry, Conn., 20
Souza, Pete, 59–60
Soviet Union, 364
Spain, 247–48
speech:
  articulation, 95, 96
  brain and, 165–66, 220
speech cortex, 180
Spencer, Dennis, 259–66, 377
spinal column, 19, 29, 141, 356
spinal surgeries, 364
spinal trauma, 153
spinsters, 86
spleen, busted, 262–63
Stamford, Conn., 12
Stanford University, 242, 358
state asylums, 148
  overcrowding of, 99–102
statues, 34, 117, 170
Stevenson, George, 147
stock market crash (1929), 21
storytelling, xiii, 199–200, 280, 281, 310, 374
  connections and, 35
Stowe, Harriet Beecher, 137
Stratton, G. M., 124
stress, neurasthenia and, 91
stroke, 221, 264, 323

suction catheters, 148, 155, 192, 195, 213, 266, 326
suicide, 54, 79, 106, 193
Sunset Hill, 105–8
superego, 86
Surgeon General, 149
surgical instruments:
  of Inca, 13
  see also tools
"Surgical Treatment of Epilepsy, The" (Price), 164–65
Sweet, Bill, 205
swimming pools, 72
Switzerland, 79–80
synapses, 7
syphilis, 118, 327

tactile memories, 186
Tan, Monsieur, 95, 132, 165, 172, 220
tantrums, 86
taste, sense of, 182
teenagers, bullfight training of, 234–36
telegraph, 186, 187
television, xiii, 44, 135, 383
templates, internal, 31
temporal, defined, 172
temporal lobe, 163, 173, 186, 239
  of Aurora, 128
  left, 177
  medial, *see* medial temporal lobes
  posterior left, 96
Terry, Norman J., 27–30
Terry (*Esquire* editor), 281–82
teslas, 334
Teuber, Hans-Lukas, 273–76
thanatology, 114
Thanksgiving dinners, 360, 376
Theaetetus, 189
Thompson, Anna, 159–61, 177, 178, 226
Thorazine, 365, 376, 377
Thornfeldt, Mrs., 55
"Thoughts on Secretaries" (Fulton), 363
Tilney, Frederick, 86
Tilton, Elizabeth, 137–38
Tilton, Theodore, 137–38
Time, 183
tin can telephones, 267
tinnitus, 323, 325
tome, 81
tools:
  of Freeman, 149, 150–51
  of Scoville, 143, 145, 146, 149, 155, 191, 192, 193, 195, 205–6, 211, 213, 263, 266, 325, 326, 370, 407–8
toros bravos, 235
torsional forces, 6
totalitarianism, 368
totem artifact, 13
touch, sense of, 269–70
tourism, in Egypt, 31, 32, 34
tracheotomy, 164
tractability, 196, 377–78
trains, xii, 55–56, 90
transorbital lobotomy, 150–51, 153, 154–55
trephines, 145, 146, 149, 191, 195, 211, 213, 214, 325, 407
trials:
  of Beecher, 137–39
  Doctors, 113–15, 118–19
  Nuremberg, 327
Tridione, 203
trigeminal nerves, 128
Trooper (Corkin's dog), 380
Truman, Harry S., 100, 225
Tulving, Endel, 276–77
tumors, brain, 82, 140, 142, 143, 153, 166, 211
Tuskegee Syphilis Experiment, 118, 327
typewriters, 22
typhus, 112

U-boats, 162
UCSD, *see* California, University of, at San Diego
uncertainty, insanity epidemic and, 99
Uncle Tom's Cabin (Stowe), 137

unconsciousness, 121, 192, 193, 206, 255, 262
induced, 15, 150
Nazi medical experiments and, 113
uncotomy, 192–95, 197
uncus, 128, 156, 173, 191–95, 206, 207, 211, 213, 224
Underwood Typewriter, 22, 203, 286
United States Nuremberg Military Tribunal, 113
see also Doctors Trial

vaccines, 112, 117
Valparaiso, 90
Vassar College, 64, 65
VE Day, 294–95
Vermont, 90
vesicovaginal fistula, 117
veterans, 132, 273
Veterans Administration, U.S., 132
videos, 276, 341–42
Vienna, 65, 107, 362
violence, 230, 316–18
visual illusion, 162
visual memories, 186
vivisections, experimental, 116, 127, 130
vocational instruction, 146
vomiting, 196, 206

Walla Walla, Wash., 51–56, 135
Walter Reed General Hospital, 52
Warren Anatomical Museum, 89–91
Warrenton, Va., 357–59
war wounds, 14
Washington, D.C., 52, 74–79, 149
congressional hearings in, 100, 367–71, 376
mental asylum in, 82
Watts, James, 82–83, 129, 148, 149–50, 377
Freeman's estrangement from, 150, 155
Wechsler, David, 225
Wechsler-Bellevue Intelligence Scale, 160, 226–27, 231
Wechsler Memory Scale (WMS), 225–26, 228
weight, 42, 53, 85
Wells, H. G., 244
Wernicke, Carl, 96
Wernicke's area, 96
Wesleyan University, 98
West Hartford, Conn., 267–68, 289, 316–18
"What, If Anything, Can Monkeys Tell Us About Human Amnesia When They Can't Say Anything at All?" (Murray), 247
Whitcomb, Ben, 262–63
Whitehorse, 339–42
Whitehorse General Hospital, 340
white matter, 81
white women, 117
Williams, Tennessee, 184
Willimantic, Conn., 20
Windham High School, 20
see also Bickford Health Care Center
Windsor Locks, Conn., 322–23
Windsor Locks Probate Court, 329, 331, 332
Wizard of Oz, The (film), 344, 348
women, 117, 409
disproportionate number of surgeries on, 196
epilepsy in, 164
word games, xii–xiii
World Federation of Neurological Societies, 364
World War II, 52, 65, 100, 120, 121, 132, 293–96, 298
in Hopkinton, 109
insanity epidemic and, 99
Milner and, 162–63

X experiments, 112
X-rays, 324
pneumoencephalogram compared with, 141–42

Yale–New Haven Hospital, 259–61
Yale School of Medicine, 259–60, 266, 364
Yale University, 23–24, 52, 101, 122–26, 132, 205, 208
  neurology experiments at, 80–81
  Pribram at, 238, 240, 357, 362–63
Yerbury, Edgar C., 99, 100, 101, 103, 154

Zehnder, Max, 122–25, 131–32
Zeus, 19, 180, 188